T0339603

Marine Structural Design Calculations

Marine Structural Design Calculations

Mohamed A. El-Reedy, Ph.D.

AMSTERDAM • BOSTON • HEIDELBERG • LONDON
NEW YORK • OXFORD • PARIS • SAN DIEGO
SAN FRANCISCO • SINGAPORE • SYDNEY • TOKYO
Butterworth-Heinemann is an imprint of Elsevier

Butterworth-Heinemann is an imprint of Elsevier
The Boulevard, Langford Lane, Kidlington, Oxford OX5 1GB, UK
225 Wyman Street, Waltham, MA 02451, USA

ISBN: 978-0-08-099987-6

British Library Cataloguing-in-Publication Data
A catalogue record for this book is available from the British Library

Library of Congress Cataloging-in-Publication Data
A catalog record for this book is available from the Library of Congress

For Information on all Butterworth-Heinemann publications
visit our website at http://store.elsevier.com/

Typeset by MPS Limited, Chennai, India
www.adi-mps.com

This book has been manufactured using Print On Demand technology.

Dedication

This book is dedicated to the spirits of my mother and my father, and to my wife and my children, Maey, Hisham, and Mayar

Contents

About the Author xiii
Preface xv

1 Introduction to Offshore Structures 1
1.1 Introduction 1
1.2 History of offshore structures 1
1.3 Overview of field development 2
1.4 Types of offshore platforms 4
1.4.1 Drilling/well protected platforms 5
1.4.2 Tender platforms 5
1.4.3 Self-contained platforms 5
1.4.4 Production platforms 5
1.4.5 Quarters platforms 6
1.4.6 Flare jackets and flare towers 6
1.4.7 Auxiliary platforms 6
1.4.8 Bridges 6
1.4.9 Helidecks 6
1.5 Types of offshore structures 6
Further reading 12

2 Engineering Management for Marine Structures 13
2.1 Overview of field development 13
2.1.1 Project cost and the life cycle 13
2.1.2 Concept and screening selection 15
2.2 FEED engineering phase 16
2.3 Detail engineering phase 17
2.4 Engineering design management 18
2.4.1 Engineering stage time and cost control 19
2.4.2 Engineering interfaces 22
2.4.3 Structural engineering quality control 23
Further reading 32

3 Offshore Structures' Loads and Strength 33
3.1 Introduction 33
3.2 Gravity load 33
3.2.1 Dead load 33
3.2.2 Live load 35

3.2.3 Impact load 38
3.2.4 Design for serviceability limit state 38
3.2.5 Crane support structures 39
3.3 Wind load 41
3.4 Offshore loads 46
3.4.1 Wave load 47
3.4.2 Current load 57
3.5 Earthquake load 59
3.5.1 Extreme level earthquake requirements 64
3.5.2 Abnormal level earthquake requirements 65
3.5.3 ALE structural and foundation modeling 65
3.6 Ice loads 68
3.7 Other loads 68
3.7.1 Marine growth 69
3.7.2 Scour 69
3.8 Design for ultimate limit state 69
3.8.1 Load factors 70
3.8.2 Partial action factors 74
3.9 Collision events 76
3.10 Material strength 78
3.11 Cement grout 79
Further reading 83

4 Offshore structures design 85
4.1 Introduction 85
4.2 Guide for preliminary design 85
4.2.1 Approximate dimensions 89
4.2.2 Bracing system 92
4.2.3 Jacket design 94
4.3 Structure analysis 95
4.3.1 Global structure analysis 96
4.3.2 The loads on the piles 99
4.3.3 Modeling techniques 102
4.4 Dynamic structure analysis 106
4.4.1 Natural frequency 107
4.5 Cylinder member strength 112
4.5.1 Cylinder member strength calculation by ISO19902 113
4.5.2 Cylinder member strength calculation by API RP2A 125
4.6 Tubular joint design 135
4.6.1 Simple joint calculation from API RP2A (2007) 136
4.6.2 Joint calculation from API RP2A (2000) 145
4.7 Fatigue analysis 150
4.7.1 Stress concentration factors 151
4.7.2 S-N curves for all members and connections, except tubular connections 154

4.7.3 S-N curves for tubular connections 155
4.7.4 Jacket fatigue design 168
4.8 Topside design 172
4.8.1 Topside structure analysis 172
4.8.2 Deck design to support vibrating machines 173
4.8.3 Grating design 174
4.8.4 Handrails, walkways, stairways, and ladders 175
4.9 Bridges 178
4.10 Vortex-induced vibration 179
Further reading 187

5 Helidecks and boat landing design **189**
5.1 Introduction 189
5.2 Helideck design 189
5.2.1 Helicopter landing loads 189
5.2.2 Safety net arms and framing 193
5.3 Design load conditions 198
5.3.1 Helideck layout design steps 201
5.3.2 Plate thickness calculation 204
5.3.3 Aluminum helideck 205
5.4 Boat landing design 205
5.4.1 Boat landing calculation 206
5.4.2 Boat landing design using a nonlinear analysis method 209
5.4.3 Boat impact methods 210
5.4.4 Tubular member denting analysis 211
5.5 Riser guard 214
5.5.1 Riser guard design calculation 214
Further reading 216

6 Geotechnical data and piles design **217**
6.1 Introduction 217
6.2 Geotechnical investigation 217
6.2.1 Performing an offshore investigation 218
6.3 Soil tests 218
6.4 In-situ testing 220
6.4.1 Cone penetration test 220
6.4.2 Field vane test 225
6.5 Soil properties 226
6.5.1 Strength 227
6.5.2 Soil characterization 230
6.6 Pile foundations 231
6.6.1 Pile capacity for axial loads 232
6.6.2 Foundation size 239
6.6.3 Axial pile performance 240
6.6.4 Pile capacity calculation methods 252

6.7 Pile wall thickness 254
6.7.1 Design pile stresses 255
6.7.2 Stresses due to the weight of the hammer during hammer placement 255
6.7.3 Minimum wall thickness 258
6.7.4 Driving shoe and head 260
6.7.5 Pile section lengths 260
6.8 Pile drivability analysis 261
6.8.1 Evaluation of soil resistance drive 261
6.8.2 Unit shaft resistance and unit end bearing for uncemented materials 261
6.8.3 Upper- and lower-bound SRD 262
6.8.4 Results of wave equation analysis 263
6.8.5 Results of drivability calculations 265
6.8.6 Recommendations for pile installation 266
6.9 Soil investigation report 267
6.10 Composite pile 269
6.11 Mud mat design 274
Further reading 278

7 Construction and installation lifting analysis 281
7.1 Introduction 281
7.2 Construction procedure 281
7.3 Engineering the execution 282
7.4 Construction process 282
7.4.1 Fabrication tolerances 283
7.4.2 Stiffener tolerances 288
7.4.3 Conductor guides and piles tolerances 290
7.4.4 Dimensional control 291
7.4.5 Jacket assembly and erection 291
7.5 Installation process 294
7.5.1 Loadout process 295
7.5.2 Transportation process 295
7.5.3 Barges 301
7.5.4 Launching and upending forces 303
7.6 Lifting analysis 306
7.6.1 Weight control 306
7.6.2 Weight calculation 307
7.6.3 Classification of weight accuracy 308
7.6.4 Loads from transportation, launch, and lifting operations 311
7.6.5 Lifting procedure and calculation 311
7.6.6 Structural calculations 330
7.6.7 Lift point design 332
7.6.8 Clearances 333
7.6.9 Lifting calculation report 337
Further reading 339

8 SACS Software **341**
8.1 Introduction **341**
8.2 In-place analysis **341**
8.3 Defining member properties **346**
8.4 Input the load data **348**
8.4.1 Joint can **355**
8.4.2 The foundation model **360**
8.5 Output data **366**
8.6 Dynamic analysis **371**
8.6.1 Eigenvalue analysis **373**
8.7 Seismic analysis **376**
8.7.1 Combination of seismic and gravity loads **378**
8.8 Collapse analysis **380**
8.9 Loadout **385**
8.10 Sea fastening **386**
8.10.1 Load combinations **388**
8.11 Fatigue analysis **394**
8.11.1 Center of damage **395**
8.11.2 Generation of the foundation superelement **398**
8.11.3 Dynamic wave response analysis **400**
8.11.4 Fatigue input data **409**
8.12 Lifting analysis **410**
8.13 Flotation and upbending **415**
8.14 On-bottom stability **417**
8.15 Launch analysis **420**
8.16 Summary **420**
8.16.1 Static analysis **420**
8.16.2 Dynamic analysis **421**
8.16.3 Seismic analysis **421**
8.17 Fatigue analysis **421**
8.17.1 Collapse analysis **421**
8.17.2 Lifting analysis **421**
8.17.3 On-bottom stability **422**
8.17.4 Tow analysis **422**

Appendix: Assignment **423**
Index **429**

About the Author

Mohamed A. El-Reedy's background is in structural engineering. His main area of research is the reliability of concrete and steel structures. He has provided consulting expertise to engineering companies and the oil and gas industry in Egypt and to international companies, such as the International Egyptian Oil Company (IEOC) and British Petroleum (BP). Moreover, he provides different concrete and steel structure design packages for residential buildings, warehouses, and telecommunication towers and electrical projects with WorleyParsons Egypt. He has participated in liquified natural gas (LNG) and natural gas liquid (NGL) projects with international engineering firms. Currently, Dr. El-Reedy is responsible for reliability, inspection, and maintenance strategies for onshore concrete structures and offshore steel structure platforms. He has performed these tasks for hundreds of structures in the Gulf of Suez in the Red Sea.

Dr. El-Reedy has consulted with and trained executives at many organizations, including the Arabian American Oil Compnay (ARAMCO), BP, Apachi, Abu Dhabi Marine Operating Company (ADMA), the Abu Dhabi National Oil Company, King Saudi's Interior ministry, Qatar Telecom, the Egyptian General Petroleum Corporation, Saudi Arabia Basic Industries Corporation (SAPIC), the Kuwait Petroleum Corporation, and Qatar Petrochemical Company (QAPCO). He has taught technical courses about repair and maintenance for reinforced concrete structures and the advanced materials in concrete industry worldwide, especially in the Middle East.

Dr. El-Reedy has written numerous publications and has presented many papers at local and international conferences sponsored by the American Society of Civil Engineers, the American Society of Mechanical Engineers, the American Concrete Institute, the American Society for Testing and Materials, and the American Petroleum Institute. Many of his research papers have been published in international technical journals, and he has authored four books about total quality management, quality management and quality assurance, economic management for engineering projects, and repair and protection of reinforced concrete structures. He received his bachelor's degree from Cairo University in 1990, his master's degree in 1995, and his PhD from Cairo University in 2000.

Preface

When the structural engineer starts work on the design of marine structures, construction, or maintenance, the assignment is like a black box. As most of engineering faculties, especially those for the structural or civil engineering aspects, focus on the design of the residential, administration, hospital, and other domestic buildings constructed from concrete or steel. On the other hand, some courses take care of the design of harbors only. Therefore, my first book about offshore structures covers their design, construction, and maintenance.

The design of offshore structure platforms combines steel structure design methods with the load that applied to the harbor, such as wave, current, and other parameters. However, the design of the platforms depends on the technical practice, which depends on the experience of the engineering company itself.

Therefore, this book focuses on the engineering design calculations for offshore structures, which are the main marine structures in the oil and gas industry. The interfaces between the engineering firm, the construction company, and the installation company, and their effect on the engineering design, is essential to address here.

Most of the readers of my first book are asking about design calculations for the marine structures. Nowadays, most engineering structure design is performed by software. Therefore, manual calculation is covered by examples in each chapter, and Chapter 8 discusses, step by step, how to use the SACS software in designing offshore structure.

Managing the engineering phase and coordination among different disciplines along the engineering life cycle is discussed. The engineering management for different stagesmust know the techniques to control time and cost while maintaining the quality of the engineering deliverables, so Chapter 2 has check lists that can be applied for different structure engineering design throughout the whole project.

On the other hand, the construction of steel structure is easily known by the structural engineer, as anyone can see in the construction of a new steel building. But, it is rare to view the construction and installation of an offshore structure platform without playing a direct role in the project. The number of offshore structures worldwide is very few whene compared with the steel structures or normal buildings. Therefore, the major design guide and roles for offshore structure depends on research and development, which is growing very fast to be match the development of worthy business all over the world.

Therefore, all the major oil and gas exploration and production companies support and sponsor more research to enhance the design and the reliability of offshore structures and to enhance the revenue from these petroleum projects and their assets.

The aim of this book is to deeply cover the design calculations of the platform with comprehensive attention to the critical issues that usually face the designer and to provide the simplest design tools, based on the most popular codes by API, ISO, and the other technical standard and practice usually used in offshore structure design.

The lifting calculation is critical, as it is done by cooperation between the engineering firm, the construction company, and the installation contractor, so a specific chapter concerns lifting with examples.

The boat landing, riser guard, and helipad require special design precautions with different approach, so these are discussed in detail with an examples for calculation.

In addition, it is also important to focus about the ways of controlling and reviewing the design. which most engineers facein the reviewing cycle, so this book covers the whole scope of the offshore structural engineer's activities whatever his or her role in the big theatre.

Further, the offshore project life cycle is identified from the viewpoint of the owner, engineering firm, and contractor, so the structural engineer or the project engineer has an overview about the relation between the structural system and its configuration from economic and engineering point of view.

This book is intended to be a guidebook to junior and senior engineers and for anyone who is working in the design, construction, repair, and maintenance of fixed offshore structure platforms. It provides guidance for an overview and practical help for the traditional and advanced technique in design of fixed offshore structure platforms.

Mohamed Abdallah El-Reedy
http://www.elreedyma.comli.com

Introduction to Offshore Structures 1

1.1 Introduction

Marine structures and specifically offshore structures have special characteristics from the economic and technical points of view. Oil and gas production depend on offshore structure platforms, which relates directly to global investment, as it has a direct effect on the oil price.

From a practical point of view, because of increasing oil prices, as happened in 2008, many offshore structure projects were begun.

From technical point of view, offshore structure platform design and construction is a merger between steel structure design and harbor design, and a limited number of engineering faculties focus on offshore structural engineering as the design of fixed offshore platforms, floating and other types. On the other hand, the number of offshore structural projects is limited relative to conventional steel structure projects, such as residences, factories, and other buildings, which depend on the continuous research and worldwide studies over a long time period. Also, no available textbook covers the calculations for these types of offshore structures, so that is the aim of this book.

All the major multinational companies working in the oil and gas industry provide a wealth of information from research and studies in offshore structures.

All major oil and gas companies continuously support research and development to enhance the capability of the engineering offices and construction contractors they deal with to support their business needs.

1.2 History of offshore structures

As early as 1909 or 1910, wells were being drilled in Louisiana. Wooden derricks were erected on hastily built wooden platforms constructed on top of wood piles.

Over the past 50 years, two majors categories of fixed platforms have been developed, the steel template type, which was pioneered in the Gulf of Mexico (GoM), and the concrete gravity type, which was first developed in the North Sea. Recently, a third type, the tension leg platform, was used due to the need to drill wells and develop gas projects in deep water. In 1976, Exxon installed a platform in the Santa Barbara Channel to water depth of 259 m (850 ft).

Marine Structural Design Calculations. DOI: http://dx.doi.org/10.1016/B978-0-08-099987-6.00001-5

There are three basic requirements in designing fixed offshore platforms:

1. Withstand all loads expected during fabrication, transportation, and installation.
2. Withstand loads resulting from severe storms and earthquakes.
3. Provide functional safety as a combined drilling, production, and housing facility.

Around 1950, while the developments were taking place in the GoM and the Santa Barbara Channel, the BP company engaged in similar exploration on the coast of Abu Dhabi in the Persian Gulf. The water depth is less than 30 m (100 ft), and the operation has grown steadily over the years.

In the 1960s, hurricanes in the GoM caused serious damage to offshore platforms, so reevaluation of platform design criteria was strongly needed. The hurricane history of the GoM follows:

- In 1964, Hurricane Hilda, had wave heights of 13 m and wind gusts up to 89 m/s. This 100-year storm destroyed 13 platforms.
- The next year, another 100-year storm, Hurricane Betsy, destroyed three platforms and damaged many others.
- For that, designers abandoned the 25-and 50-year conditions and began to design for a storm recurrence interval of 100 years.

1.3 Overview of field development

Estimating future oil reserves in different areas of the world is based on geological and geophysical studies and oil and gas discoveries. As of January 1996, about 53% of these reserves were in the Middle East, which could be a reason for the political troubles in that area. Noting that, 60% of all reserves were controlled by the Organization of Petroleum Exporting Countries (OPEC). This explains why OPEC and the Middle East are so important for the world's current energy needs

In reality, all the companies or countries have a good assessment of the undiscovered reserves in the Middle East and the former Soviet republics. Most researchers believe that the major land-based hydrocarbon reserves are already discovered and most significant future discoveries are expected to be in offshore, arctic, and other difficult-to-reach and -produce areas of the world

Geological research found that North America, northwest Europe, the coastal areas of West Africa, and eastern South America appear to have similar potential for deepwater production. During an early stage in geological history, the sediments were deposited in basins with restricted circulation, which were later converted to the super source rocks found in the coastal regions of these areas. The presence of these geological formations gives us the initial indication for the discovery of hydrocarbons. Before feasible alternatives for producing oil and gas from an onshore field are identified and the most desirable production scheme is selected, exploratory work defining the reservoir characteristics have to be completed. First, a decision has to be made whether an offshore location has the potential for hydrocarbon reserves. This assessment is usually performed through a study of geological formations by geologists and geophysicists

The geologists and geophysicists must decide that this field could be economically viable and further exploratory activities are warranted. This decision involves preparing cost, schedule, and financial return estimates for selected exploration and production schemes. After that, they must compare several of these alternatives and identify the most beneficial one.

Due to the absence of detailed information with respect to the reservoir characteristics, future market conditions, and field development alternatives, the experts have to make judgments based on their past experience. The total cost and schedule estimates are based on the data available to the company from its previous history regarding this type of project. The power of the oil and gas companies depends on their expertise in these decisions, so most of these companies keep such expertise within the company and compete with each other to steal them. Sometimes, these data do not help enough, so the decision can result from brainstorming sessions attended by experts and management and be greatly affected by the company culture and past experience.

The reservoir management plan is affected by the reservoir and produced fluid characteristics, Reservoir uncertainty may regard size and topography, regional as well as national politics, company and partner culture, and the economics of the entire field development scheme. Well system and completion design are affected by the same factors that affect the reservoir management plan, except perhaps political factors. Platforms and facilities for process and production, storage and export are affected by all these factors.

The following factors affect the decision on the field development:

- Reservoir characteristics.
- Production composition (oil, gas, water , H_2S, and other).
- Reservoir uncertainty.
- Environment at the water depth.
- Regional development status.
- Local technologies available.
- Politics.
- National politics.
- Partners.
- Company culture.
- Schedule.
- Equipment availability.
- Construction facilities availability.
- Market availability.
- Economics.

If the preliminary economic studies in this feasibility study are positive, seismic data generation and evaluation by geophysicists follow. These result in reasonable information with respect to the reservoir characteristics, such as its depth, spread, faults, domes, other factors and an approximate estimate for recoverable reserves of hydrocarbons.

If the seismic indications are positive and the decision is to explore further, exploratory drilling activities commence. Depending on the water depth, environment, and a suitable exploration scheme is selected. A jack-up exploratory unit is suitable for shallow water depths. In water depths exceeding 120 m (400 ft), ships or semisubmersible drilling units are utilized. In the case of 300 m (1000 ft) depths, floating drilling units require special mooring arrangements or a dynamic positioning system. Noting that, a floating semisubmersible drilling rig capable of operating in 900 to 1200 m (3000 to 4000 ft) water depths is needed.

Delineation of the exploratory drilling work follows the discovery well. This generally requires three to six wells drilled at selected points of a reservoir. These activities and production testing of the wells where oil and gas are encountered give reasonably detailed information about the size, depth, extent, and topography of a reservoir, such as the fault lines, impermeable layers and their recoverable reserves, viscosity (API grade), liquid properties (such as oil/water ratio), and other impurities (such as sulfur or other critical components).

Reservoir information enables the geologists and geophysicists to suggest the best location and number of wells required to produce a field and volumes of oil, gas, and water production. This information is needed to estimate the type of production equipment, facilities, and the transport system necessary to produce the field.

So, it is obvious that the accuracy of reservoir data has a major impact on selecting a field. In marginal or complex reservoirs, the reliability of the reservoir data and the flexibility of a production system to accommodate changes in reservoir appraisal becomes a very desirable feature. In summary, the selection of the type of the structures and the number of wells and the platform locations depend on the reservoir.

Offshore structure design in engineering firms has two categories, green field (new construction) and brown filed (existing structure). Nowadays, the oil and gas platforms face the problem of aging, as many platforms were constructed over 40 years ago. Note that the API was started in 1969 as a design guide for a fixed offshore structure platforms. So, all the old platforms were designed according to the engineering office experience and now they need some strengthening or monitoring. Around 50% of workhours worldwide are spent on strengthen or assessing existing platforms, with major rehabilitation projects run worldwide.

It is worth remembering the world of 40 years ago. Companies were not using computers as they do these days. All the reports and drawings were only hard copies, which were destroyed over time, and there was no management of change policy as we apply it now. Based on that, you can find a drawing that does not match the actual condition, in other words, no as-built drawings are available. This book covers green field and brown field design calculations.

1.4 Types of offshore platforms

There are different types of fixed offshore platform as follow which are depending on the function of this platform:

- Drilling/Well-protector platforms
- Tender platforms

- Self-contained platforms (template and tower)
- Production platforms
- Flare jacket and flare tower
- Auxiliary platforms

Every kind of these types of platforms has its characteristics from functionality point of view and these platforms classifications will be as follow:

1.4.1 Drilling/well protected platforms

Oil and gas wells are drilled from this type of platform, so the rig approaches the platform to drill the new wells or perform work throughout the platform's life.

Platforms built to protect the risers on producing wells in shallow water are called *well-protectors* or *well jackets*. Usually a well jacket serves from one to four wells.

1.4.2 Tender platforms

The tender platform is not used as commonly now as it was 40 years ago. This platform functions as the drilling platform, but in this case, the drilling equipment rests on the platform topside to perform the job, but now, the jack up rig is normally used.

In tender platforms, the derrick and substructure, drilling mud, primary power supply, and mud pumps are placed on the platform.

As already mentioned, these types of platform are not seen in new designs or new projects but you can find old platforms like these still in use, so you must have information on them in case of assessing the drilling platform.

On the main deck of a tender platformare two rail beams in parallel, facing in the long direction of the platform, used as a railway to the tender tower above the deck for drilling activity. So, if you perform an integrity assessment for an existing platform, it is important to define whether the is railway beam exists physically and check the original drawings, as it may be removed over time.

1.4.3 Self-contained platforms

The self-contained platform is a large and usually has multiple decks, which provided adequate strength and space to support the entire drilling rig with its auxiliary equipment, crew quarters, and enough supplies and materials to last through the longest anticipated period of bad weather, when supplies could not be brought in. There are two types: the template type and the tower type.

1.4.4 Production platforms

Production platforms support control rooms, compressors, storage tanks, treating equipment, and other facilities.

1.4.5 Quarters platforms

The living accommodations platform for offshore workers is commonly called a *quarters platform.*

1.4.6 Flare jackets and flare towers

A flare jacket is a tubular steel truss structure that extends from the mud line to approximately 3–4.2 m (10–13 ft) above the mean water line (MWL). It is secured to the bottom by driving tubular piles through its three legs.

1.4.7 Auxiliary platforms

Sometimes, small platforms are built adjacent to larger platforms to increase the available space or to permit carrying heavier equipment loads on the principal platforms. Such auxiliary platforms have been used for pumping or compressor stations, oil storage, quarters platforms, or production platforms. Sometimes, they are free standing; other times, they are connected by bracing to the older structure.

1.4.8 Bridges

A bridge 30–49 m (100–160 ft) in length that connects two neighboring offshore structures is called a *catwalk*. The catwalk supports pipelines, pedestrian movement, or a bridge of materials handling. The different geometries of bridges is shown in Figures 1.1, and Figure 1.2 presents photo of a bridge between two platforms.

1.4.9 Helidecks

A helideck is the landing area of the helicopter, as shown in Figure 1.2, so it must be larger enough to handle loading and unloading operations. The square heliport has a side length from 1.5 to 2.0 times the expected length of the largest helicopter. The heliport landing surface should designed for a concentrated load of 75% of the gross weight. The impact load is twice the gross weight of the largest helicopter, this load must be sustained area 24" × 24" anywhere on the helideck surface.

1.5 Types of offshore structures

The types of offshore platforms differ from structure system point of view, which developed over time from the requirement to obtain the oil and gas in locations with greater water depth.

A concrete gravity platform was constructed for Shell in 1997. Most of concrete gravity platforms are in the North Sea. Some have been developed in Canada, the last one, with depth water 109 m, is under construction for Exxon Mobil.

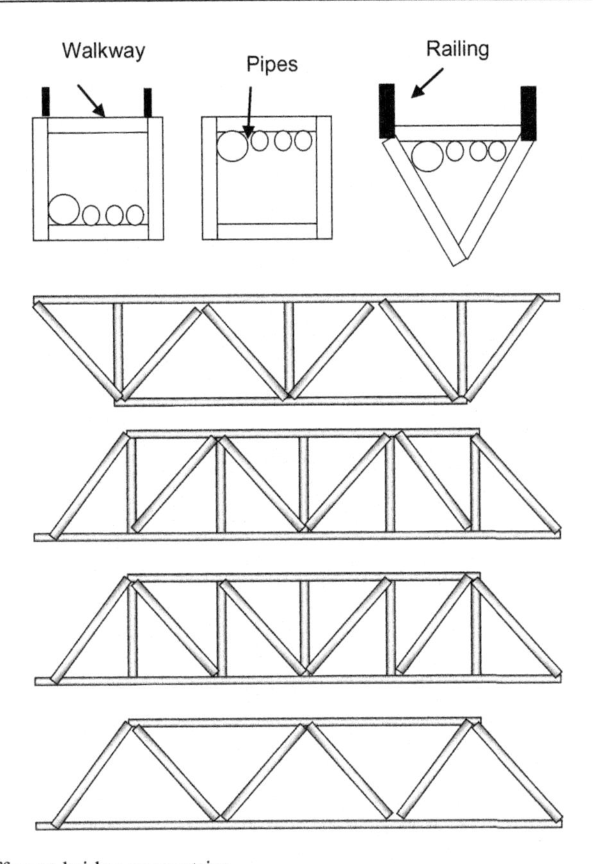

Figure 1.1 Different bridge geometries.

Figure 1.3 shows a well protector platform with four legs, which is a satellite platform. The complex platform, which consists mainly of a production, drilling platform and an auxiliary platform, connected by a bridges.

In some areas, the reserves are low, so only one well needs to be drilled. Many solutions were developed to solve this problem and obtain the business target. One is to have a subsea well connected to the nearest platform by a pipeline. This solution is costly but it is now used widely in case of a deep water.

Another solution is to use a minimum offshore structure, as shown in Figure 1.4. The platform uses the conductor itself as the main support to the small deck. Also, two diagonal pipes are connected to the soil by piles, as shown. The shape of the topside for this three-legged platform as shown in Figure 1.5.

The first oil floating production, storage, and offloading (FPSO) vessel was the Shell Castellon, built in Spain in 1977. The first ever LNG carrier (Golar LNG–owned Moss-type LNG carrier) converted into an LNG floating storage and regasification unit was carried out in 2007 by the Keppel shipyard in Singapore.

Figure 1.2 Bridges connecting a helideck platform.

In the last few years, LNG FPSOs have also been launched. An LNG FPSO works under the same principles as an oil FPSO, but it produces only natural gas, condensate, or LPG, which is stored and offloaded.

Floating production, storage, and offloading vessels are particularly effective in remote or deepwater locations, where seabed pipelines are not cost effective. FPSOs eliminate the need to lay expensive long-distance pipelines from the oil well to an onshore terminal. They can also be used economically in smaller oil fields, which can be exhausted in a few years and do not justify the expense of installing a fixed oil platform. Once the field is depleted, the FPSO can be moved to a new location. In areas of the world subject to cyclones, such as northwest Australia, or icebergs, as in Canada, some FPSOs are able to release their mooring or riser turret and steam away to safety in an emergency. The turret sinks beneath the waves and can be reconnected later.

The FPSO operating in the deepest water depth is the Espirito Santo FPSO from Shell America, operated by SBM Offshore. The FPSO is moored in a water depth of 1800 m in the Campos Basin—Brazil and is rated for 100,000 bpd. The EPCI contract was awarded in November 2006 and the first oil delivered in 2009. The FPSO conversions and internal turret were done at the Keppel Shipyard Tuas in Singapore, and the topsides were fabricated in modules at Dynamac and BTE in Singapore

Figure 1.3 A satellite platform.

A tension-leg platform is a vertically moored floating structure normally used for the offshore production of oil or gas and is particularly suited for water depths greater than 300 m (about 1000 ft).

The first tension leg platform (TLP) was built in the north sea for Conoco's Hutton field in 1980s. The magnolia TLP is considered the deepest TLP worldwide. It was constructed in water depth around 1425 m (4674 ft). The hull consists of four circular columns connected at the bottom by rectangular pontoons. Conoco Phillips owns this platform, which is located in the Gulf of Mexico.

Figure 1.6 presents a brief summary of different types of platform structures and their range depending on the water depth and function. Noting that, these ranges are changing with time as there are many researches and development to construct a structure in a deep water.

Nowadays, there is a major trend to use a gas that was rejected over 40 years and was flared but due to its bad impact on environment and a new technology in gas processing and it is economic due to its wide use in industries; there are many projects seek to discover gas for production, so the exploration is extended into deep water, which is not a good match with the conventional steel structure platform. Researchers and the engineering office utilize a leg with a tension wire platform.

On the other hand, many shapes of platform are constructed as a minimum offshore structure, which is one pile by using the conductor itself as a support with two inclined members to support with the pile. The subsea structure is used widely

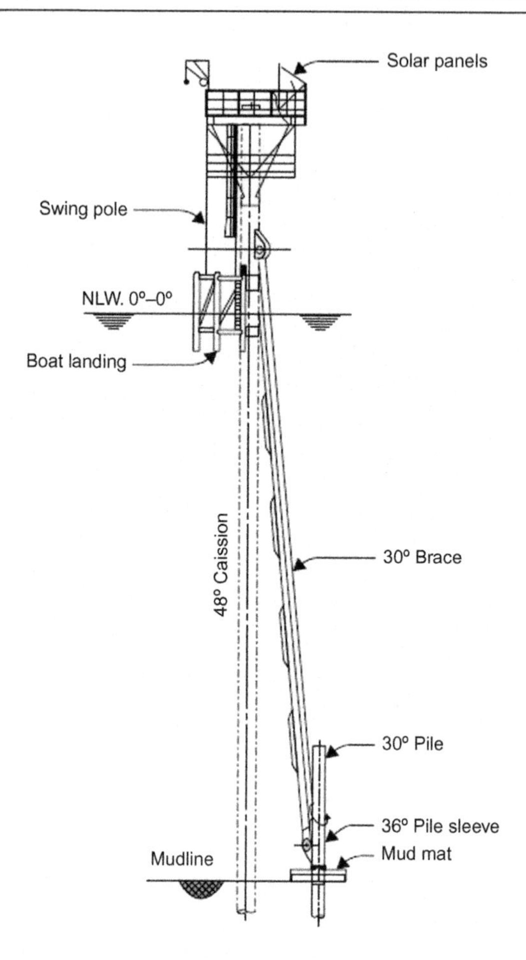

Figure 1.4 Elevation view of a three-legged platform.

for most deepwater drilling as a less-expensive alternative. In addition to that, for over 14 years, projects have constructed concrete gravity platforms using a reinforced concrete structure, and researches has been done on these types of the platform. Conoco Phillips in 1973 started the first concrete gravity platform in the North Sea in the Ekofisk field.

Worldwide, multinational companies in the petroleum industry and the countries that have oil and gas reserves have come to an agreement over control of the reserves. It is important to recognize that most tructural engineers focus on stresses, strain, structural analysis, codes, and design standards, which are the main elements of their job but really not all their job. It is important to be challenged to create a solution, select the structural configuration, and satisfy the owner's requirements and expectations. As contractors, the engineering firms and the engineering staff of the owner's organization should be on the same page to achieve the owner's target and goals.

Figure 1.5 Three-legged platform of Figure 1.4.

The company targets and policies are based on its business targets and profit, the expected oil and gas reserves, expected oil and gas prices, and the last important factor the country that owns the land and this reserve. Therefore, the terms and condition and the political situation of this country direct the investment of a project. Any engineer working on this type of project should have a helicopter view of all the constraints around the project, as these constraints are a guide to the engineering solutions, options, and alternative design; and it is very important that this overview be known and communicated by senior management to the junior staff.

The design and installation of the platform requires an interface among different engineering disciplines and the construction contractor, which is addressed in Chapter 2. The project life cycle and the cost estimate in each stage is discussed.

The load calculation is the first step in any structural analysis. The loads on the platforms are usually defined by the owner specifications. Some studies govern the loads and other technical practices, which are discussed in detail in Chapter 3, with a practical example for calculating these loads. The wave theory and the wave, wind, and seismic loads are addressed in that chapter in detail with examples.

To start design of the steel structure platform, an engineer must understand the main principles of the steel structure design and the effect of the environment and the major loads on the offshore structures, all these design principles are illustrated within the scope of the API, ISO19902, and the ISO in Chapter 4 with a solved

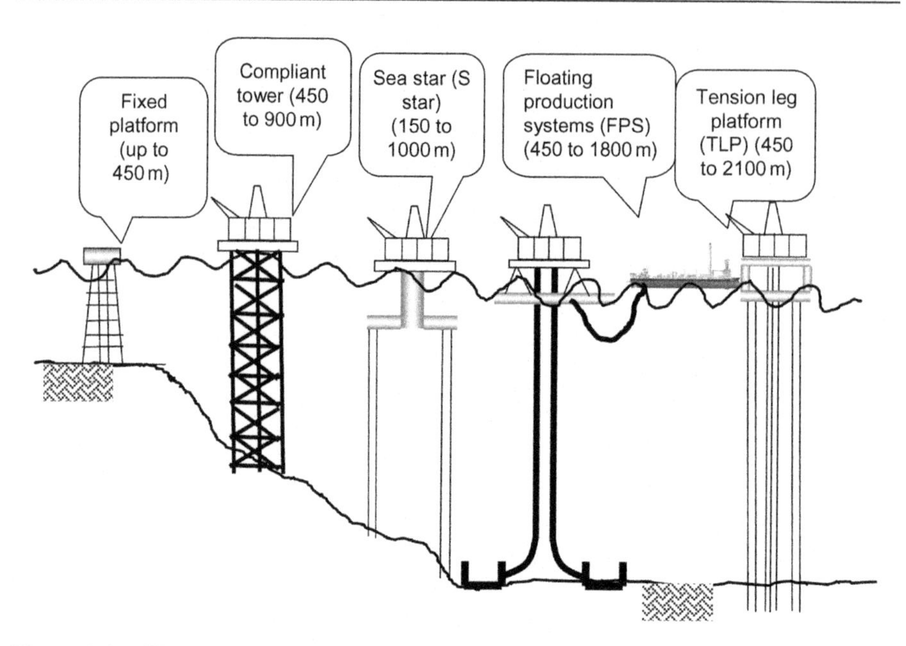

Figure 1.6 Different types of offshore structures.

example of designing different platform members. Chapter 5 covers the design of the helidecks, boat landings, and riser guards.

This foundation is a critical element of structure safety and includes many variations and uncertainties. Therefore, Chapter 6 presents research and studies performed to estimate the pile capacity in offshore structure on sand and clay soil. Chapter 6 illustrates the geotechnical investigation and presents many example of calculating the pile capacity, discussing all the tools and data required to estimate the pile capacity by the different researchers.

During construction and installation, lifting is the main factor that affects the design and needs interface among the design office, the construction crew, and the installation contractor. Also, a marine wire survey is needed for insurance guarantees, so Chapter 7 discusses the lifting process and calculation with examples.

The last chapter presents, step by step, the SACS software, which is the most traditional software use in the market for offshore structure. This chapter is designed to assist the structural engineer in using this software and understanding its main features. The appendix contains an assessment to to test the reader's knowledge all the subjects covered. The answers are found on the website www.elreedyma.comli.com.

Further reading

International Organization for Standardization, Petroleum and Natural Gas Industries. 2004. Offshore Structures. Part 2. Fixed Steel Structures. ISO/DIS 19902.

Engineering Management for Marine Structures

2

2.1 Overview of field development

Offshore structures have special economic and technical characteristics. So managing this type of project needs some special skills and an overview of all projects phases. Economically, offshore structures are dependent on oil and gas production, which is directly related to global investment, which is in turn affected by the price of oil. For example, in 2008, oil prices increased worldwide, as a result, many offshore structure projects were started during that time period.

During this phase, due to the absence of detailed information about reservoir characteristics, future market conditions, and field-development alternatives, experts rely on their past experience and schedule estimates based on data available from previous history. The success of oil and gas projects depends on this expertise, so most of companies keep experts on hand and compete with each other to recruit them.

2.1.1 Project cost and the life cycle

Field development for a new project or for the extension of existing facilities is a multistep process. The first step involves gathering input data, such as the reservoir and environmental data; the selection and design of major system components, such as the production drilling and the wells, facilities, and offtake system; in addition to the decision criteria, such as the economics. The next step is evaluating the different field-development options that satisfy the input requirements and establishing their relative merits with respect to the decision criteria. In this design process loop, not only alternatives for field-development systems but also alternatives for each system need to be taken into account.

For the next stage, a conceptual design for the selected system is required. In this phase, the selection activity focuses on the system components and detail elements. During this phase, design iterations are generated until all the engineering team member disciplines and operation teams are satisfied from a technical point of view. All the system components and construction activities must be well defined. Once the design is completed, few changes to the system and its components can be made without suffering delays and cost overruns.

All project activities that precede the start of the detailed design phase are considered the FEED (front-end engineering design) phase. FEED is the most important phase of a project life cycle.

Marine Structural Design Calculations. DOI: http://dx.doi.org/10.1016/B978-0-08-099987-6.00002-7

An ideal field-development schedule should allow sufficient lead time for the performance of all FEED work before detailed design starts.

Experience shows that the FEED phase will identify viable options; develop, evaluate, and select concepts; and provide a conceptual design. The FEED phase usually consumes only about 2–3% of the total installed cost (TIC) of field development and has the highest impact on cost, schedule, quality, and success. It is common to observe major cost overruns when a full FEED phase is not performed. Reanalysis of projects that did not have a satisfactory FEED phase because of political factors indicates that a 50% TIC reduction could have been achieved if a satisfactory FEED phase had been performed.

Figure 2.1 shows the variation in the accuracy of the TIC estimates in different phases of a project. In general, understanding the economics and other features of a field-development system improve as we move along the project life cycle. At the start of the FEED phase, a number of options are available and identifying the right field-development concept profoundly influences a project's success. During the conceptual design phase before FEED, each system component, such as well systems, platform(s), topside facilities, transportation, and their subcomponents, such as hull, the mooring system, tethers, living quarters, processes, utility systems, pipelines, storage, and risers, are defined; also a cost and schedule estimate is prepared.

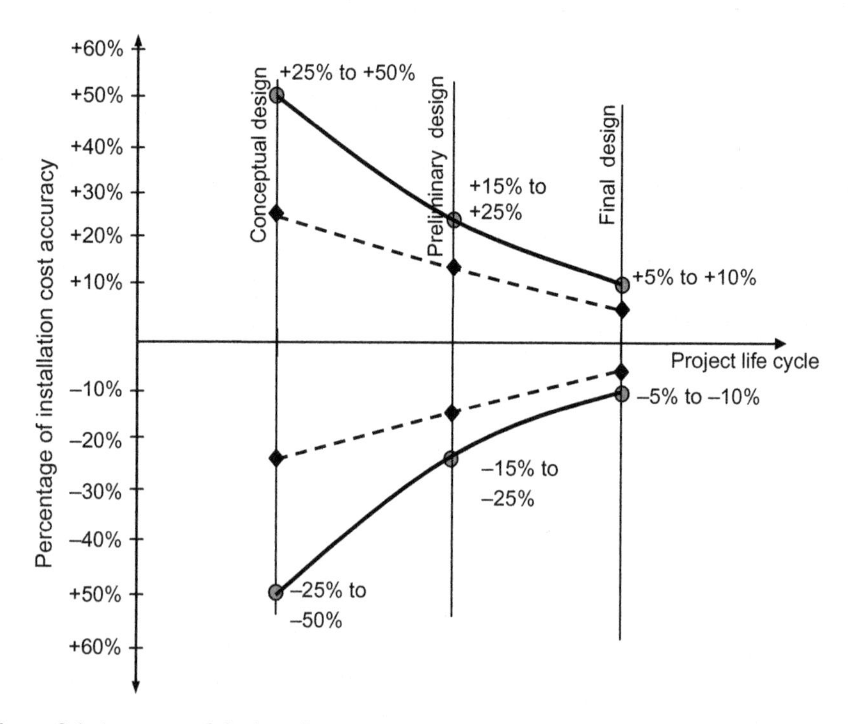

Figure 2.1 Accuracy of the installation cost at every project stage.

Selection and definition of the system components and subcomponents also have a significant impact on costs and the schedule. At this phase, the accuracy of the TIC estimates is approximately ±15% to ±25%.

The ability to influence cost and savings decreases with progress along the project life cycle. At the concept development stage, the selection and development of the right concept have a major impact on the TIC. Savings in the detailed design and construction phases generally stem from good project control and execution, which result in TIC optimization. The final design phase provides detailed engineering analysis and design, approved-for-construction (AFC) drawings and fabrication, transportation, installation, precommissioning, and the hookup and commissioning of the field by the contractor. Efficient project management in the execution of the plan; cost, schedule, and quality control; verification; quality, safety assurance; purchasing; and documentation have some effect on the TIC but not as profound an effect as the FEED phase. Past experience indicates that, at the start of the construction phase, the accuracy of the TIC estimates is approximately ±5% to ±10%.

2.1.2 Concept and screening selection

At the concept stage, external factors, such as country requirements and characteristics, technology transfer, and environmental pollution potential, as well as the culture, politics, economics, and infrastructure of the country that owns the land and the operating oil company and its partners may have major influences on the concept of selection. Not easily comparable criteria, such as the economics, design completeness maturity, and external factors, have to be weighed against each other and used for ranking and selection. In a screening process, first of all, the goal of the exercise is defined. Then, the viable field-development options are identified. This is followed by the identification of a multitude of selection criteria, which are grouped and ordered in a hierarchy, after which experts determine the importance of each criterion by comparing them. The comparisons are then passed through an analytical process to obtain rankings for each comparison and its alternatives.

The basic design defines the platform, production facility, structural configurations, and dimensions in enough detail to allow the detailed design to start. Basic design results enable reliable cost and schedule estimates and the ordering of long-lead major item and structural components.

The conceptual drawings show major component configurations for platforms, topside facility layouts, well locations and well systems, reservoir maps and production profiles, storage (if needed) and offloading systems, pipelines to shore, and preliminary sizes.

A general platform structure configuration is defined by conceptual drawings that show side elevations, plans for legs, and major bracing. Preliminary process flow diagrams (PFDs) and major equipment lists may also be available. A concept cost is estimated in average (±40%) as an order of magnitude.

A conceptual design package that includes this information is prepared by the owner or specialized engineering team (this called the *appraise phase*) and is given to the design contractor as input to the preliminary design phase to start the FEED stage.

2.2 FEED engineering phase

In the FEED phase, after the facilities required on the topside of the deck have been defined, the geometrical shape of the platform is defined. A preliminary structure analysis run through structure analysis software to identify the sections that match the loads and the geometrical shape with appropriate deck dimensions that serve the required facilities on the deck. The structure system of the platform subsea structure, which is called a *jacket*, and a construction method based on the water depth are selected in this phase. In management of a new project, this is called the *select phase*.

The FEED process ends with the completion of the preliminary engineering design. At this point, the FEED phase has provided the following deliverables:

- Basic design drawings for all major platform and deck structures and components (i.e., jacket, deck, piles and conductors). These should contain enough detail to enable reliable field- development cost and schedule estimates. This information is particularly important if owners wish to solicit bids and to enter into a lump-sum EPC (engineering procurement and construction) agreement with a contractor.
- A basis of detailed design (BOD) document for the detailed design phase. The BOD specifies the design requirements, including the platform configuration and environmental parameters (e.g., metocean, seismic, ice).
- Site-specific information (e.g., water depth, temperature, soil characteristics, mudslides, shallow gas pockets).
- Definition of nongenerated loads (e.g., equipment, wet or dry supplies and operating loads, such as dynamic vibrations from rotating machinery, mud pumps and operations). The design life of the structure should be determined, considering the operation requirements and fatigue effect along its lifetime.
- Definition of accidental loads (e.g., collisions with boats, dropped objects, fire and explosions).
- Load combinations (extreme environmental, operational, serviceability, transportation, lift and launch). Damage stability and redundancy requirements (e.g., missing member or flooded leg or compartment).
- Preferred material classes.
- Design regulations, codes, and recommended practices.
- Definition of appurtenances and their locations (e.g., escape and evacuation equipment, escape routes, stairs, boat landings, barge bumpers, conductors, and mud-mats) with different discipline engineers.
- Corrosion protection requirements (e.g., sacrificial anode or impressed current, and types of anodes).
- A narrative of the construction methods and procedures that will affect the platform configuration and size (e.g., skidding and load-out procedures, pulling points, lifting eyes, launch skids) and the verification and certification requirements.

- Any other owner requirements that will affect the detailed design (e.g., jack-up drilling unit clearances, tender rig sizes and weights).

This document, which contains most of the information just listed, is issued to all qualified bidders. It is very important to note that most contractors will not be interested in bidding a lump-sum EPC contract if a basic design package is not available.

The following tasks should also be completed in this phase:

- Preparation of detailed and final process flow diagrams. If a detailed design is attempted without final PFDs, process changes during the final design phase may cause significant rework as well as schedule and cost overruns.
- Preparation of P&IDs (piping and instrumentation diagrams).
- Preparation of final deck and facility layouts, providing adequate space and clearance for all equipment and operations.
- Preparation of equipment and material lists, data sheets, and specifications.
- Preparation of detailed engineering, transportation, precommissioning, and commissioning scopes of work.
- Preparation of detailed design schedules and cost estimates.
- Performance of global in-place analyses to confirm that major structure members and equipment will fit.

2.3 Detail engineering phase

At the end of this phase, the engineering office delivers the full construction drawings and specifications for the whole project, which contain all the details that enable the contractor to execute its function. A huge number of engineering hours make up this phase, so there must be good coordination among the different disciplines. This happens if there is good organization in the engineering office and the client provides a free mode of communication between the different parties through a system channel with continuous coordination.

The complexity of this phase is such that it needs a high-quality system. A system of quality assurance is important at this stage because it organizes the work. The target of the project and each team member's responsibility is clear. The concept of quality is defined by supporting documents.

The documents are regarded as the executive arm of the quality assurance process. For example, any amendment or correction in the drawings should be made through the agreed procedure and system. Moreover, the drawings should be sent within a specified time to the client for review and discussion, with an official transmittal letter to control the process time. Any comments or inquiries should be handled through agreement between the two technical parties; then, the modification should be done by the engineering firm and resent to the client through the same communication procedure.

The development of a system to avoid older copies of the drawings becoming confused with the current copies can prevent human error. The most current set of drawings may be assured by the establishment of a system for continuous amendment of the date and number of the drawings and engineering reviews until the final stage of the project. Approval of the final set of drawings should be sealed with a stamp ("Approved for Construction") indicating that these are the final drawings approved for the construction.

After completion of the detailed engineering phase, the specifications and drawings are ready to be used in the execution phase. You can imagine that, in some projects, the documents may reach hundreds of volumes, especially the specifications, operation manual, and volumes of maintenance and repairs.

2.4 Engineering design management

The goal of engineering management is to control the design stage to provide high quality at a better price on time. The design input comprises all technical information necessary for the design process. The basis of this information is the owner through the SOR (statement of requirements) document, so the engineering firm should review this document carefully and, if there is any confusion or misunderstanding, it should be clarified in the document and through meetings. The main item in the SOR is the scope of work, so work required must be described in detail and the borders for this scope of work provided. For example, the available crane capacity and the way of lifting must be specified, as the lifting design during construction or installation depends on them, as we discuss in later chapters. On the other hand, in case of jacket installation, the maximum barge crane capacity information should be transferred to the engineering firm and the launching runner in case of jacket launching technique. In addition to that, the sea fastening analysis depends on the barge data, so who do what is very important to be defined early and the data available to the engineering firm in the design process. The engineering firm needs the following data:

- Instructions for control of the design, whether the client controls the whole process or requests some specific actions (e.g., a representative from the audit during the design phase) are often provided in the contract.
- The designer must take into account the materials available in the local market in relation to the project and its location and match these with the capabilities of the owner. The designers must have a realistic view as well as full and up-to-date knowledge of the best equipment, machinery, and available materials.
- The design must be matched to the project specifications, the permissible deviations and tolerances should be in accordance with the specifications and requirements of the owner.
- Health, safety, and the environment are t critical subjects these days, so every design should conform to health, safety, and environmental regulations.
- The computer is one of the basic tools used now in the design process. As well as in the recording and storage of information, computers offers the ability to change the design of the work easily, by using computer-aided design (CAD) software to obtain more precise information with the access to various tables and diagrams.

- The design output must be compatible with all design requirements, and the design should be reviewed through an internal audit. The new design should be compared with old designs that have been approved before for similar projects. This is a simple procedure for checking designs. Any engineering firm should have a procedural checklist for reviewing designs.

Audits or reviews of the design are intended to be conducted on a regular basis at important stages in the design. The audit must require complete documentation and can take analytical forms, such as the analysis of collapse with an assessment of the risk of failure.

In addition to that, the owner is responsible for collecting all the expectations and the requirements in detail from all the stakeholders and specifically from the representatives of the operations phase, as they are the main client to the project or engineering project managers. All their requirements should be discussed with them and defined in the engineering phase. It is important also to have an operations representative review the drawings specific for process, piping, and mechanical parts to ensure they agree with their expectations and requirements, as that person represents the end user for the project.

The main challenge for the engineering firm and its engineering manager or project manager in the design phase is make sure the work of the different disciplines match the project objective and target. The project manager has two meetings that are very important, the first meeting is after obtaining the bidding document form the client to put price. This meeting is with all the discipline departments or leaders and clearly presents the scope of the project to them. All the discipline leaders must attend the meeting, so they know the scope and borders of the project to minimize errors estimating workhours. The second meeting occurs after winning the contract, to start the work and raise the prospect of any risk during implementation.

2.4.1 Engineering stage time and cost control

Most of engineering contracts depend on cost time and resources (CTR) sheets, which is specifies the cost of time and resources to be delivered from the engineering firm to the client to review and approve. This sheet is very essential even for small projects, as it defines the deliverables that will be received and the cost and time for each deliverable. A sample CTR form is shown in Table 2.1.

2.4.1.1 Time schedule control

We use a real project as an example: a four-legged platform in a water depth around 80 m. The percentage of workhours of different disciplines could be as in Table 2.2.

The approximate workhours for structural engineering activity is about 5600–6300 hours, whereas the total project workhours is about 38,000–44,000 hours. Noting that workhours reflect money, the cost is around for this project is \$2 million. If a firm has a high fee rate per workhour, the workhours may be less

Table 2.1 **Sample CTR form**

Date	Project name	CTR	Project number	Discipline
Deliverables:				
Input Data				
Comments				

Item	Activity	Discipline	Workhours	Cost/hr	Cost
Total cost					

Table 2.2 **Percentage of workhours in the engineering phase**

Discipline activity	Percentage of workhours
Project management	25%
Process	12%
Mechanical	7%
Piping and layout	16
Electrical	8%
Control and Instrument	3%
Structural engineering	17%
Pipeline	5%
HSSE	7%

Table 2.3 Engineering report progress measurement guideline

Activity	Progress percentage
Issue draft to the client	30%
Client reviews completed document	50%
Incorporate client comments	80%
Approve final report	100%

Table 2.4 Engineering preparation for a P&ID

Activity	Progress percentage
Study and prepare draft or sketches	20%
Completed document reviewed by the client	60%
Issue for design	75%
Issued for construction finally without hold	100%

Table 2.5 Procurement progress measurement

Activity	Progress percentage
Issue RFQ to bidders	20%
Bid evaluation and selection	30%
Issue purchase order to vendor	40%
Approval of vendor prints	50%
Materials successfully passes tests or inspection	85%
Materials ready for FOB	90%
Materials received on site	100%

than in the table. As it depends on the engineering firm's approach in its proposal, some engineering firms reduce the number of workhours and increase the cost per workhour and vice versa, according to the company's strategy and the market share in this time period.

The progress of engineering activity must be identified if it is to be controlled, so some measurement of progress should be identified and agreed upon by the client and the engineering firm that will do the work before starting the project. Table 2.3 is an example for the progress measured in study reports.

The measured progress should be defined for different engineering discipline activities. In an oil and gas project, preparation of a P&ID is critical, so the most practical progress measurement will be as in Table 2.4. For procurement services, Table 2.5 presents the traditional progress measuring percentage.

Table 2.6 **Manufacturing delivery progress measuring guideline**

Manufacturing and delivery milestones	Equipment	Bulk materials
All vendor drawings and vendor data approved	10%	
All sub ordered materials received by vendor	25%	10%
Materials successfully inspected, tested, and accepted	50%	75%
All materials delivered on site	15%	15%

Table 2.7 **Overall project progress measuring guideline**

Activity	Percentage
Detailed engineering	10%
Procurement services	5%
Delivery of equipment materials	25%
Construction	25%
Installation	25%
Professional services	10%

The requisition of bulk materials, like structural steel, piping, or electrical cables is based on the materials takeoffs (MTOs), and in most cases, three MTOs are issued, the following weights are to be applied for bulk MTOs:

First MTO requisition 65%.
Second MTO requisition 25%.
Third and final MTO requisitions 10%.

For equipment delivery, the milestone progress will be as in Table 2.6, which defines the percentage progress after finishing the milestones. For overall project progress, the measurements in Table 2.7 can be considered a guideline.

2.4.1.2 *Engineering cost control*

The most expensive project engineering phase accounts for 10% of the total project cost. It is worth mentioning that most of the cost is in design and drafting, which is around 57% of the total engineering cost. Table 2.8 presents a breakdown of the engineering phase costs. In addition, the engineering contractor's fee or the administration overhead and profit cost usually average 5%. Table 2.8 shows that design and drafting are the most significant items of technical workhour cost, so in most cases, it is best to calculate the required design and drafting cost, then develop the other items as a percentage.

2.4.2 *Engineering interfaces*

The greatest potential risk in any project is managing the interface among all the stakeholders and all the departments work in this project. The main goals of project

Table 2.8 **Breakdown of engineering phase costs**

Activity	Percentage of the engineering cost
Design and drafting	57%
Proposal	1%
Project management	19%
Procurement	12
Structural engineering	4%
Project control, estimating, and planning	7%

management are to reduce miscommunication and manage the interfaces in an appropriate manner. In case of engineering procurement and construction, the contract will reduce some of these risks to the owner. In most mean construction and installation for marine projects, the EPC comes after the FEED engineering phase, so it is very important to have an interface between EPC contractor and the engineering firm during the FEED phase, and it is necessary that this phase begin only after an execution strategy plan is established, noting that this document is dynamic throughout the project life.

The interface here is managed by the contractor project manager, who coordinates the interface among different departments in the company. In case different contractors are used for detailed engineering, construction, and installation, the company's owner defines exactly who does what and when in the tender package. This should be done through an assignment matrix after discussion with potential construction and installation contractors; but the detailed engineering in most cases will be done by the same people who manage the FEED phase, so the interface between those handling two phases is not a problem. If the detailed engineering is done by another company, this should be decided early on. The best solution is to let the company that will manage the detailed phase receive the documents ordering both the FEED and detail phases. Any money spent on the additional review will be saved during the FEED stage and will avoid the hassles of arguments, redoing the work, and changing orders.

2.4.3 Structural engineering quality control

An offshore structure is complicated and not easy sto design, so it requires many quality checks. One method is to use checklists to ensure that all the factors are taken into account in the design. Tables 2.9 through 2.12 are examples of such checklists.

The calculation report delivered to the client should include, as a minimum, the following data:

- Calculation cover sheet.
- Contents page.
- Description of the analysis methodology and techniques.
- Explanation of the model geometry and axis system.

Table 2.9 Checklist for jacket in-place analysis

Items	Check point	Yes or no
Computer model		
1.	Framing dimensions according to the drawings or sketches	
2.	Framing elevations according to the drawings or sketches	
3.	Water depth and mud line elevation match with the basis of design	
4.	Member group properties: E, G, density match with the basis of design	
	a. Section details, segment lengths, member offsets	
	b. Corrosion allowance in splash zone	
	c. Zero density for wishbone elements in ungrouted jackets	
	d. Grout density corrections for grouted leg and pile	
5.	Member properties: Kx, Ky, Lx, Ly	
6.	Member end releases where applicable	
7.	Flooded members (legs, risers, J tubes, caissons, etc.)	
8.	Dummy members (relevant joints kept, rest deleted)	
9.	Plate and membranes modeled correctly	
10.	Drag and inertia coefficients (smooth, rough members)	
11.	Marine growth data as per basis of design	
12.	Member and group overrides	
	a. No wave load and marine growth on piles and wishbones in jacket legs	
	b. Enhancement of Cd, Cm for anode supported members	
	c. Enhancement of Cd, Cm for jacket walkway members	
	d. Enhancement of Cd, Cm	
13.	Hydrostatic collapse check selected with redesign option	
14.	Allowable stress modifiers for extreme storm load cases	
15.	Unity check ranges in the analysis input file	

(Continued)

Table 2.9 **(Continued)**

Items	Check point	Yes or no
Loads		
16.	Load description, calculations and distribution	
17.	Wave theory, wave, current, and wind directions	
18.	Equivalent Cd, Cm calculations for items mentioned above	
19.	Load contingencies match with the basis of design	
20.	Load combinations for operating and extreme cases	
21.	Load summations	
22.	Load summation verification against weight control data	
PSI data and input file		
23.	Units for T-Z, Q-Z, and P-Y data from the geotechnical data	
24.	Pile segmentation data, end bearing area	
25.	With reference to the input format	
	With reference to the input format, check T, Q, and P factors	
Analysis results		
26.	Enclose sea-state summary to be checked against load summation	
27.	Enclose member check summary: review for overstressed members	
28.	Enclose joint check summary: review overstressed joints (check Fy = 2/3Fu for chords of high-strength members)	
29.	Check maximum pile compression and tension	
30.	Enclose model plots: joints/group/section names, Kx, Ky, Lx, Ly, Fy, and loading	
31.	Enclose deflection plots, member unity check ratio plots	
32.	Enclose hydrostatic collapse check reports and check for need of rings	
33.	Review pile factor of safety calculations	
34.	Review permissible deflection calculations	
35.	Review plot plan and latest structural drawings	
36.	Review relevant sections of weight control report	

(*Continued*)

Table 2.9 (Continued)

Items	Check point	Yes or no
Dynamic in-place analysis (if required)		
37.	Determine dynamic amplification factor (DAF) based on single Degree of Freedom (DOF) concept and apply on wave load cards.	
38.	Determine DAF based on Inertia load distribution and apply on total structure	
General		
39.	Joint name range identified for each framing level and sequential	
40.	Member group name specific to each framing level and sequential	
41.	Check for future loads, loads due to specific requirement, such as rigless interventions	

- Explanation of boundary conditions.
- Explanation of the input loads and any supporting calculations showing the development of the loads.
- Load combination matrix.
- Explanation of the stress analysis assumptions, including member effective lengths philosophy.
- Discussion of the results, including deflections, member stresses, joint stresses, reactions, pile capacities, and the like.
- Additional calculations supporting the results, including calculations showing the justification of overstressed members and so forth.
- Attachments should contain the following:
  - Any reference information used for the model.
  - Drawings.
  - Input files (including the model, pile-soil interaction [PSI] input file, and joint-can input).
  - Model plots showing joint numbers and member groups. Member effective lengths and the like may also be presented if required.
  - Selected results, including reactions, summary of maximum member unity check (UC), summary of maximum joint (UC), summary of maximum joint deflections, pile and safety factors, and other relevant results.
  - Calculation checklist and any other checklists that ought to be included, a summary report checklist to be used by the lead engineer, analyst engineer, and checking engineer. This document should be included in the project document.

Table 2.9 is a checklist for jacket in-place analysis, and Table 2.10 is a checklist for topside in place analysis. The checklist for the on-bottom stability analysis is in Table 2.11.

Table 2.13 presents a check list to verify the topside lifting analysis, as the quality assurance to the engineering process in this phase. Table 2.14 is a report checklist.

Table 2.10 **Checklist for topside in-place analysis**

Items	Check points	Yes or no
Computer model		
1.	Framing dimensions and elevations match with the drawings	
2.	Member properties: Kx, Ky, Lx, Ly	
3.	Member end releases where applicable	
4.	Plate and membranes modeled correctly.	
5.	Review the loads in the analysis input file	
6.	Allowable stress modifiers for extreme storm load cases	
7.	Unity check ranges in the analysis input file	
Loads		
8.	Load description and calculations	
9.	Secondary structural item dead load calculations	
10.	Equipment, piping operating and dry load, Electrical and Instrument bulk load calculations	
11.	Wind load calculations, wind area considered.	
12.	Load contingencies	
13.	Load combinations for operating and extreme cases	
14.	Load combinations for local checks.	
15.	Load summations	
	Installation and preservice loads may be applied as a separate load case	
	Crane load cases may be magnified and applied for local checks	
16.	Check for future loads, loads due to a specific requirement, such as rigless interventions	
Analysis results		
17.	Enclose member check summary: review for overstressed members	
18.	Enclose joint check summary: review overstressed joints (check $Fy = 2/3\ F_u$ for chords of high-strength members)	
19.	Enclose model plots: joint/group/section names, K_x, K_y, L_x, L_y, F_y	
20.	Enclose deflection plots, member unity check ratio plots	

Table 2.11 **Checklist for jacket on-bottom stability analysis**

Project:		
Client:		
Items	**Items**	**Yes or no**
I. Computer model		
1.	It is assumed that the model is checked for dimensions, elevations, member group, and section properties in the in-service analysis and is upgraded to suit the current analysis	
	a. Remove pile, appurtenances	
	b. Revise Cd, Cm with member and group overrides as for clean members in the entire structure	
	c. Remove marine growth card from the input file	
	d. Check for flooded members and support conditions	
II. Loads		
1.	It is assumed that the load calculations are verified in the in-service analysis and relevant load cases are picked only for the current analysis	
2.	Wave parameters: installation wave conditions and directions	
3.	Weight of lead and add-on pile sections (before driving in)	
4.	Load contingencies	
5.	Load combinations	
6.	a. Load combination without contingencies + environmental forces	
	b. Load combinations with lead piles in	
	1. Each leg, one at a time + environmental forces	
	2. All legs at the same time + environmental forces	
	c. Load combination with lead + add-on pile in	
	1. Each leg, one at a time + environmental forces	
	2. Two diagonally opposite legs at the same time + environmental forces	
	3. Two opposite legs at a time + environmental forces	

(*Continued*)

Table 2.11 (Continued)

Items	Items	Yes or no
III. Analysis results		
1.	Determine factor of safety (FoS) for sliding and overturning (translate model origin to match load Center of Gravity (COG); basic load case summary is calculated)	
2.	Determine FoS for bearing (translate model origin to match the mudmat COG: combined load case summary is considered)	
3.	Analyze bearing pressure check on the jacket structure and mudmat	
4.	Enclose sea-state summary (basic/combined load case summary)	
5.	Enclose member check report: review overstressed members	
6.	Enclose joint check summary: review overstressed joints (check Fy = 2/3Fu for chords of high-strength members)	
IV. Factor of safety and mudmat design		
1.	Factor of safety against sliding (>1.50)	
2.	Factor of safety against overturning (>1.50)	
3.	Factor of safety against bearing (>1.50 without environmental forces)	
4.	Factor of safety against bearing (>2.00 with environmental forces)	

Table 2.12 Checklist for jacket and topsides transportation analysis

S/N	Items to check	Yes or no
Computer model		
	It is assumed that the model has been checked for dimensions, elevations, and member group and section properties in the in-place /load-out analysis and was upgraded to suit the current analysis	
1.	Member end releases for sea fastening	
2.	Sea fastening material yield strength	
3.	Define conditions for dead and tow load cases	

(*Continued*)

Table 2.12 **(Continued)**

S/N	Items to check	Yes or no
Loads		
	It is assumed that the load calculations have been verified in the in-service and load-out analysis and relevant load cases are picked only for the current analysis	
4.	Remove future loads, operating weights, and live loads; add rigging loads	
5.	Load contingencies in static and tow analysis	
6.	Load combinations in static analysis	
7.	Center of rotation data	
8.	Roll and pitch direction representation	
9.	Consideration of self-weight during inertia load generation.	
10.	Roll and pitch acceleration data	
11.	Coefficients for the lateral load components due to combinations	
12.	Primary and secondary load case identification	
13.	Load combinations	
14.	Chord strength reduction in the joint can Input file (check Fy = 2/3 Fu for chords of high-strength members)	
15.	Allowable stress modifiers	
Analysis results		
16.	Enclose load case summary for dead load case	
17.	Enclose reaction summary from combined analysis	
18.	Cross-check reaction summary values from basic motion equations	
19.	Enclose member check report: review overstressed members	
20.	Enclose joint check summary: review overstressed joints	
21.	Enclose model plots: joint/group/section names, Kx, Ky, Lx, Ly, Fy	
22.	Enclose deflection plots, member unity check ratio plots	
Sea fastening design		
23.	Check for adequacy of base plate connections for sea fastenings (weld size less than the barge deck plate thickness, gusset connections for uplift forces preferred)	
24.	Check for adequacy of doubler plate connection (weld, doubler plate) on the topside members	

Table 2.13 **Checklist for topside lift analysis**

Items	Check point	Yes or no
Computer model		
1.	It is assumed that the model has been checked for dimensions, elevations, and member group and section properties in the in-service analysis and is upgraded to suit the current analysis	
2.	Check the input file and focus on the member group properties for slings as E, G, and density.	
3.	Member end releases for slings	
4.	Member end offsets for slings at pad eye locations	
5.	Member properties: Kx, Ky, Lx, Ly for changed support conditions	
6.	Support conditions for hook point(s)	
7.	Adequate analytical springs are provided	
8.	Modeling of sling and hook for 75–25% sling load distribution (sling mismatch)	
Loads		
1.	It is assumed that the load calculations have been verified in the in-service analysis and relevant load cases are picked only for the current analysis. Remove future loads, postinstalled items, operating weights, and live loads	
2.	Load contingencies	
3.	DAF, skew load factors, consequence factors	
4.	Proportionate distribution of sling loads (75–25%)	
5.	Calculations for COG shift	
6.	Check load combinations, include rigging loads	
Analysis results		
1.	Enclose sea-state summary	
2.	Enclose member check report: review overstressed members	
3.	Enclose joint check summary: review overstressed joints (check Fy = 2/3 Fu for chords of high-strength members)	
4.	Enclose sling forces to allow zero moment and shear and only axial forces	
5.	Enclose deflection plots, member unity check ratio plots	
6.	Weight control report extract enclosed and the weight comparison carried out	

Table 2.14 Report checklist

Task		LE	AE	CE
Model set-up	Latest data and information used			
	Unique model and run name used			
	Geometry			
	Support conditions			
	Member effective lengths			
	Loads			
	Load combinations and AMOD			
	Line-by-line check of input file			
Self-checking	Review model geometry plots			
	Check load sums against load combination matrix			
	Program generated loads reasonable (e.g., dead, wave, wind)			
Postrun verification	Review errors and warning messages			
	Check that reaction totals match load combination matrix			
	Check that deflections are consistent with expected loadings, storm directions, and the like			
	Check pile convergence, stresses, and safety factors			
	Review member stresses			
	Review tubular joint stresses			
	Model and results saved on the server			
	All checks complete and satisfactory			

LE = Lead engineer.
AE = Senior engineer.
CE = Checking engineer.

Further reading

Elreedy, M.A., 2011. Construction Management for Industrial Projects. Wiley-Scrivener, New Jersy, USA.

Project Management Institute Standards Committee, 2013. A Guide to the Project Management Body of Knowledge. Project Management Institute, Upper Darby, PA.

Tricker, R., 1997. ISO 9000 for Small Business: A Guide to Cost-Effective Compliance. Butterworth-Heinemann, New York, USA.

Offshore Structures' Loads and Strength

3.1 Introduction

Fixed offshore platforms are unique structures, as they located in the sea or ocean and their main function is to carry industrial equipment that services oil and gas production and drilling.

Note that the robust design of the fixed offshore structure is dependent on defining all the applied load very well. Most of loads that affect the platform laterally, such as wind and waves, are variable, so we depend on metocean environmental data for the location of the platform.

The structure design depends on the strength of materials from which it was built and the applied load. In general the loads that act on the platform are as follows:

- Gravity loads
- Wind loads
- Wave loads
- Current loads
- Earthquake loads
- Installation Loads
- Other loads

3.2 Gravity load

The gravity load consists of the dead load and the live (imposed) load.

3.2.1 *Dead load*

The dead load is the weight of the overall platform structure, including the piling, superstructure, jacket, stiffeners, piping and conductors, corrosion anodes, decking, railing, grout, and other appurtenances.

Sealed tubular members are to be considered either buoyant or flooded, whichever produces the maximum stress in the structure analysis.

The main function of the topside is to carry the load from the facilities and drilling equipment. The percentage between the weights of the topside component is as

Marine Structural Design Calculations. DOI: http://dx.doi.org/10.1016/B978-0-08-099987-6.00003-9

Table 3.1 Example of the weight and weight percentages of an eight- legged drilling/production platform

	Weight (tons)	Percentage
1. Deck		
Drilling deck		
Plate	72	11
Production deck		
Plate	52	7.8
Grating	1.0	0.16
Subtotal	125	18.8
2. Deck beams		
Drilling deck	174	26.3
Production deck	56	8.5
Subtotal	230	34.8
3. Tubular trusses	**146**	**22.1**
4. Legs	**105**	**15.9**
5. Appurtances		
Vent stack	6	0.9
Stairs	12	1.8
Handrails	4	0.6
Lifting eyes	2	0.3
Drains	6	0.9
Fire wall	4	1.7
Stiffeners	14	2.2
Total	661	100

shown in Table 3.1. From this table one can compute the self-weight of the topside for an eight-legged platform in 90 m water depth.

In calculating the platform's self-weight, a contingency allowance of 5% is considered to cover the variables in these loads.

The function of the jacket is to surround the piles and hold the pile extensions in position all the way from the mud line to the deck substructure. Moreover, the jacket provides support for boat landings, mooring bits, barges bumpers, a corrosion protection system, and many other platform components; virtually all the decisions about the design depend on the number of jacket legs. The soil conditions and foundation requirements often tend to control the leg size. The golden rule in design is to minimize the projected area of members near the water surface in a high-wave zone to minimize the load on the structure and reduce the foundation requirements. Table 3.2 shows the jacket weight for the eight-legged drilling /production platform at a water depth of 91 m.

Table 3.2 **Jacket weight for eight-legged drilling/production platform drilling/production for a 91 m water depth**

ID	Component	Weight (tons)	Percent of total weight	Percent weight of main element
1.	**Legs**			40
	Joint can In between tubular and others	177 309	14.6 25.4	
2.	**Braces**			40.7
	Diagonal in vertical plan horizontal Diagonal in horizontal plan	232 163 100	19.1 13.4 8.2	
3.	**Other framing**			9.8
	Conductor framing Launch trusses and runners Miscellaneous framing	35 82 2	2.9 6.7 0.2	
4.	**Appurtenances**			9.5
	Boat landing Barge bumpers Corrosion anodes walkways Mud mats Lifting eyes Closure plates Flooding system miscellaneous *Total*	28 29 22 16 5 2 2 7 4 1215	2.3 2.4 1.8 1.3 0.4 0.2 0.2 0.6 0.3	100

3.2.2 Live load

Live loads are the loads imposed on the platform during its use, and these may change during a mode of operation or from one mode of operation to another and should include the following:

1. The weight of drilling and production equipment.
2. The weight of living quarters, heliport, and other life support equipment.
3. The weight of liquid in storage tanks.
4. The forces due to deck crane usage.

The owner defines the live load, and normally, it is be included in the statement of requirements (SOR) or basis of design (BOD) documents approved by the owner. Table 3.3 lists some guidelines.

Table 3.3 Guidelines for defining a live load

	Uniform load beams and decking kN/m^2 [lbs/ft^2]	Concentrated line load on decking kN/m′[lbs/ft]	Concentrated load on beams kN[kips]
Walkways and stairs	4.79 kN/m^2 [100]	4.378 kN/m′ [300]	4.44 kN [1]
Areas over 400 ft^2	3.11 kN/m^2 [65]		
Areas of unspecified light use	11.97 kN/m^2 [250]	10.95 KN/m′ [750]	267 KN [60]
Areas where specified loads are supported directly by beams		7.3 KN/m′ [500]	

Table 3.4 Guidelines for decks carrying a live load

Area	Loading (kN/m^2)				Point load (kN)
	Deck plate grating and stringers	Deck beams	Main framing	Jacket and foundation	
Laydown areas	12	10	c		30
Open deck areas and access hatches	12	10	c	d	15
Mechanical handling routes	10	5	c	d	30
Stairs and landings	2.5	2.5	b	—	1.5
Walkways and access platforms	5	2.5	(Note)	(Note)	5, a

Note: Member categories are as follows:

a. Point load for access platform beam design is 10 kN and 5 kN for deck grating and stringers, respectively.

b. Loading for deck plate, grating, and stringers are combined with the structural dead loads and designed for the most onerous of the following:

- Loading over entire contributory deck area.
- A point load (applied over a 300 mm × 300 mm footprint).
- Functional loads plus design load on clear areas.

c. For the design of the main framing two cases are considered:

- Maximum operating condition: All equipment, including future items and helicopter, together with 2.5 kN/m^2 on the laydown area.
- Live load condition: All equipment loads but no future equipment, together with 2.5 kN/m^2 on the laydown areas, and a total additional live load of 50 tonnes. This live load is applied as a constant uniformly distributed load over the open areas of the deck.

d. Deck loading on clear areas for extreme storm conditions for substructure design may be reduced to zero in view of the not normally manned status of the platform during storm conditions. A total live load of 200 kN at the topside center of gravity is assumed for the design of the jacket and foundations

For general deck area design loading, the topside deck structure is designed for the imposed loads specified in Table 3.4. These are applied to open areas of the deck, where the equipment load intensity is less than the values shown.

DvV (2008) defines the variable functional loads on deck areas of the topside structure based on the guidelines in Table 3.5. These values are considered guidelines, so the values should be defined in the basis of design or the design criteria approved by

Table 3.5 Variable functional loads on deck areas

	Local load design		Primary design	Global design
	Distributed load (kN/m^2)	**Point load (kN)**	**Apply factor for distributed load**	**Apply factor to primary design load**
Storage areas	q	$1.5q$	1.0	1.0
Lay-down areas	q	$1.5q$	f	f
Lifeboat platforms	9.0	9.0	1.0	May be ignored
Area between equipment	5.0	5.0	f	May be ignored
Walkways, staircases, and platforms crew spaces	4.0	4.0	f	May be ignored
Walkways and staircases for inspection only	3.0	3.0	f	May be ignored
Areas not exposed to other functional loads	2.5	2.5	1.0	May be ignored

the owner. If the owner needs to increase the load, over that mentioned in the code, it should be stated on the basis of design and the detail drawings should include the load on the deck. In Table 3.5, the loads are identified for the local design considered in designing the plates, stiffeners, beams, and brackets. The loads in the primary design should be used in design of girders and columns. The loads determine the design of the deck main structure and substructure, which constitute the global design.

Using Table 3.5, the wheel loads are added to distributed loads where relevant. (Wheel loads can be considered acting on an area of 300×300 mm.)

Point loads are applied on an area 100×100 mm at the most severe position but are not added to wheel loads or distributed loads.

The value of q is evaluated for each case. Lay down areas should not be designed for less than 15 kN/m^2.

The value of f is from the following :

$$f = \min\left\{1.0, \left(0.5 + 3/\sqrt{A}\right)\right\} \tag{3.1}$$

where A is the loaded area in m^2.

Global load cases are established based on a worst case scenario, characteristic load combinations, and complying with the limiting global criteria to the structure. For buoyant structures, these criteria are established by requirements for the floating position in still water and intact and damage stability requirements, as documented in the operational manual, considering variable load on the deck and in tanks.

In calculating the dry weight of piping, valves, and other structure s,support and increase 20% for contingency to all estimates of piping weight. As in most cases, there are changes in piping dimensions and location over the structure lifetime. In addition to that, all the piping and fittings are calculating in the operating condition by assuming pipes are full of water with a specific gravity equal to 1 and a 20% contingency.

Table 3.6 **Minimum uniform loads from industrial practice**

Platform deck	Uniform load, kN/m^2 (Psf)
Helideck	
Without helicopter	14 (350)
With bell 212	2.0 (40)
Mezzanine deck	12 (250)
Production deck	17 (350)
Access platforms	12 (250)
Stairs and walkways	4.7 (100)
Open area used in conjunction with the equipment operating and piping loads for operating and storm conditions	2.4 (50)

Table 3.7 **Impact load factor**

Structural item	Load direction	
	Vertical	Horizontal
Rated load of cranes	100%	100%
Supports of light machinery	20%	0
Support of reciprocating machinery	50%	50%
Boat landings	200 kips (890 KN)	200 kips (890 KN)

If calculating the dry weight of all equipment, equipment skid , storage, and helicopters, the contingency allowance of 10% is included in the equipment.

From practical point of view, Table 3.6 presents the live load values from industrial practice.

3.2.3 Impact load

For the structural component carrying live loads that include impact, the live load must be increased to account for the impact effect, as shown in Table 3.7.

3.2.4 Design for serviceability limit state

The serviceability of the topside structures can be affected by excessive relative displacement or vibration in the vertical or horizontal direction. Limits for either can be dictated by

1. Discomfort to personnel.
2. Integrity and operability of equipment or connected pipework.
3. Limits to control deflection of supported structures, as in flare structures.
4. Damage to architectural finishes.
5. Operational requirements for drainage (free surface or piped fluids).

Table 3.8 **Maximum vertical deflection based on ISO 9001**

Structural element	Δ_{max}	Δ_2
Floor beams	$L/200$	$L/350$
Cantilever beams	$L/100$	$L/150$
Deck plate		$2t$ or $b/150$

Note: L is the span.
t is deck thickness.
b is the stiffener spacing.

All sources of vibration should be considered in the design of the structure. The following is the main sources of the vibration on the platforms:

- Operating mechanical equipment, including that used in drilling operations.
- Vibrations from variations in fluid flow in piping systems, in particular slugging.
- Oscillations from vortex shedding on slender tubular structures.
- Global motions from the effect of environmental actions on the total platform structure.
- Vibrations due to earthquake and accidental events.

Large cantilevers, whether formed by simple beams or trusses forming an integral part of the topsides but excluding masts or booms, normally are proportioned to have a natural period of less than 1 second in the operating condition.

The final deflected shape, Δ_{max}, of any element or structure comprises three components, as follows:-

$$\Delta_{max} = \Delta_1 + \Delta_2 - \Delta_0 \tag{3.2}$$

where

Δ_0 is any precamber part of a beam or element in the unloaded state, if it exists.
Δ_1 is the deflection from the permanent loads (actions) immediately after loading.
Δ_2 is the deflection from the variable loading and any time-dependent deformations from permanent loads.

The maximum values for vertical deflections are given in Table 3.8, based on ISO 9001.

Lower limits can be necessary to limit ponding of surface fluids and ensure that drainage systems function correctly.

Horizontal deflections generally are limited to 0.3% of the height between floors. For multifloor structures, the total horizontal deflection should not exceed 0.2% of the total height of the topsides structure. Limits can be defined to limit pipe stresses, so as to avoid risers or conductors over tresses or failure. In most cases, some designers allow higher deflections, which may be acceptable for structural elements where serviceability is not compromised by deflection.

3.2.5 Crane support structures

The crane support structure comprises the crane pedestal and its connections to the topside primary steelwork. It does not include the slew ring or its equivalent nor the connections between the slew ring and the pedestal.

Crane support structures, where practical, should be attached at the intersection of topside primary trusses and connected at main deck elevations with minimal eccentricities. The pedestal should be included in the analytical model of the primary structure, as its stiffness can have a significant effect on load distribution. The maximum rotation at the top of the pedestal or in the plane of the effective point of support should not exceed the manufacturer's recommended requirements and in no case exceed 1° for the most onerous case of loading. Where this criterion cannot be met, the dynamic response should be checked.

A number of separate situations shall be considered for the design of the crane support structures as follows.

1. A crane working in calm conditions.
2. A crane working at maximum operating wind conditions. The maximum operating wind may be different for platform lifts and sea lifts, as lifts to or from an adjacent vessel, and may also vary depending on the weight being lifted.
3. A crane at rest, extreme wind conditions. For this situation, the crane boom may be considered resting in a boom support in the case of a fitted crane.
4. Crane collapse. This situation is included to ensure that, in event of a gross overload of the crane, causing collapse of any part of the crane structure, in most cases, the boom will suffer the greatest damage, no damage to the crane support structure will be suffered, and progressive collapse will be resisted.

The cases of loading shall be checked as follows based on ISO 9001.

1. Crane working without wind:

$$F_G + fF_L + F_H \tag{3.3}$$

where
F_G is the vertical load due to dead weight of components.
F_L is the vertical load due to the suspended load, including sheave blocks, hooks, and others.
F_H is the horizontal loads due to offlead and sidelead.
f is a dynamic coefficient, taken as 2.0 for sea lifts and 1.3 for platform lifts.

2. Crane working with wind:

$$F_G + fF_L + F_H + F_W \tag{3.4}$$

where F_G, F_L, F_H, and f are as previously.
F_W is the operating wind action.
f is taken as 2.0 for sea lifts and 1.3 for platform lifts for a maximum crane operating wind.

3. Crane not working, extreme wind:

$$F_G + F_{W,\max} \tag{3.5}$$

where F_G is as previously.
$F_{W,\max}$ is the extreme wind action.

The action factors used with each of these should be those for normal operating conditions.

For cases 1 and 2, F_L is selected to check the lifted load applicable to both maximum and minimum crane radius, for sea and platform lifts. For cases 2 and 3, the most onerous wind directions are checked.

It should be demonstrated that the crane pedestal and its components are designed to safely resist the forces and moments from the most onerous loading condition applicable to the prevailing sea state, together with associated offlead and sidelead forces. These values are obtained from the crane manufacturer, and the angles used should be no less than the following:-

- Offlead angle, 6 degrees.
- Sidelead angle, 3 degrees.

The crane support structure should be designed such that its failure load exceeds the collapse capacity of the crane.

The crane manufacturer's failure curves, for all crane conditions, are used to determine the worst loading on the pedestal. It is assumed that the maximum lower bound failure moment of the weakest component places an upper bound on the forces and moments to which the pedestal can be subjected. The design moment for the crane failure condition is taken as the lower bound failure moment just described, multiplied by a safety factor of 1.3.

3.3 Wind load

The wind data is provided by the owner according to the metocean study, which defines the prevailing wind direction and the maximum wind speed on the basis of 1 year, 50 years, and 100 years. Figures 3.1 and 3.2 is a sample of the wind data in the metocean study report.

The most important design considerations for an offshore platform are the storm wind and storm wave loadings it will be subjected to during its service life. These data usually are available to the owner and submitted to the engineering firm officially. If such data are not available, information should be delivered from an authority with experience in the country or international specific offices.

The wind speed at any elevation above a water surface is presented as

$$V_z = V_{10}(z/10)^{1.7} \tag{3.6}$$

From the code equation, which is the same for the different codes, we can calculate the wind force on the structure

V_{10} = wind speed at height 10 m.
10 = reference height, m.
z = desired elevation, m.

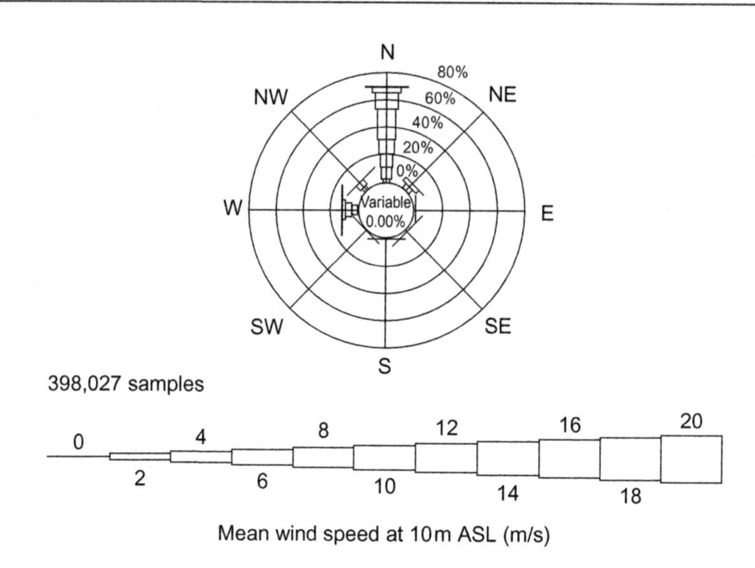

Figure 3.1 Sample of wind rose.

Mean wind speed at 10m ASL (m/s)	N	NE	E	SE	S	SW	W	NW	Total	CFD
0.0–2.0	0.34	0.28	0.24	0.24	0.19	0.22	0.39	0.43	2.3	100.0
2.0–4.0	1.9	1.2	0.61	0.60	0.32	0.34	2.0	2.9	9.8	97.7
4.0–6.0	8.2	1.9	0.16	0.47	0.22	0.31	4.8	3.6	19.8	87.9
6.0–8.0	19.3	1.7	0.01	0.35	0.18	0.25	3.6	0.93	26.2	68.1
8.0–10.0	21.6	0.74	<0.01	0.23	0.13	0.15	1.4	0.13	24.4	41.8
10.0–12.0	12.6	0.18	<0.01	0.11	0.05	0.03	0.62	0.02	13.6	17.4
12.0–14.0	3.2	0.02	<0.01	0.03	0.01	<0.01	0.18	<0.01	3.4	3.9
14.0–16.0	0.32	<0.01		<0.01	0.01	<0.01	0.04	<0.01	0.38	0.42
16.0–18.0	<0.01			<0.01	<0.01	<0.01	0.01		0.03	0.04
18.0–20.0	<0.01			<0.01			<0.01	<0.01	<0.01	<0.01
20.0–22.0							<0.01		<0.01	<0.01
Total	67.4	6.0	1.0	2.0	1.1	1.3	13.0	8.0	100.0	

Figure 3.2 Tabulated data for mean wind speed from different directions.

The wind velocity, vertical wind profile, and time averaging duration in relation to the dimensions and dynamic sensitivity of the structure's components should be determined. In special cases, a dynamic response to wind action can be significant and must be taken into account.

Table 3.9 Shape coefficient, C_s, for perpendicular wind approach angles

Component	Shape coefficients, C_s
Flat walls of buildings	1.5
Overall projected area of structure	1.0
Beams	1.5
Cylinder	
Smooth $R_e > 5 \times 10^5$	0.65
Smooth $R_e \leq 5 \times 10^5$	1.2
Rough all R_e	1.05
Covered with ice, all R_e	1.2

For all directions and angles of wind approach to the structure, wind actions on vertical cylindrical objects may be assumed to act in the direction of the wind. Actions on cylindrical objects that are not in a vertical attitude should be calculated using appropriate formulas that take into account the direction of the wind in relation to the attitude of the object. Actions on walls and other flat surfaces that are not perpendicular to the direction of the wind also should be calculated using appropriate formulas that account for the skewness between the direction of the wind and the plane of the surface. Where appropriate, loads caused by wind should be calculated taking into account increased exposure area and surface roughness due to ice. Local wind effects, such as pressure concentrations and internal pressures, should be considered by the designer where applicable.

However, in accordance with these observations, due attention should be given to which velocity (in magnitude and direction) and which area are to be used. Similarly, the direction of the resulting action vector should be carefully considered. The basic relationship between the wind velocity and the wind load on an object is presented by the following equation:

$$F = 1/2\ \rho_a V_w\ C_s A \tag{3.7}$$

where

F is the wind action on the object.
ρ_a is the mass density of air (at standard temperature and pressure).
V_w is the wind speed.
C_s is the shape coefficient.
A is the area of the object.

In the absence of data indicating otherwise, the shape coefficients in Table 3.9 are recommended for perpendicular wind approach angles with respect to each projected area. The shape conditions factor for different structure element is presented in Table 3.10

Wind loads on downstream components can be reduced due to shielding by upstream components.

Table 3.10 **Shape coefficient, C_s, by API RP2A**

Beams	1.5
Sides of buildings	1.5
Cylindrical section	0.5
Overall projected are of platform	1.0

Table 3.11 **Design wind pressures at 125 mph for 100 years**

Structure member	Pressure kN/m^2 (psf)
Flat surface such as wide flange beams, gusset plates, sides of building	2.9 (60)
Cylindrical structural members	2.3 (48)
Cylindrical deck equipment (L = 4*D*)	1.4 (30)
Tanks standing on end ($H \le D$)	(25)

Table 3.12 **Design wind pressures at 125 mph for 100 years**

Component	Shielding factor
Second in a series of trusses	0.75
Third or more in a series of trusses	0.50
Second in a series of beams	0.50
Third or more in a series of beams	0.00
Second in a series of tanks	1.00
Short objects behind tall objects	0.00

The extreme quasi-static global action caused by wind is calculated as the vector sum of the preceding wind loads on all objects.

When wind loads are important for structural design, wind pressures and resulting local loads should be determined from wind tunnel tests on a representative model or from a computational model representing the structure and considering the range and variation of wind velocities. Computational models should be validated against wind tunnel tests **or** *full-scale measurements* of similar structures.

Table 3.11 lists some design wind pressures that have been used in conjunction with a 100 year sustained wind velocity of 125 mph.

The shielding coefficient is applied by API using of these factors in Table 3.12 are based on the judgment of the designer, and the following factors are usually used in the case of applying the lateral wind load to the topside APIRP2A advice on shielding reduction.

Based on the DNV, the wind pressure acting on the surface of helidecks may be calculated using a pressure coefficient $C_p = 2.0$ at the leading edge of the helideck, linearly reducing to $C_p = 0$ at the trailing edge, taken in the direction of the wind. The pressure may act both upward and downward.

Table 3.13 **Live load on the fixed platform from technical practice**

Area	Loadings (psf) by member category		
	Deck plate, grating, and stringers	**Deck beams**	**Main truss framing, girders, jacket, and foundation**
Cellar and main decks	14.4	9.6	9.6
Walkway, stairs, and access decks	4.8	2.4	7.2
Laydown areas	19	14.4	150

3.3.1.1.1 Example 3.1

For stairs design the first bay of the stair that connects to the main deck at level 13.53 meter to the helideck is the most critical stair bay.

Gravity loads

Uniform gravity loads/one stair channel = (dead load + live load) × ½ stair width + handrail weight/meter + channel's own weight/meter

Dead load = 0.5 kN/m′

Live load = 50 psf = 2.5 kN.m' (extracted from the Table 3.13, as per project specifications.)

- Handrail weight/meter = 0.4 kN/m′
- Channel own weight/meter = 0.379 kN/m′

Uniform gravity loads (ton/m)/one stair channel = (0.5 + 2.5) × 0.5 × 1.2 + 0.4 + 0.379 = 2.6 kN/m′

Moment = (uniform load / m) × inclined length × projected length/8 = 2.6 × 6.22 × 4.4/8 = 8.9 m. kN

Actual stress = moment/section modulus = 8.9 (10^6)/371 = 24000 kN/m^2

Unsupported length = 6.22 m

Allowable stress (ksi) = 12,000 C_b/(member length × depth/area of flange) = 12000 × 1/[622 × 26/(9 × 1.4)] = 9.35 ksi = 64466 kN/m^2 > actual stress

3.3.1.1.2 Wind loads

Wind loads/one stair channel = $C_d \times ½ \times \rho \times v^2 \times ½$ stair width

- C_d (drag coefficient) = 2
- ρ (air intensity) = 1.3 kg/m^3
- V (wind velocity) = 29 m/sec.

Wind loads/one stair channel = 2 × ½ × 1.3 × 29^2 × 0.6 = 655.98 N/m = 0.66 kN/m^2

Moment = (wind load/m) × inclined length × projected length/8 = 0.66 × 6.22 × 4.4/8 = 0.23 m. kN

Actual stress (gravity + wind) = moment/section modulus = (8.9 + 0.23)(10^6)/371 = 24,609 kN/m^2 < allowable stress × 1.33

Note that wind loads on a channel in the minor axis direction is neglected, since horizontal bracing is provided

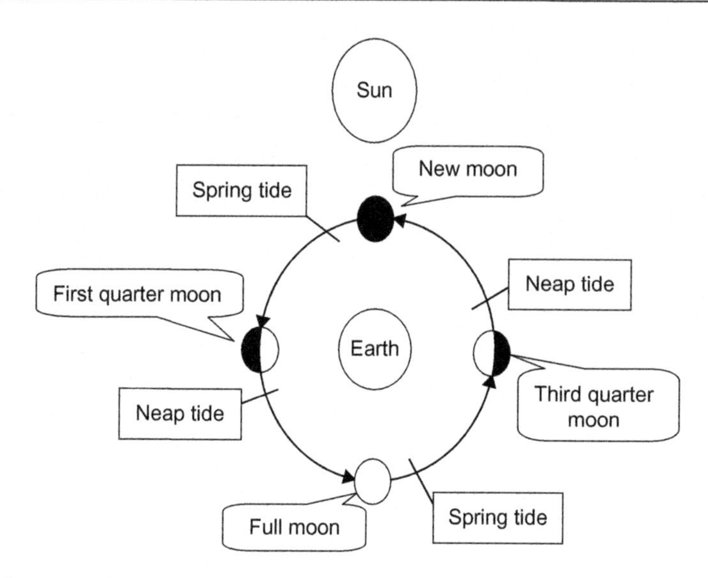

Figure 3.3 Times of neap and spring tide.

3.4 Offshore loads

Offshore loads affect the platforms as waves, tide, and current. The terminology that follows relates to the water level in the sea and ocean:

- High high water level (HHWL).
- Mean high water level (MHWL).
- Mean water level (MWL), equal to mean sea level (MSL) and still water level (SWL).
- Low low water level (LLWL) which is equal to the chart datum (CD).
- Mean low water level (MLWL).

The main low water lever is less than the mean high water level by 304 mm (1 ft).

Tides are the rise and fall of the sea level caused by the combined effects of the gravitational forces exerted by the moon and the sun and the rotation of the earth.

Most places in the ocean usually experience two high tides and two low tides each day (semidiurnal tide), but some locations experience only one high and one low tide each day (diurnal tide). The times and amplitude of the tides at the coast are influenced by the alignment of the sun and moon, by the pattern of tides in the deep ocean as shown in Figure 3.3, and by the shape of the coastline and near-shore bathymetry.

Most coastal areas experience two high and two low tides per day. The gravitational effect of the moon on the surface of the earth is the same when it is directly overhead as when it is directly underneath. The moon orbits the earth in the same direction as the earth rotates on its axis, so it takes slightly more than a day, about 24 hours and 50 minutes, for the moon to return to the same location in the sky. During this time, it has passed overhead once and underneath once, so in many places the period of strongest tidal force is 12 hours and 25 minutes. The high tides do not necessarily occur when the moon is overhead or underneath, but the period of the force still determines the time between high tides.

Storm surge is an offshore rise of water associated with a low-pressure weather system, typically a tropical cyclone. Storm surges are caused primarily by high winds pushing on the ocean's surface. The wind causes the water to pile up higher than the ordinary sea level. Low pressure at the center of a weather system also has a small secondary effect, as can the bathymetry of the body of water. This combined effect of low pressure and persistent wind over a shallow water body is the most common cause of storm surge flooding problems. The term *storm surge* in casual (nonscientific) use is storm tide; that is, it refers to the rise of water associated with the storm, plus tide, wave run-up, and freshwater flooding. When referencing storm surge height, it is important to clarify the usage, as well as the reference point.

3.4.1 Wave load

Large forces result when waves strike a platform's deck and equipment. Where an insufficient air gap exists, all actions resulting from waves, including buoyancy, inertia, drag, and slam, should be taken into account, see ISO 19901-1 and ISO 19902.

Waves characteristic to the ocean often appear as a confused and constantly changing sea of crests and troughs on the water surface because of the irregularity of wave shape and the variability in the direction of propagation. The direction of wave propagation can be assessed as an average of the direction of individual waves.

In general, actual water-wave phenomena are complex and difficult to describe mathematically because of nonlinearities, three-dimensional characteristics, and apparent random behavior. However, two classical theories, one developed by Airy in 1845 and the other by Stokes in 1880, describe simple waves. The Airy and Stokes theories generally predict wave behavior better where water depth relative to wavelength is not too small. For shallow water regions, conidial wave theory, originally developed by Korteweg and DeVries in 1895, provides a rather reliable prediction of the wave form and associates motions for some conditions. Recently, the work involved in using conidial wave theory has been substantially reduced by the introduction of graphical and tubular forms of function by Wiegel in 1960 and Masc and Wiegel in 1961; however, application of the theory is still complex.

Stokes in 1880 developed a finite amplitude theory that is more satisfactory. Only the second-order Stokes equation is presented, but the use of higher-order approximations is sometimes justified for the solution of a practical problems.

Another widely used theory, known as the *stream function theory*, is a nonlinear solution similar to the Stokes' fifth-order theory, both use summations of sine and cosine wave forms to develop a solution to the original differential equation.

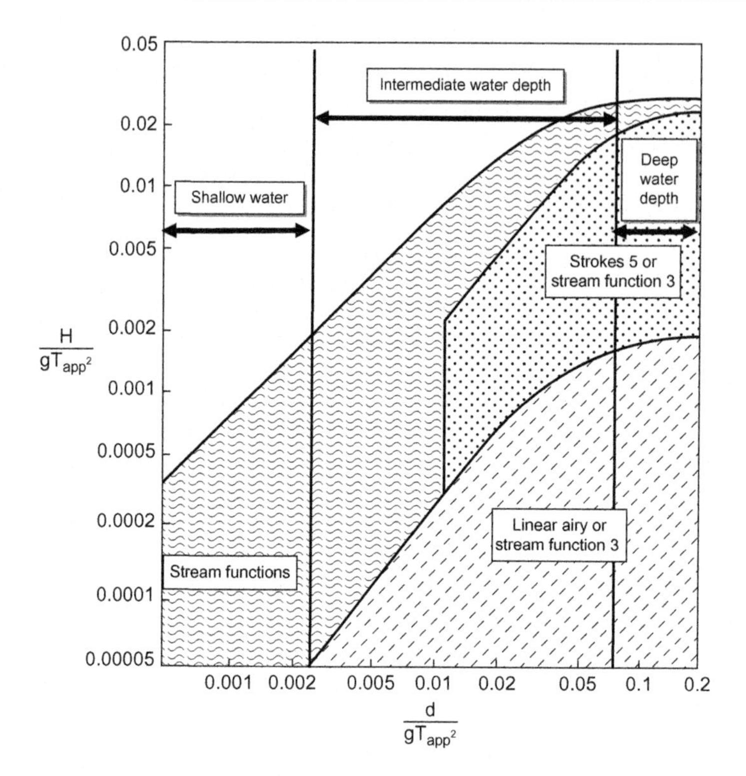

Figure 3.4 Regions of applying wave theories.

The theory to be used for a particular offshore design is determined by the policy under which the designing engineers are working. The selection of the best method is defined by the curve in Figure 3.4, from APIRP2A, where

H/gT_{app}^2 is dimensionless wave sleepiness.
d/gT_{app}^2 is dimensionless relative depth.
d is mean water depth.
T_{app} is the wave period.
H is the wave height.
g is the acceleration of gravity.

Metocean data provides the Table 3.14 data as an example of a platform location.

The sea state (E) is calculated by dividing a wave form into small slides noting that, for each slide, the height H_i is squared, the heights added together and averaged. Then, E is calculated as

$$E = 2\frac{\sum_{i=1}^{N} H_i}{N}$$

Table 3.14 **Example metocean data**

Return period	W_s (m/s)	H_s (m)	T_z (s)	T_p (s)	H_{max} (m)	T_{ass} (s)	H_c (m)	U (cm/s)
1 year	16.1	2.2	4.7	6.2	3.2	6.3	1.9	1
10 years	17.7	2.8	5.4	7.1	5.3	7.2	3.2	5
50 years	18.7	3.2	5.9	7.7	6.6	7.7	4.0	11
100 years	19.2	3.3	6.0	7.9	7.2	8.0	4.3	15
10,000 years	22.0	4.5	7.1	9.4	10.9	9.4	6.6	51

Note:
W_s is the 1-hour mean wind speed at 10 m above sea level, m/s.
H_s is the significant wave height, estimated from the wave energy spectrum, Equivalent to the mean height of the highest one third of the wave in a sea state.
H_{max} is the maximum wave height, highest individual zero crossing wave height in a storm of 24 hours duration, m.
T_z is the mean zero crossing wave period; the average period of the zero-crossing wave heights in a sea state. estimated from the wave energy spectrum, s.
T_p is the peak wave period, the period associated with the peak in the wave energy spectrum, estimated from the wave energy spectrum, s.
T_{ass} is the wave period associated with maximum wave height, s.
H_c is the crest height, highest crest to mean-level height of an individual wave in a storm of 24 hours duration, m.
U is the horizontal wave orbital velocity at 3 m above the seabed, estimated from H_{max} and T_{ass} using stream function wave theory, cm/s.

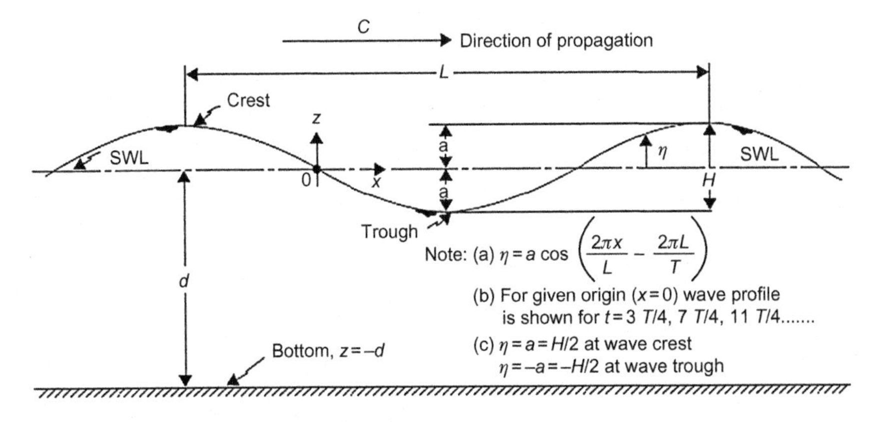

Figure 3.5 Definitions of sinusoidal progressive curve.

Figure 3.5 shows a two-dimensional, simple progressive wave propagation in the positive x-direction, the symbol η denotes the displacement of the water surface relative to the still water level, which is a function of x and time, t, at the wave crest, η, equal to one half of the wave height.

Water particle displacement is presented in Figure 3.6 for deep and shallow water. Water particle displacement is an important factor of linear wave mechanics, dealing with the displacement of individual water particles within the wave. Water particles generally move in elliptical paths in shallow water or to transitional water and in circular path shapes in deep water, as is clearly presented in Figure 3.6. If the mean particle position is considered to be at the center of the ellipse or circle,

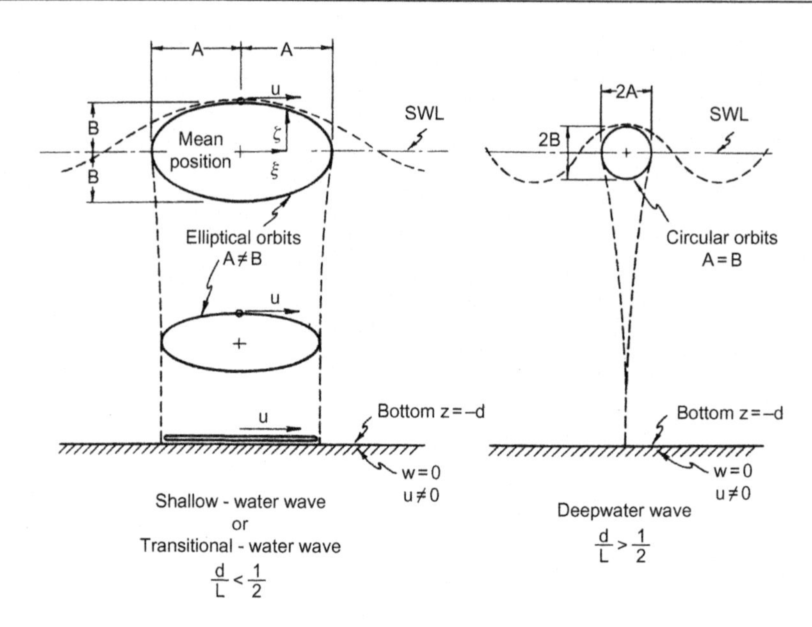

Figure 3.6 Water particle displacement for shallow and deepwater waves.

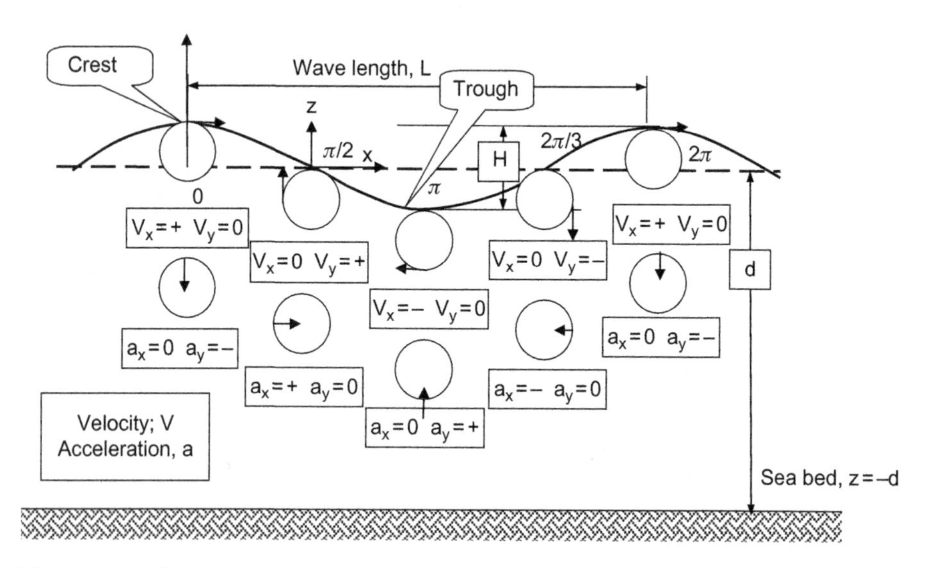

Figure 3.7 Fluid particles velocities and acceleration directions at every location.

then vertical particle displacement with respect to the mean position cannot exceed one half the wave height. Therefore, the displacement of the fluid particle is small if the wave height is small.

Figure 3.7 presents the horizontal and vertical velocities and acceleration for various locations of the particles, which is very important in calculating the wave forces in any structure member in subsea, as the drag force and inertia force are functions of the particle velocity and acceleration, respectively. The following equations calculate the wave velocity and acceleration:

$$F1 = [2\pi\,(z + d)/L] \tag{3.8}$$

$$F2 = (2\pi d/L) \tag{3.9}$$

$$F3 = (2\pi x/L) - (2\pi t/T) \tag{3.10}$$

For velocity,

$$U = [(H/2)\,(gT/L)\,\cosh F1/\cosh F2]\,\cos F3 \tag{3.11}$$

$$W = [(H/2)\,(gT/L)\,\sinh F1/\cosh F2]\,\sin F3 \tag{3.12}$$

For acceleration,

$$a_x = [+gpH/L\,\cosh F1/\cosh F2]\,\sin F3 \tag{3.13}$$

$$a_z = [-gpH/L\,\sinh F1/\cosh F2]\,\cos F3 \tag{3.14}$$

The theories concerning the modeling of ocean waves were developed in the 19th century. Practical wave force theories concerning actual offshore platforms were not developed until 1950, when Morison equation was presented:

$$F = F_D + F_I \tag{3.15}$$

where

F_D is the drag force.
F_I is the inertia force.

3.4.1.1 Drag force

The drag force due to a wave acting on an object (Figure 3.8) can be found by

$$F_D = ½\rho C_d V^2 A \tag{3.16}$$

where

F_D = drag force (N).
C_d = drag coefficient (no units).
V = velocity of object (m/s).
A = projected area (m^2).
ρ = density of water (kg/m^3).

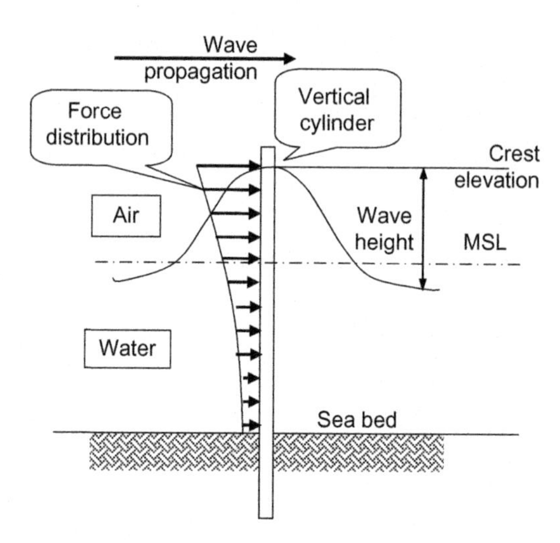

Figure 3.8 Wave force distribution on a vertical pipe.

3.4.1.2 *Inertia force*

The inertia force due to wave acting on an object can be found by

$$F_I = \pi \rho a C_m . D^2/4 \tag{3.17}$$

where

F_I = inertia force (N)
C_m = mass coefficient (no units)
a = horizontal water particle acceleration (m²/s)
D = diameter of the cylinder (m)
ρ = density of water (kg/m³)

3.4.1.3 *Wave load calculation*

- The values of C_d and C_m are dimensionless and those most often used in the Morison equation are 0.7 and 2.0, respectively. The API recommends 0.65 and 1.6, respectively, for smooth or 1.05 and 1.2 for rough surfaces in case of existing marine growth.
- Water particle velocity and acceleration are functions of wave height, wave period, water depth, distance above the bottom, and time.
- The most elementary wave theory was presented by Airy in 1845.
- Another widely used theory, known as the *stream function theory*, is a nonlinear solution similar to the Stokes fifth-order theory.

Waves are omnidirectional. API RP2A mentions that the maximum wave height for a 1-year or 100-year storm is applied on the eight directions by the same values applied to the platform.

3.4.1.4 Comparison between wind and wave calculation

When we calculate the force affect structure due to wind we take the drag force into consideration and neglect the inertia force but in case of wave we should consider drag force and inertia force and the following example provide us the reason to neglect the inertia force in case of wind load.

3.4.1.4.1 Example 3.2

Calculate the forces in tubular member for wind and wave

Pipe Dia. = 0.4 m
$V_{air} = 25$ m/s $V_{water} = 1$ m/s
$a_{air} = 1$ m^2/s $a_{water} = 1$ m^2/s
$\rho_{air} = 1.3$ kg/m^3 $\rho_{water} = 1000$ kg/m^3
$F_d = (1/2).Cd.\ \rho.V^2.\ A$
$F_m = Cm.\ \rho.\ \pi.\ (D^2/4).\ a$

For the air,

$F_d = (1/2)\ (0.8)(1.3)(25)^2(0.4) = 130$ Newton
$F_m = 2\ (1.3)(p)(0.4)^2/4\ (1) = 0.33$ Newton

For the water,

$F_d = (1/2)(0.8)(1000)(1)^2(0.4) = 160$ Newton
$F_m = 2\ (1000)(\pi)(0.4)^2/4\ (1) = 251$ Newton

3.4.1.4.2 Example 3.3

Calculate the maximum bending moment and shearing force applied on a vertical tubular member in the water and resting on the soil bed.

The input parameters are

Density 1025 kg/m^3
Kinatic velocity = 0.0000018 m^2/s
Tubular member length $h = 138$ m
The level of the top of tubular member = −48 m, as the mean water level is zero level
$C_m = 2$
Wave time period $T = 9.8$ s
Wave length $L = g/(2\pi) \times T^2$
Wave height; $H = 10.40$ m
$K = 2\pi/L0.04$ m^{-1}

The calculation in this example based on the following instructions:

- Calculate the drag for and inertia force every 6.9 m at 20 sections.
- Calculate every phase angle = $\pi/12 = 0.262$.
- The calculation of every section for phase angle 6.28 is presented in Table 3.15.
- Calculate the maximum bending moment and shear for every phase angle and choose the maximum, as shown in Table 3.16.

Table 3.15 **Calculating the force in every segement for case of phase angles 6.28 and 5.76**

z (m)	Horizontal		Drag force (kN)	Inertia force (kN)	Resultant force (kN)	Bending moment (kNm)
	Velocity (m/s)	Acceleration (m/s^2)				
Phase angle (rad) = 6.28						
−48.00	0.00	−0.29	0.00	−2.74	−2.74	−246.49
−52.50	0.00	−0.24	0.00	−2.27	−2.27	−193.97
−57.00	0.00	−0.20	0.00	−1.88	−1.88	−152.24
−61.50	0.00	−0.16	0.00	−1.56	−1.56	−119.13
−66.00	0.00	−0.13	0.00	−1.29	−1.29	−92.92
−70.50	0.00	−0.11	0.00	−1.07	−1.07	−72.23
−75.00	0.00	−0.09	0.00	−0.89	−0.89	−55.91
−79.50	0.00	−0.08	0.00	−0.74	−0.74	−43.10
−84.00	0.00	−0.06	0.00	−0.61	−0.61	−33.06
−88.50	0.00	−0.05	0.00	−0.51	−0.51	−25.22
−93.00	0.00	−0.04	0.00	−0.42	−0.42	−19.12
−97.50	0.00	−0.04	0.00	−0.36	−0.36	−14.40
−102.00	0.00	−0.03	0.00	−0.30	−0.30	−10.76
−106.50	0.00	−0.03	0.00	−0.25	−0.25	−7.96
−111.00	0.00	−0.02	0.00	−0.22	−0.22	−5.82
−115.50	0.00	−0.02	0.00	−0.19	−0.19	−4.19
−120.00	0.00	−0.02	0.00	−0.16	−0.16	−2.95
−124.50	0.00	−0.02	0.00	−0.15	−0.15	−1.98
−129.00	0.00	−0.01	0.00	−0.14	−0.14	−1.22
−133.50	0.00	−0.01	0.00	−0.13	−0.13	−0.58
−138.00	0.00	−0.01	0.00	−0.13	−0.13	0.00
					−15.98	−1103.24

Phase angle (rad) = 5.76						
−48.00	−0.22	−0.25	−0.06	−2.37	−2.44	−219.34
−52.50	−0.18	−0.21	−0.05	−1.97	−2.01	−172.07
−57.00	−0.15	−0.17	−0.03	−1.63	−1.66	−134.68
−61.50	−0.13	−0.14	−0.02	−1.35	−1.37	−105.04
−66.00	−0.11	−0.12	−0.02	−1.12	−1.13	−81.69
−70.50	−0.09	−0.10	−0.01	−0.93	−0.94	−63.33
−75.00	−0.07	−0.08	−0.01	−0.77	−0.78	−48.93
−79.50	−0.06	−0.07	−0.01	−0.64	−0.64	−37.65
−84.00	−0.05	−0.06	0.00	−0.53	−0.53	−28.84
−88.50	−0.04	−0.05	0.00	−0.44	−0.44	−21.97
−93.00	−0.03	−0.04	0.00	−0.37	−0.37	−16.64
−97.50	−0.03	−0.03	0.00	−0.31	−0.31	−12.52
−102.00	−0.02	−0.03	0.00	−0.26	−0.26	−9.35
−106.50	−0.02	−0.02	0.00	−0.22	−0.22	−6.92
−111.00	−0.02	−0.02	0.00	−0.19	−0.19	−5.06
−115.50	−0.02	−0.02	0.00	−0.16	−0.16	−3.64
−120.00	−0.01	−0.01	0.00	−0.14	−0.14	−2.56
−124.50	−0.01	−0.01	0.00	−0.13	−0.13	−1.72
−129.00	−0.01	−0.01	0.00	−0.12	−0.12	−1.06
−133.50	−0.01	−0.01	0.00	−0.11	−0.11	−0.50
−138.00	−0.01	−0.01	0.00	−0.11	−0.11	0.00
					−14.07	−973.50

Table 3.16 Maximum force and bending moment in every angle phase

Phase angle (rad)	Maximum force (kN)	Bending moment (kNm)
0.000	−15.98	−1103.24
0.262	−15.38	−1060.52
0.524	−13.62	−937.57
0.785	−10.89	−747.50
1.047	−7.41	−505.60
1.309	−3.43	−230.22
1.571	0.75	58.77
1.833	4.84	340.86
2.094	8.58	597.64
2.356	11.72	812.72
2.618	14.07	973.29
2.880	15.50	1070.77
3.142	15.98	1103.24
3.403	15.38	1060.52
3.665	13.62	937.57
3.927	10.89	747.50
4.189	7.41	505.60
4.451	3.43	230.22
4.712	−0.75	−58.77
4.974	−4.84	−340.86
5.236	−8.58	−597.64
5.498	−11.72	−812.72
5.760	−14.07	−973.29
6.021	−15.50	−1070.77
6.283	−15.98	−1103.24
Maximum	15.98	1103.24

3.4.1.5 Conductor shielding factor

Determining the conductor shield factor (SF) depends on the configuration of the structure and the number of well conductors, which can be a significant portion of the total wave forces.

If the conductors are closely spaced, the forces on them may be reduced due to hydrodynamic shield.

$$\begin{aligned} SF &= 0.25(S/D)1.5 < S/D < 4.0 \\ SF &= 1.0\ S/D \geq 1.0 \end{aligned} \tag{3.18}$$

where S is the spacing in the wave direction and D is the conductor diameter.

3.4.1.5.1 Example 3.4

In case of a conductor diameter of 24 in. and the spacing between the conductors is 71 in., find the shielding factor.

Solution

Shielding factor = 0.75

3.4.2 Current load

Based on ISO 9002, the most common categories of ocean currents are

- Wind generated currents
- Tidal currents
- Circulational currents
- Loop and eddy currents
- Soliton currents
- Longshore currents

Wind-generated currents are caused by wind stress and atmospheric pressure gradient throughout a storm.

Tidal currents are regular, following the harmonic astronomical motion of the planets. Maximum tidal current precedes or follows the highest and lowest astronomical tides (HAT and LAT). Tidal currents are generally weak in deep water, but are strengthened by shoreline configurations. Strong tidal currents exist in inlets and straights in coastal regions.

Circulational currents are steady, large-scale currents of the general oceanic circulation (e.g., the Gulf Stream in the Atlantic Ocean). Parts of the circulation currents may break off from the main circulation to form large-scale eddies. Current velocities in such eddies (*loop and eddy currents*) can exceed that of the main circulation current (e.g., Loop Current in the Gulf of Mexico).

Soliton currents are due to internal waves generated by density gradients. Loop and eddy currents and soliton currents penetrate deeply in the water column.

Longshore currents in coastal regions run parallel to the shore as a result of waves breaking at an angle on the shore. They are also referred to as *littoral* currents.

Earthquakes can cause unstable deposits to run down continental slopes and thereby set up gravity driven flows. Such flows are called *turbidity currents*. Sediments give the flow a higher density than the ambient water. Such currents should be accounted for in the design of pipelines crossing a continental slope with unstable sediments. Strong underwater earthquakes can also lead to generation of *tsunamis*, which in coastal regions behave like a long shallow water wave similar to a strong horizontal current.

The effects of currents should be considered in the design of ships and offshore structures, as well as their construction and operation.

The following items should be considered in design of offshore structures:

- Currents can cause large steady excursions and slow drift motions of moored platforms.
- Currents give rise to drag and lift forces on submerged structures.
- Currents can give rise to vortex induced vibrations of slender structural elements and vortex induced motions of large volume structures.
- Interaction between strong currents and waves leads to change in wave height and wave period.
- Currents can create seabed scouring around bottom-mounted structures.

Information on the statistical distribution of currents and their velocity profiles is generally scarce for most areas of the world. Currents measurement campaigns are recommended during early phases of an offshore oil exploration development.

Site specific measurements should extend over the water profile and over a period that captures several major storm events. Some general regional information on current conditions are given in ISO 19901-1 (2005) "Metocean Design and Operating Considerations."

If sufficient joint current-wave data are available, joint distributions of parameters and corresponding contour curves (or surfaces) for given exceedance probability levels can be established. Otherwise conservative values, using combined events should be applied.(NORSOK N-003, DNV-OS-C101).

The presence of current in the water produces some minor effects, the most variable of which is the current velocity. This should be added vectorially to the horizontal water particle velocity.

From a practical point of view, some designer increase the maximum wave height is by 5–10% to account for the current effect, and the current is neglected from calculation.

3.4.2.1 Design current profiles

When detailed field measurements are not available the variation in shallow of tidal current velocity water with depth may be modeled as a simple power law, assuming unidirectional current:

$$V_{c,\text{tide}}(Z) = V_{c,\text{tide}}(0)\left(\frac{d+z}{d}\right)^{\alpha} \text{ for } Z \leq 0 \tag{3.19}$$

The variation of *wind* generated current can be taken as either a linear profile from $z = -d_0$ to still water level,

$$V_{c,\text{wind}}(Z) = V_{c,\text{wind}}(0)\left(\frac{d_o+z}{d_o}\right) \text{ for } -d_o \leq Z \leq 0 \tag{3.20}$$

or a slab profile,

$$V_{c,\text{wind}}(Z) = V_{c,\text{wind}}(0) \text{ for } -d_o < Z < 0 \tag{3.21}$$

The profile giving the highest loads for the specific application should be applied.

Wind-generated current can be assumed to vanish at a distance below the still water level,

$$V_{c,\text{wind}}(Z) = 0 \text{ for } z < -d_o \tag{3.22}$$

where

$v_c(z)$ = total current velocity at level z.
z = distance from still water level, positive upward.
$v_{c,\text{tide}}(0)$ = tidal current velocity at the still water level.

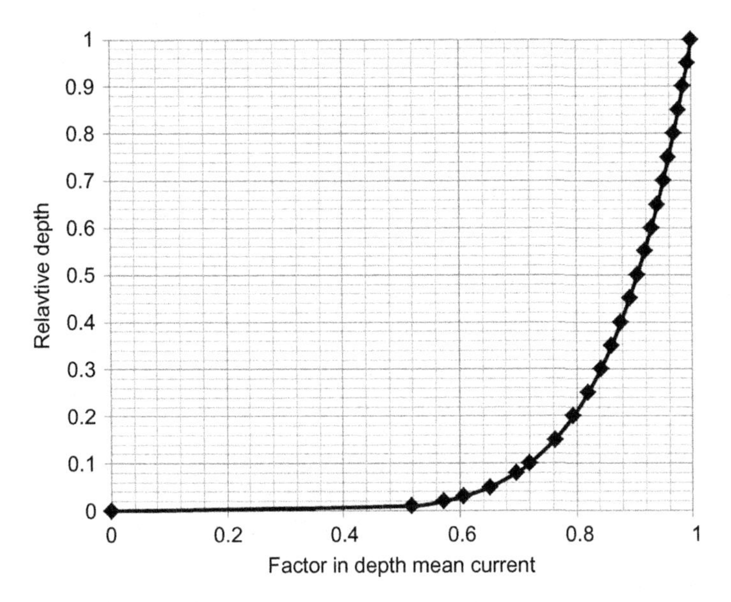

Figure 3.9 Current profile.

$v_{c,\mathrm{wind}}(0)$ = wind-generated current velocity at the still water level.
d = water depth to still water level (taken as positive).
$d0$ = reference depth for wind generated current.
$d0 = 50$ m.
α = exponent, typically $\alpha = 1/7$.

In deep water along an open coastline, wind-generated current velocities at the still water level may, if statistical data are not available, be taken as follows:

V_c,wind(0) = k U_1 hour, 10 m, where $k = 0.015$–0.03

U_1 hour, 10 m is the 1 hour sustained wind speed at a height 10 m above sea level.

The variation in current velocity over depth depends on the local oceanographic climate, the vertical density distribution, and the flow of water into or out of the area. This may vary from season to season. Deepwater profiles may be complex. Current direction can change 180° with depth.

API mentions that the minimum speed is 0.31 km/hr (0.2 knots), as shown in the current profile figure.

A sample of a current profile is illustrated in Figure 3.9, and Table 3.17 gives the factors to be applied to the depth-mean current speed, to give current speeds at different depths.

3.5 Earthquake load

Approximately more than a hundred conventional steel pile supported platforms have been installed in high-activity earthquake regions, such as offshore California,

Table 3.17 **Relation between relative depth and the factors**

Relative depth	Factors
0	0
0.01	0.517947
0.02	0.57186
0.03	0.605963
0.05	0.651836
0.08	0.697106
0.1	0.719686
0.15	0.762603
0.2	0.794597
0.25	0.820335
0.3	0.841982
0.35	0.86073
0.4	0.877307
0.45	0.892193
0.5	0.905724
0.55	0.91814
0.6	0.929624
0.65	0.940315
0.7	0.950323
0.75	0.959736
0.8	0.968625
0.85	0.97705
0.9	0.985061
0.95	0.992699
1	1

Alaska, New Zealand , Japan, China, and Indonesia and new areas in high-activity earthquake regions such as offshore structures in Venezuela, Trinidad, and the Caspian Sea.

As a first step, the designer should have a specific a seismic study about the area on which the platform will be constructed to define the seismic activity. Evaluation of the intensity and characteristics of ground motion should be considered in the study.

It is worth mentioning that fixed offshore structures are usually long-term structures that likely will spend 1–5 seconds under in lateral, flexural, and torsion stress. The first vertical mode periods are frequently in the range of 0.3 to 0.5 seconds. The ductility requirements are intended to ensure that the platform has sufficient reserve capacity to prevent its collapse during rare intense earthquake motion, although structuraol damage may occur.

In regions of low seismic activity, less than 0.05 g the design of environmental loading other than earthquake will provide sufficient resistance against earthquake.

Table 3.18 Relation between seismic zone and effective horizontal ground acceleration

Zone or relative seismicity factor, Z	0	1	2	3	4	5
Ratio of effective horizontal ground acceleration to gravitational acceleration, G	0	0.05	0.10	0.20	0.25	0.40

In case of seismic activity of 0.05–0.1 g, the important item is the deck constituents, such as piping and facilities.

The platform should be checked against earthquake by using a dynamic analysis procedure, such as spectrum analysis or time history analysis.

The uniform structure system is based on the following principles: the structure's vertical elements are continues down to the foundation and do not experience sudden changes in stiffness, the building height is no more than 100 m, and the maximum ratio of height to the horizontal dimension is equal to or more than 5 in the direction of the earthquake force. The seismic load is calculated from the following general equation:

$$V = Z.I.K.C.S.W \quad (3.23)$$

where Z is the earthquake intensity factor and its value is 0.1 in zone 1 and 0.2 in zone 2, and 0.3 in zone 3, as shown in Table 3.18, based on API RP2A (2007); I is the importance factor of the building, which in critical structures is considered to be 1.25. The value of S depends on the soil type, K is the factor for the structure system, ad W is the structure weight. From this equation , the structure weight is the major factor that affects the earthquake load.

The first step in the seismic analysis is dynamic analysis to obtain the eigenvalue and eigenvector for the structure. The mass used in the dynamic analysis should consist of the mass of the platform associated with gravity loading, the mass of the fluids enclosed in the structureand the equipment on it, and the added mass. The added mass may be estimated as the mass of the displaced water for motion transverse to the longitudinal axis of the individual structural framing and appurtenances. The damping ratio is considered to be 5% in the dynamic analysis.

Based on API RP2A(2007), by using the response spectrum method and one design spectrum is applied equally in both horizontal directions, the complete quadratic combination (CQC) method may be used to combine modal responses, and the square root of the sum of the squares (SRSS) may be used for combining the directional responses.

The spectrum analysis method is for the structure with a uniform shape and structure system and a structure height between 100 and 150 m or the ratio between the heights to the horizontal dimension more than 5 in the earthquake load direction.

Table 3.19 Response spectra relation between time period and normalizes spectral acceleration to gravity

Time period, T, Sec.	Spectral acceleration/effective ground acceleration(Sa/G) for different soil type		
	A	B	C
0.04–0.05	1.0		
0.05–0.13	20T		
0.125–0.72	2.5		
0.125–0.48	2.5		
0.125–0.32	2.5		
0.32–5.0	1.8/T		
0.5–5.0		1.2/T	
0.8–5.0			0.8/T

The effect of the seismic force on the structure in this item as a static lateral load effect on the floor level of the structure, and its values are calculated from the dynamic properties as the natural period and natural mode that is calculated by the modal analysis.

API RP2A presents a response spectral soil types in the Table 3.19 are A, B, and C, the classification of which follows: Soil A is a rock crystalline, conglomerate, or shalelike material generally having shear wave velocities in excess of 914 m/s (3000 ft/s). Soil type B is shallow strong alluvium competent sands, silts, and stiff clays with shear strengths in excess of about 72 kPa (1500 psf), limited to depths of less than about 200 ft (61 m), and overlying rocklike materials. Soil Type C is deep strong alluvium competent sands, silts, and stiff clays with thicknesses in excess of about 61 m (200 ft) and overlying rocklike materials.

For the strength requirement, the basic AISC allowable amounts may be increased by 70%. The procedures in ISO 19901-2 for design against seismic events should be followed. The design loads, combinations of loads, and action effects resulting from ground motion, and the requirements that a fixed steel offshore structure subjected to seismic actions should satisfy the procedures.

Based on API RP2A, in areas of low seismic activity, platform design normally is controlled by storm or other environmental loading rather than earthquakes. For areas where the strength level design horizontal ground acceleration is less than 0.05 g, such as the Gulf of Mexico, no earthquake analysis is required, since the design for environmental loading other than earthquake provides sufficient resistance against potential effects from seismically active zones.

As outlined in ISO 19901-2, a two-level seismic design procedure should be followed. The structure should be designed to the ultimate limit state (ULS) for strength and stiffness, when it is subjected to an extreme level earthquake (ELE). Under the ELE, it should sustain little or no damage. The structure is then checked when it is subjected to an abnormal level earthquake (ALE) to ensure that it meets

Table 3.20 Applicability of ductile design recommendation

SRC	Recommendations for ductile design
1	Not applicable
2	Optional
3	Recommended
4	Recommended

reserve strength and energy dissipation requirements. Under the ALE, the structure may sustain considerable damage, but structural failures causing loss of life or major environmental damage are not expected to occur.

In accordance with API RP2A, earthquake loading should be determined by response spectrum analysis. Both the strength level (SLE) and ductility level (DLE) earthquakes should be considered in the design. The following return periods are used in design: SLE for 200 years and DLE for 500 years.

ISO 19901-2 gives alternative procedures for determining seismic loads and alternative methods of evaluation of seismic activity. The selection of the procedure and the method of evaluating the activity depend on a structure's seismic risk category (SRC). The SRC depends on the platform's exposure level and the seismic zone in which it stands, which is given in ISO 19901-2. Its requirements are

1. Determine seismic actions using either the simplified seismic action procedure or the detailed seismic action procedure, as specified in ISO 19901-2.
2. Evaluate seismic activity and the associated response spectra for the design of a structure against excitation of its base by ground motions using either ISO maps, regional maps, or a site-specific seismic hazard analysis, as specified in ISO 19901-2.
3. Demonstrate the ALE performance of the structure, which can require a nonlinear analysis.

Specific recommendations for the ductile design of fixed steel offshore structures are dependent on the structure's SRC, as given in Table 3.20.

Both simplified and detailed seismic action procedures require an estimate of the seismic reserve capacity factor, C_r. This factor represents a structure's ability to sustain ground motions due to earthquakes beyond the ELE. It is defined as the ratio of spectral acceleration that causes structural collapse or catastrophic system failure to the ELE spectral acceleration. For fixed steel offshore structures, the representative value of C_r may be estimated from the general characteristics of a structure's design; see Table 3.21.

Where the values of C_r in Table 3.21 are not used a value may be assumed. In such cases, both of the following conditions apply:

- If the simplified seismic action procedure is followed, the assumed value of C_r should not exceed 2.8 for manned unevacuated platforms, 2.4 for manned evacuated platforms, and 2.0 for unmanned platforms.
- A nonlinear time history analysis should be performed to ensure survival in the event of an ALE. As an alternative, a static pushover analysis may be performed to confirm that that C_r is equal to or higher than that assumed.

Table 3.21 Representative values of seismic reserve capacity factor, C_r

Characteristics of structure design	C_r
The recommendations for ductile design are followed and a nonlinear static pushover analysis is performed to verify the global performance of the structure under ALE conditions	2.80
The recommendations for ductile design are followed, but a nonlinear static pushover analysis to verify ALE performance is not performed	2.00
The structure has a minimum of three legs A bracing pattern consisting of leg-to leg diagonals with horizontals, or X-braces without horizontals The slenderness ratio $(K \cdot L/r)$ of diagonal bracing in vertical frames is limited to no more than 80 and $F_y \cdot D/E \cdot t \leq 0.069$. For X-bracing in vertical frames the same restrictions apply, where the length L to be used depends on the loading pattern of the X-bracing. A nonlinear analysis to verify ALE performance is not performed	1.40
If none of these characterizations apply	1.1

3.5.1 Extreme level earthquake requirements

The mass used in the dynamic analysis consists of the mass of the structure associated with

- The permanent loads, G_1 and G_2.
- 75% of the variable loads, Q_1.
- The mass of entrapped water.
- The added mass.

The added mass may be estimated as the mass of the displaced water for motion transverse to the longitudinal axis of individual structural members and appurtenances. For motions along the longitudinal axis of the structural members and appurtenances, the added mass may be neglected.

The structural model includes the three-dimensional distribution of the stiffness and mass of the structure.

Joints in the model may be treated as rigid. Asymmetry in the distribution of the stiffness and mass of the structure can lead to significant torsion and should be considered in design.

In computing the dynamic characteristics of braced, pile-supported fixed steel offshore structures, a modal damping ratio of up to **5%** may be used in the dynamic analysis of the ELE event. Additional damping, including hydrodynamic or soil-induced damping, should be substantiated by special studies.

Pile-soil performance and pile design requirements should be determined on the basis of studies that consider the design loads, installation procedures, cyclic, and strain rate effects on soil properties and the characteristics of soils as appropriate for the axial or lateral capacity algorithm being used. The stiffness and capacity of the pile foundation should be addressed in a manner compatible with the calculation of the axial and lateral response.

For the design of piles for an ELE event, a partial resistance factor of 1.25 is used to determine the axial pile capacity, and a partial resistance factor for the $p-y$ curves of 1.0 is used to determine the lateral pile capacity.

3.5.2 Abnormal level earthquake requirements

Based on ISO 19002, the structure foundation system should be analyzed to demonstrate an ability to withstand the rare, intense ALE earthquake. This analysis should establish that

- The structure does not globally collapse during the earthquake.
- The structural integrity of the topside is maintained.

The characteristics of the rare, intense ALE event are developed according to ISO 19901-2. The stability of the structure foundation system during the ALE event is demonstrated by analytical procedures that are rational and reasonably representative of the expected response of the structural and soil components to intense ground shaking.

The designer should develop a thorough insight into the performance of the structure and its foundation during the ALE event. The expected nonlinear effects, including material yielding, buckling of structural components, and pile failures, should be adequately modeled and captured. The time history method of analysis is recommended; however, a static pushover analysis may be used. While an analysis for the extreme level of earthquake event focuses on internal forces and stresses, the focus of an analysis for the abnormal level of earthquake event is on strains and displacements.

3.5.3 ALE structural and foundation modeling

Analysis of an abnormal level earthquake should be based on representative, best-estimate values of parameters such as steel yield strength, member slenderness, member strength, and soil strength, including the axial and lateral capacities of piles. The partial resistance factors are equal to 1.0 for axial and lateral pile capacities under ALE. Models of the structural and soil elements should include, as appropriate, the representative degradation of strength and stiffness under abnormal action reversals, the interaction of axial forces and bending moments in structural members, hydrostatic pressures, local inertial effects caused by members vibrating out of plane, and the $P-\Delta$ effect of earthquake actions.

In an analysis of the abnormal earthquake load event, the yield stress of the material and the nonlinear behavior of the joints, including any design strength bias, should be represented. Local joint deformations greater than 5% of the chord diameter should be specifically investigated; otherwise, the joint should be assumed to fracture at this deformation level.

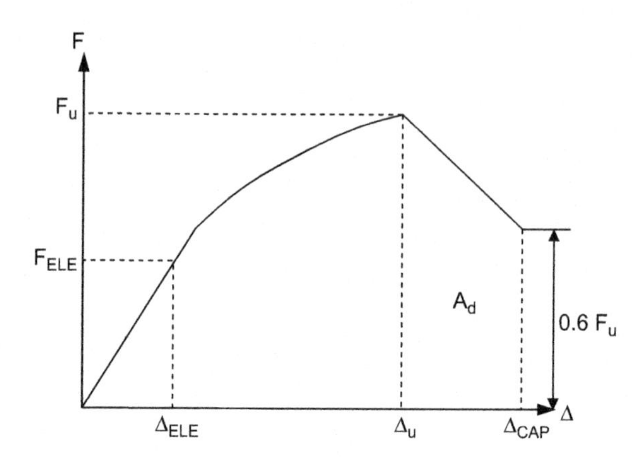

Figure 3.10 Deformation curve for seismic action.

The objective of nonlinear static pushover analysis is to verify that the seismic reserve capacity factor, C_r, of the structure as designed is greater than that initially estimated for the design. The actions used in a static pushover analysis should represent the pattern of ALE seismic actions on the structure and foundation. Action patterns in a pushover analysis may be constructed to match the shear and moment distributions determined from an ALE response spectrum analysis along the height of the structure. Pushover analyses should be performed in several directions, as given next, to identify the structure's weakest direction:

- The seismic load effect in the direction aligned with the longitudinal (end-on) axis of the structure.
- The seismic load effect in the direction aligned with the transverse (broadside) axis of the structure.
- The seismic load effect in the direction aligned with one or more diagonal axes of the structure.

Diagonal direction(s) can be the weakest direction(s), especially with regard to foundation performance.

Yielding of structural members or piles should not occur at global action levels lower than or equal to the global ELE action F_{ELE}, Figure 3.10.

The seismic reserve capacity factor may be estimated from the global seismic load deformation curve obtained in a static pushover analysis, specifically from global shear vs. deck displacement, as shown in Figure 3.10:

$$C_r = C_{\text{sr}} C_{\text{dr}} \tag{3.24}$$

C_{sr} is a factor corresponding to the strengthening regime of the action-deformation curve and is estimated as

$$C_{\text{sr}} = \Delta_u / \Delta_{\text{ELE}} \tag{3.25}$$

where Δ_{ELE} is the deformation caused by the global ELE action F_{ELE}, and Δ_u is the deformation corresponding to F_u, as presented in Figure 3.10, the ultimate action where the slope of the action-deformation curve becomes negative.

C_{dr} is a factor corresponding to the degrading regime of the load-deformation curve. It is a measure of the energy dissipation capacity of the structure beyond the ultimate seismic action, and the corresponding deformation. C_{dr} is estimated as

$$C_{dr} = [1 + (A_d/F_u\ \Delta_u)]^{0.5} \tag{3.26}$$

where A_d is the area under the action-deformation curve starting from Δ_u and ending with Δ_{CAP}, the deformation capacity of the structure. For the purpose of nonlinear static pushover analysis, the deformation capacity is assumed to be the deformation where the global action falls to 60% of F_u.

The preceding determination of C_r presupposes that the primary sources of degradation of the global resistance have been properly modeled in the static pushover analysis, such as soil degradation, buckling of compression members, and local buckling of members due to rotations at the member end, which reduces the plastic moment capacity.

Alternatively, Δ_u is set as the deformation where the slope of the action-deformation curve is reduced to 5% of the initial elastic slope, and C_{dr} is assumed equal to 1.0.

To ensure the seismic design process is conservative, the lower of the two values of C_r thus determined should be adopted.

3.5.3.1 Topside appurtenances and equipment

Topside design accelerations include the effects of the global dynamic response of the structure and, if appropriate, local dynamic response of the topside and equipment itself.

It is recommended that topside response spectra or other in-structure response spectra be obtained from time history analyses of the complete structure. The topside response spectra are the average values from at least four time history analyses. Direct spectra-to-spectra generation techniques may also be used; however, such methods should be calibrated against the time history method. The topside response spectra should be broadened to account for uncertainties in structure frequencies and soil-structure interaction.

Seismic actions on topside equipment, piping, and appurtenances should be derived by dynamic analysis using either (1) an uncoupled analysis with deck-level floor response spectra as input or (2) a coupled analysis that properly includes a simple dynamic model of the relevant part of the topside or the appurtenance in the global structural model.

Equipment, piping, and other topside appurtenances are designed and supported such that they can resist extreme level earthquake loads. On the other hand, displacements and deformations of the topside inthe case of an extreme earthquake load should be limited or designed against to avoid damage to the equipment, piping, appurtenances, and supporting structures.

Safety critical systems and structures on or in the topside should be designed such that they are functional during and after an abnormal level of earthquake event. Hazardous systems should be designed such that they do not fail catastrophically or rupture during an abnormal earthquake level event. In lieu of performing an ALE analysis of deck-supported structures, topside equipment, and equipment tie-downs, they should be designed with an increased partial action factor on E of 1.15 rather than 0.9.

3.6 Ice loads

Ice applies an impact force to the offshore structure especially in polar areas of the sea such as Alaska. To imagine the effect of the ice on the platforms, it is very important to know that the drifting ice travels at a speed about 1–7% of the wind speed. A typical ice island in Alaska is about 1 km in diameter with 1 km thickness and travels at a speed of 3 knots.

Note that the ratio of the thickness of the ice above water to that underwater is about 1:2, although in general it is in the range from 1:1 to 1:7.

The effect of ice load depends on the spacing of individual members. Generally, the following rules are used:

1. **Spacing ≥ 6 Diameters.** Ice will crush against the tubular members and pass through and around the platform if the tubular spacing is greater than six times the tubular diameter. For groups of tubular members of different size, the average tubular diameter should be used to determine the spacing.
2. **Spacing ≤ 4 Diameters.** As the tubular spacing decreases, interference effects may occur that influence both the load on the tubular members and the failure mode of the ice. With a tubular spacing less than four diameters or with closely spaced conductors between platform legs, ice blocks may wedge inside the structure and the effective contact area becomes the out-to-out dimension across the closely spaced tubular members in the direction of the ice movement. In this case, the total ice load on the structure should be calculated with D taken as the out-to-out dimension across the closely spaced tubular members.
3. **4 < Spacing < 6.** Ice forces are determined by linear interpolation between loads for spacing of 4 and 6.

Note that shielding occurs when the tubular members are located in the lee of other structure members. The loads on these piles may be considerably less as the ice may fail in another mode or simply be cleared away under pack ice pressures. These clearing forces are estimated as the product of the pack ice pressure (estimated at 2 ton/m width based on floe area, floe profile, wind speed, and current velocity) and pile diameter.

3.7 Other loads

These loads are presented by the configuration of the platform and the environmental condition, as follows:

1. Marine growth
2. Scour

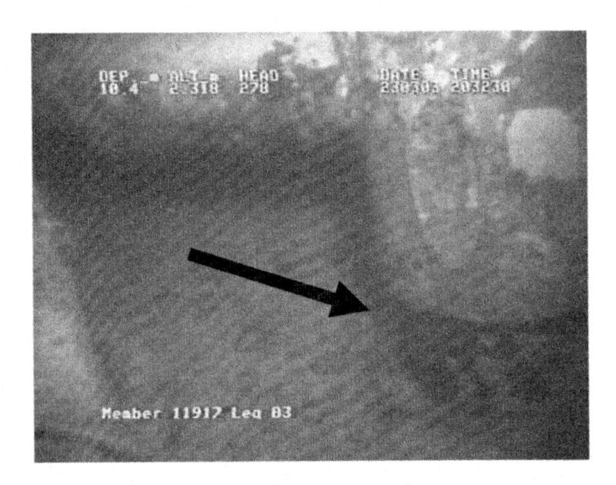

Figure 3.11 Scour on a platform leg.

3. Material selection and corrosion, stress analysis, welding, structure analysis, design for fabrication and installation, welding
4. Marine civil engineering as the installation equipment, installation methods, navigation safety instrument are discussed in Chapter 7
5. Naval architecture as the floating and buoyancy, towing, launching, and controlled flooding also are discussed in Chapter 7.

3.7.1 Marine growth

The marine growth increases the diameter of the jacket member so the drag force increases. Based on API 1.5, from MHHW to −150 ft. MHHW is 300 mm (1 ft) higher than MLLW. Smaller or larger values of thickness may be used from a site-specific study. Structural members are considered smooth if above MHHW or lower than 45 m (150 ft), where marine growth is light enough to ignore. The zone between MHHW and 45 m is considered rough.

3.7.2 Scour

Seabed scour (Figure 3.11) affects both lateral and axial pile performance and capacity. Scour prediction remains an uncertain art. Sediment transport studies may assist in defining scour design criteria, but local experience is the best guide. In practical design, scour is assumed to be 1.5 pile diameter.

3.8 Design for ultimate limit state

An action factor should be applied to each of the nominal external actions in the load combination. The action factors depend on the national or regional building code in use. This is to ensure that similar levels of reliability for topside design are achieved to that implied in other ISO 19900 series international standards.

The combination of factored nominal actions causes amplified internal forces, S. A resistance factor is applied to the nominal strength of each component to determine its factored strength. Each component should be proportioned to have sufficient factored strength to resist S. The appropriate strength and stability criteria are taken from the appropriate national or international building code. These criteria are the formulas for the nominal strength of the component and the associated resistance factors.

Under some conditions, particularly during construction and installation, the internal forces should be computed from unfactored nominal actions and the action factors applied to the internal forces to arrive at S, as discussed in ISO 19902.

Deformation actions can arise from the effects of fabrication tolerance, foundation settlement, or the indeterminate effects of transportation and lift. For the primary structure supported by a multicolumn gravity-base structure (GBS), the movement of the column tops can also cause significant deformation actions. They can occur from operational or accidental thermal effects. All such actions should be considered in combination with operating actions to ensure that serviceability and ultimate limit states are not exceeded.

3.8.1 Load factors

The partial action factors to be used when AISC-LRFD, EC 3 , NS 3472 , BS 5950, or BS 5400 Part 3 is the chosen code follow. The action factors cover maximum gravity and extreme environmental and operating environmental combinations. Other relevant codes or standards may be used. In such cases, appropriate action factors should be evaluated to achieve a similar level of reliability to that implied in the international standard. The procedure should be followed to derive appropriate sets of action factors, as necessary.

The internal force, S, resulting from the design action, F_d, ise calculated using Equation (3.27):

$$F_d = \gamma_G(G_1 + G_2) + \gamma_Q(Q_1 + Q_2) \tag{3.27}$$

where

G_1 = the permanent load on the structure such as the weight of the structure with associated equipment and other objects.

G_2 = the permanent load on the structure such as the weight of equipment and other objects that remain constant for long periods but can change be changed during a mode of operation.

Q_1 = the variable load on the structure by the weight of consumable supplies and fluids in pipes, tanks, and stores; the weight of transportable vessels and containers used for delivering supplies; and the weight of personnel and their personal effects on the structure.

Q_2 = the short-duration variable load imposed on the structure from operations, such as lifting of drill string, lifting by cranes, machine operations, vessel mooring, and helicopter loadings.

Table 3.22 Partial actions factors in different international standards and specifications for maximum gravity

Code	Permanent, γ_G	Variable, γ_Q
ISO 19902	1.30	1.50
AISC–LRFD	1.25	1.40
NS 3472	1.25	1.45
EC3	1.30	1.50
BS5950	1.45	1.65
BS 5950 Part3	1.25	1.45

Table 3.23 Partial load factors for in-place situation

Design situation	Partial load factors[a]					
	γ_{G1}	γ_{G2}	γ_{Q1}	γ_{Q2}	γ_{Eo}	γ_{Ee}
Permanent and variable actions only	1.3	1.3	1.5	1.5	0.0	0.0
Operating situation with corresponding wind, wave, and current conditions[b]	1.3	1.3	1.5	1.5	$0.9\gamma_{f,E}$	0.0
Extreme conditions when the action effects due to permanent and variable actions are additive[c]	1.1	1.1	1.1	0.0	0.0	γ_E
Extreme conditions when the action effects due to permanent and variable actions opposed[d]	0.9	0.9	0.8	0.0	0.0	γ_E

[a] A value of 0 for a partial action factor means that the action is not applicable to the design situation.
[b] For this check, G_2, Q_1, and Q_2 are the maximum values for each mode of operation.
[c] For this check, G_1, G_2, and Q_1 include those parts of each mode of operation that can reasonably be present during extreme conditions.
[d] For this check, G_2 and Q_1 exclude any parts associated with the mode of operation considered that cannot be ensured of being present during extreme conditions.

The partial action factors for selected codes and standards are given in Table 3.22.

3.8.1.1 In-place analysis by ISO19902

In performing in-place analysis through load resistance factor design based on ISO 19902, the general equation for determining the design load (action) (F_d) for in-place situations is given in Equation (3.28), the appropriate partial action factors for each design situation are given in Table 3.23:

$$F_d = \gamma_{G1} G_1 + \gamma_{G2} G_2 + \gamma_{f,Q1} Q_1 + \gamma_{Q2} Q_2 + \gamma_{,\mathrm{Eo}}(E_o + \gamma_{f,D} D_o) + \gamma_{\mathrm{Ee}}(E_e + \gamma_D D_e) \tag{3.28}$$

where

G_1 and G_2 are the permanent loads.
Q_1 and Q_2 are the variable loads.
E_o is the environmental load defined by the owner as the operating wind, wave, and current parameters.
D_o is the equivalent quasi-static action representing dynamic response but caused by the wave condition that corresponds with that for E_o.
E_e is the extreme action due to wind, waves, and current.
D_e is the equivalent quasi-static action representing dynamic response.
γ_{G1}, γ_{G2}, γ_{Q1} and γ_{Q2} are the partial action factors for the various permanent and variable actions and for which values for different design situations are given in Table 3.23.

On the other hand, γ_E and γ_D are the partial action factors for the environmental loads the values of which are defined by the owner through the statement of requirement (SOR) document; γ_{Eo} and γ_{Ee} are partial action factors applied to the total quasi-static environmental action plus equivalent quasi-static action representing dynamic response for operating and extreme environmental conditions, respectively, and for which values for different design situations are given in Table 3.23.

3.8.1.2 *Extreme environmental situation for fixed offshore platforms*

The internal force, S, resulting from the design action, F_d, is calculated using Equation (3.29):

$$F_d = \gamma_G (G_1 + G_2) + \gamma_Q Q_1 + \gamma_E (E_e + 1.25 D_e) \quad (3.29)$$

When the internal forces due to gravity forces oppose those due to wind, wave, and current, the internal force, S, resulting from the design action, F_d, is calculated using reduced partial action factors as:

$$F_d = (1/\gamma_G)(G_1 + G_2) + (1/\gamma_Q) Q_1 + \gamma_E (E_e + 1.25 D_e) \quad (3.30)$$

For this combination, G_2 and Q_1 exclude any actions that cannot be assured to be present during an extreme storm to maximize the difference between the opposing action effects.

The appropriate partial action factors for the environmental load depends on the location of the installation. ISO 19902 allows a value of γ_E of 1.35 where no other information is available. The partial action factors for selected codes and standards are given in Table 3.24.

3.8.1.2.1 Example 3.6

In the case of a live load equal to 10 kN/m^2, a self-weight is equal to 12 kN/m^2, and a wave load of 20 kN/m^2, what is the design load for different codes for the wave load effect in the direction of a live load and a dead load? Table 3.25 has the solution.

Table 3.24 **Partial actions factors in different international standards and specifications in extreme environmental condition**

Code	Partial load factor		
	Permanent, γ_G	Variable, γ_Q	Environmental, γ_E
ISO 19902	1.10	1.10	1.00 γ_{ELs}
AISC -LRFD	1.05	1.05	0.96 γ_{ELs}
NS 3472	1.05	1.05	0.96 γ_{ELs}
EC3	1.10	1.10	1.00 γ_{ELs}
BS5950	1.20	1.20	1.11 γ_{ELs}
BS 5950 Part3	1.05	1.05	0.96 γ_{ELs}

Note: γ_{ELs} is the appropriate partial factor for the substructure.

Table 3.25 **Solution to Example 3.6**

Code	Design load, kN/m^2
ISO 19902	44.2
AISC LRFD	42.3
EC3	44.2
BS5950	48.6
BS 5950 part3	42.3

3.8.1.2.2 Example 3.7

In the case of a live load equal to 10 kN/m^2, a self-weight is equal to 12 kN/m^2, and a wave load of 20 kN/m^2, what is the design load for different codes for the wave load effect in the opposite direction to the live load and dead load? Table 3.26 has the solution.

3.8.1.3 Operating environmental situations for fixed platforms

Platform operations are often limited by environmental conditions and differing limits may be set for different operations. Examples of operations that might be limited by environmental conditions include

- Drilling and workover.
- Crane transfer to and from supply vessels.
- Crane operations on deck.
- Deck and overside working.
- Deck access.

Each operating situation that might be restricted by environmental conditions should be assessed as followsw, in which E_o and D_o represents the environmental action limiting the operations. The value of Q_2 is that associated with the particular operating situation being considered.

Table 3.26 **Solution to Example 3.7**

Code	Design load, kN/m^2
ISO 19902	−20
AISC LRFD	−19.0476
EC3	−20
BS5950	−21.6667
BS 5950 Part3	−19.0476

Table 3.27 **Partial actions factors in different international standards and specifications in operating environmental condition**

Code	Partial load factor		
	Permanent, γ_G	Variable ,γ_G	Environmental, γ_E
ISO 19902	1.30	1.50	1.20
AISC LRFD	1.25	1.40	1.15
NS 3472	1.25	1.45	1.15
EC3	1.30	1.50	1.20
BS5950	1.45	1.65	1.35
BS 5950 Part3	1.25	1.45	1.15

The internal force, S, resulting from the design action, F_d, is calculated using Equation (3.31):

$$F_d = (1/\gamma_G)(G_1 + G_2) + 1/\gamma_Q Q_1 + \gamma_E(E_o + D_o) \tag{3.31}$$

The action factors for selected codes and standards are given in Table 3.27.

3.8.2 *Partial action factors*

Each member, joint, and foundation component should be checked for strength using the internal force due to load effect (S) resulting from the design action F_d calculated by equations (3.32) and (3.33):

$$Q = 1.1G_1 + 1.1G_2 + 1.1Q_1 + 0.9E \tag{3.32}$$

where E is the inertia action induced by the ELE ground motion and determined using dynamic analysis procedures, such as response spectrum analysis or time history analysis. G_1, G_2, and Q_1 include loads that are likely to be present during an earthquake.

Table 3.28 Case for 1-year storm conditions

Load case	Load condition	Combination
1	1	Dead load + buoyancy
2	2	Unmodeled dead load (jacket and deck)
3	3	Blanket live load on main deck
4	4	Blanket live load on helideck
5	11	Wind + wave + current hitting 0.0°
6	12	Wind + wave + current hitting 45°
7	13	Wind + wave + current hitting 90°
8	14	Wind + wave + current hitting 135°
9	15	Wind + wave + current hitting 180°
10	16	Wind + wave + current hitting 225°
11	17	Wind + wave + current hitting 270°
12	18	Wind + wave + current hitting 315°

Table 3.29 Case for 100-year storm conditions

Load case	Load condition	Combination
1	1	Dead load + buoyancy
2	2	Unmodeled dead load (jacket and deck)
3	3	Blanket live load on main deck
4	4	Blanket live load on helideck
5	21	Storm wind + wave + current hitting 0.0°
6	22	Storm wind + wave + current hitting 45°
7	23	Storm wind + wave + current hitting 90°
8	24	Storm wind + wave + current hitting 135°
9	25	Storm wind + wave + current hitting 180°
10	26	Storm wind + wave + current hitting 225°
11	27	Storm wind + wave + current hitting 270°
12	28	Storm wind + wave + current hitting 315°

When contributions to the internal forces due to weight oppose the inertia actions due to the earthquake, the partial load factors for permanent and variable actions are reduced such that

$$Q = 0.9G_1 + 0.9G_2 + 0.8Q_1 + 0.9E \quad (3.33)$$

where G_1, G_2, and Q_1 include only loads that are reasonably certain to be present during an earthquake.

For global assessment of the offshore structure platform Tables 3.28 and 3.29 presents a matrix for load combination. This is a traditional load combination used

Table 3.30 **Load combination in 1-year storm condition factors**

Load combination	Load condition										
	1	2	3	4	11	12	13	14	15	16	17	18
30	1.1	1.0	1.0	1.0	1.0							
31	1.1	1.0	1.0	1.0		1.0						
32	1.1	1.0	1.0	1.0			1.0					
33	1.1	1.0	1.0	1.0				1.0				
34	1.1	1.0	1.0	1.0					1.0			
35	1.1	1.0	1.0	1.0						1.0		
36	1.1	1.0	1.0	1.0							1.0	
37	1.1	1.0	1.0	1.0								

Table 3.31 **Load combination in 100-year storm condition factors**

Load combination	Load condition											
	1	2	3	4	21	22	23	24	25	26	27	28
40	1.1	1.0	1.0	1.0	1.0							
41	1.1	1.0	1.0	1.0		1.0						
42	1.1	1.0	1.0	1.0			1.0					
43	1.1	1.0	1.0	1.0				1.0				
44	1.1	1.0	1.0	1.0					1.0			
45	1.1	1.0	1.0	1.0						1.0		
46	1.1	1.0	1.0	1.0							1.0	
47	1.1	1.0	1.0	1.0								1.0

in the input for the design or assessment of fixed offshore platforms. Table 3.30 presents the load combination for 1-year storm conditions, and Table 3.31 presents the load combination for 100 = year storm condition. Tables 3.30 and 3.31 use a matrix for the load combination versus the applied load in Working Stress Design (WSD) concept for 1- and 100-year storm conditions, respectively.

The load combinations for the maximum pile tension condition are in Table 3.32.

3.9 Collision events

In a rigorous impact analysis, if required, accidental design situations are established representing bow, stern, and beam-on impacts on all exposed components.

Table 3.32 Load combination factors for maximum pile tension conditions

Load combination	Load condition											
	1	2	3	4	11	12	13	14	15	16	17	18
40	0.9	0.9	0.75		1.0							
41	0.9	0.9	0.75			1.0						
42	0.9	0.9	0.75				1.0					
43	0.9	0.9	0.75					1.0				
44	0.9	0.9	0.75						1.0			
45	0.9	0.9	0.75							1.0		
46	0.9	0.9	0.75								1.0	
47	0.9	0.9	0.75									1.0

The collision events represent both a fairly frequent condition, during which the structure should suffer only insignificant damage, and a rare event, where the emphasis is on avoiding a complete loss of integrity of the structure.

Two energy levels are considered:

1. A low energy level, representing the frequent condition, based on the type of vessel that would routinely approach alongside the platform, such as a supply boat with a velocity representing normal maneuvering of the vessel approaching, leaving, or standing alongside the platform.
2. A high energy level, representing a rare condition, based on the type of vessel that would operate in the platform vicinity, drifting out of control in the worst sea state in which it is allowed to operate close to the platform.

The accidental design situations for both energy levels are considered. Level 1 represents a serviceability limit state to which the owner can set the requirements based on practical and economic considerations. Level 2 represents an ultimate limit state, in which the structure is damaged but progressive collapse does not occur.

In both cases, the analysis accounts for the vessel's mass as well as its added mass, orientation, and velocity. Effective operational restrictions on vessel approach sectors can limit the exposure to impacts in some areas of the structure.

The vertical height of the impact zone is established based on the dimensions and geometry of the structure and the vessel and account for tidal ranges, operational sea state restrictions, vessel draft, and motions of the vessel.

Accidental damage for a vessel collision should be considered for all exposed elements of an Installation in the collision zone.

The vertical extent of the collision zone should be assessed on the basis of visiting vessel draft, maximum operational wave height, and tidal elevation.

The total kinetic energy, E, involved in accidental collisions can be expressed as

$$E = \frac{1}{2} a m V^2$$

Table 3.33 Steel mechanical properties

Steel type	Yield strength, N/mm^2	Tensile strength, N/mm^2
API5LX52	359	455
API 5L 'B'	241	414

where

m = vessel displacement (kg)
a = vessel added mass coefficient
= 1.4 for sideways collision
= 1.1 for bow or stern collision
V = impact speed (m/s)

The total kinetic energy, E, should be taken to be at least

- 14 MJ for sideways collision
- 11 MJ for bow or stern collisions

which corresponds to a vessel of 5000 ton displacement with an impact speed of 2 m/s. A reduced impact energy may be acceptable in cases where the size of visiting vessels or their operations near the Installation are restricted. In this instance. a reduced vessel size or impact speed may be considered.

The reduced impact speed, V, in meters per second may be estimated numerically from the empirical relation

$V = \frac{1}{2}H_s$ (m/s)

where H_s = maximum permissible significant wave height in meters for vessel operations near the Installation, as in OTI (1988).

The energy absorbing mechanisms during the collision should be evaluated. Typically, local member denting, elastic and plastic deflection of the impacted member, and global elastic and plastic response of the whole structure and denting of the ship are the main mechanisms.

In a rigorous impact analysis, the collision actions should be evaluated based on a dynamic time simulation. The duration of the simulation should be sufficient to cover all relevant phases of the collision and the energy dissipation process.

3.10 Material strength

Steel should conform to a definite specification and minimum strength level, group, and class specified by the designer. Certified mill test reports or certified reports of tests made by the fabricator or a testing laboratory in accordance with ASTM A6 or A20, as applicable to the specification listed in Table 3.33, constitutes evidence of conformity with the specification. Unidentified steel should not be used.

Steel may be grouped according to strength level and welding characteristics as follows:

- Group I designates mild steels with specified minimum yield strengths of 40 ksi (280 MPa) or less. The carbon equivalent is generally 0.40% or less, and these steels may be welded by any of the welding processes as described in AWS D1.1.
- Group II designates intermediate strength steels with specified minimum yield strengths of over 40 ksi (280 MPa) through 52 ksi (360 MPa). Carbon equivalent ranges are up to 0.45% and higher, and these steels require the use of low hydrogen welding processes.
- Group III designates high strength steels with specified minimum yield strengths in excess of 52 ksi (360 MPa).

In most offshore structures using ASTM A572 Grade 50 for this grade, the material should conform to the requirements of ASTM A572 and ASTM A6 except as noted later. All steel should be supplied in the normalized condition.

If the steel required by ASTM A36 in an offshore structure topside, the material conforms to the requirements of ASTM A36 and ASTM A6 except as noted later. All steel should be supplied in the normalized condition.

Tubular members should not contain any dents greater than 3 mm or 1% × OD, whichever is the lesser. The length of the dent should not exceed 25% × OD.

Unless otherwise specified by the designer, plates should conform to one of the specifications listed in Table 3.34. The structural shape specifications are listed in Table 3.35. Steels above the thickness limits stated may be used and are considered by the designer.

Structural pipe should be fabricated in accordance with API Spec. 2B, ASTM A139, ASTM A252, ASTM A381, or ASTM A671 using grades of structural plate listed in Table 3.36 except that hydrostatic testing may be omitted.

3.11 Cement grout

If required by the design, the space between the piles and the surrounding structure should be carefully filled with grout. Prior to installation, the compressive strength of the grout mix design should be confirmed on a representative number of laboratory specimens cured under conditions that simulate the field conditions. Laboratory test procedures should be in accordance with ASTM C109. The unconfined compressive strength of 28-day-old grout specimens computed as described in ACI 214-77 but equating f'_c to f_{cu}, should not be less than either 2500 psi (17.25 MPa) or the specified design strength.

A representative number of specimens taken from random batches during grouting operations should be tested to confirm that the design grout strength has been achieved. Test procedures should be in accordance with ASTM 109. The specimens taken from the field should be subjected, until test, to a curing regime representative of the in-situ curing conditions, that is, underwater and with appropriate seawater salinity

Table 3.34 **Mechanical properties for structural steel plates**

Group	Class	Specifications and grade	Yield strength, MPa (ksi)	Tensile strength, MPa (ksi)
I	C	ASTM A36 (to 50 mm thickness)	250 (36)	400–550 (58–80)
		ASTM A131 Grade A (to 12 mm thick)	235 (34)	400–490 (58–71)
		ASTM A285 Grade C (to 19 mm)	205 (30)	380–515 (55–75)
I	B	ASTM A131 Grades B,D	235 (34)	400–490 (58–71)
		ASTM A516 Grade 65	240 (35)	450–585 (65–85)
		ASTM A573 Grade 65	240 (35)	450–530 (65–77)
		ASTM A709 Grade 36T2	250 (36)	400–550 (58–80)
I	A	ASTM A131 Grades CS,E	235 (34)	400–490 (58–71)
II	C	ASTM A572 Grade 42 (to 50 mm thick)	290 (42)	415 min (60 min)
		ASTM A572 Grade 50 (to 50 mm thick; S91 required over 12 mm)	345(50)	450 min (65 min)
II	B	API Spec 2 MT1	345 (50)	483–620 (70–90)
		ASTM A709 Grades 50T2, 50T3	345 (50)	450 min (65 min)
		ASTM A131 Grade AH32	315 (45.5)	470–585 (68–85)
		ASTM Grade AH36	350 (51)	490–620 (71–90)
II	A	API spec 2H Grade 42	290 (42)	430–550 (62–80)
		Grade 50 (to 62 mm thick)	345 (50)	483–620 (70–90)
		(Over 62 mm thick)	325 (47)	483–620 (70–90)
		API Spec 2W Grade 50 (to 25 mm thick)	345–517 (50–75)	448 min (65 min)
		(Over 25 mm thick)	345–483 (50–70)	448 min (65 min)
		API Spec 2Y Grade 50 (to 25 mm thick)	345–517 (50–75)	448 min (65 min)
		(Over 25 mm thick)	345–483 (50–70)	448 min (65 min)
		ASTM A131 Grades DH32, EH32	315 (45.5)	470–585 (68–85)
		Grades DH36, EH36	350 (51)	490–620 (71–90)
		ASTM A537 class I (to 62 mm thick)	345 (50)	485–620 (70–90)

		ASTM A633 Grade A	290 (42)	435–570 (63–83)
		Grades C, D	345 (50)	485–620 (70–90)
		ASTM A678 Grade A	345 (50)	485–620 (70–90)
III	A	ASTM A537 Class II (to 62 mm thick)	415 (60)	550–690 (80–100)
		ASTM A678 Grade B	415 (60)	550–690 (80–100)
		API Spec 2 W Grade 60 (to 25 mm thick)	414–621 (60–90)	517 min. (75 min)
		(Over 25 mm thick.)	414–586 (60–85)	517 min. (75 min)
		API Spec2Y Grade 60 (to 25 mm thick)	414–621 (60–90)	517 min. (75 min)
		(Over 25 mm thick.)	414–586 (60–85)	517 min. (75 min)
		ASTM A710 Grade A Class 3 (quenched and precipitation heat treated)		
		To 50 mm	515(75)	585 (85)
		50 mm to 100 mm	450(65)	515(75)
		Over 100 mm	415(60)	485 (70)

Table 3.35 **Mechanical properties for structural steel shapes**

Group	Class	Specifications and grade	Yield strength, MPa (ksi)	Tensile strength, MPa (ksi)
I	C	ASTM A36 (to 50 mm thickness)	250 (36)	400–550 (58–80)
		ASTM A131 Grade A (to 12 mm thick)	235 (34)	400–550 (58–80)
I	B	ASTM A709 Grade 36T2	250 (36)	400–550 (58–80)
II	C	API Spec 2 MT2 class C	345 (50)	450–620 (65–90)
		ASTM A572 Grade 42 (to 50 mm thick)	290 (42)	415 min (60 min)
		ASTM A572 Grade 50 (to 50 mm thick; S91 required over 12 mm)	345 (50)	450 min (65 min)
		ASTM A992	345–450 (50–65)	450 min (65 min)
II	B	API Spec 2 MT2 class B	345 (50)	450–620 (65–90)
		ASTM A709 Grades 50T2, 50T3	345 (50)	450 min (65 min)
		ASTM A131 Grade AH32	315 (45.5)	470–585 (68–85)
		ASTM Grade AH36	350 (51)	490–620 (71–90)
II	A	API spec 2MT2 Class A	345 (50)	450–620(65–90)
		ASTM A913 Grade 50	345 (50)	450 min (65 min)

Table 3.36 Mechanical properties for structural steel pipes

Group	Class	Specifications and grade	Yield strength, MPa (ksi)	Tensile strength, MPa (ksi)
I	C	API 5L Grade B	240 (35)	415 min (60 min)
		ASTM A53 Grade B	240 (35)	415 min (60 min)
		ASTM A135 Grade B	240 (35)	415 min (60 min)
		ASTM A 139 Grade B	240 (35)	415 min (60 min)
		ASTM A500 Grade A (round)	230 (33)	310 min (45 min)
		(Shaped)	270 (39)	310 min (45 min)
		ASTM A501	250 (36)	400 min (58 min)
I	B	ASTM A106 Grade B (normalized)	240 (35)	415 min (60 min)
		ASTM A524 Grade I (to 10 mm thick)	240 (35)	415 min (60 min)
		Grade II (over 10 mm thick)	205 (30)	380–550(55–80)
I	A	ASTM A333 Grade 6	240 (35)	415 min (60 min)
		ASTM A334 Grade 6	240 (35)	415 min (60 min)
II	C	API 5L Grade X42 2% max. cold expansion	290 (42)	415 min (60 min)
		API 5L Grade X52 2% max. cold expansion	360 (52)	455 min (66 min)
		ASTM A500 Grade B (round)	290 (42)	400 min. (58 min)
		(Shaped)	320 (46)	400 min. (58 min)
		ASTM A618	345 (50)	485 min. (70 min)
II	B	API 5L Grade X52 with SR5 or SR6	360 (52)	455 min. (66 min)

Further reading

American Institute of Steel Construction, 1989. *Specification for Structural Steel Buildings, Allowable Stress Design and Plastic Design*. AISC, Chicago, USA.

American Institute of Steel Construction, 1999. *Load and Resistance Factor Design Specification for Structural Steel Buildings*. AISC LRFD. AISC, Chicago, USA.

American Petroleum Institute, 2007. Recommended Practice for Planning, Designing, and Constructing Fixed Offshore Platforms – Working Stress Design, API RP2A-WSD. twentieth ed. API, Washington, DC.

British Standard, 2000 Structural use of Steelwork in Buildings, Code of Practice for Design. BS 5950, London.

British Standard, 2000. Steel, Concrete and Composite Bridges, Code of Practice for Design of Steel Bridges. BS 5400 Part 3, London.

DNV, 2008. Design of Offshore Steel Structures, General (LRFD Method).

ENV 1993-1-1:1992, Eurocode 3. 1992. Design of Steel Structures, General Rules and Rules for Buildings, cen, Brussels.

International Organization for Standardization, 2005. *Metocean Design and Operating Considerations*. ISO 19901-1. IOS, Amsterdam, petroleum and natural gas industry, specific requirement for offshore structure, part2: seismic design procedures and criteria, cen, Brussels.

International Organization for Standardization ISO 19901-2.

DNV (Det Norske Veritas), Offshore Standard DNV-OS-C101: Design of Offshore Steel Structures, General (LRFD Method), October 2008. http://dc140.4shared.com/doc/Pflypj7r/preview.html Design and Plastic Design, June 1989.

Norwegian Standard, 1988. Steel Structures, Design Rules. NS 3472,OTI.

Offshore structures design

4

4.1 Introduction

Offshore fixed platform design consists mainly of three parts, the first one is the design of the deck that carries the topside facility. The dimension of the deck depends on the function of the platform and the facilities located on it.

The second part is the design of the jacket, which depends mainly on the water depth, the wave and current loads, and other load effects, as described in Chapter 3. The configuration of the jacket structure system is chosen based on the water depth and the designer's experience.

The third phase of the design is to check the deck and jacket design for cases of lifting, pullout, transportation, launch, and installation.

The three main components to a steel template platform are

1. Topside facilities (decks)
2. Jacket
3. Piles

Topside facilities frequently comprise three decks:

1. Drilling deck
2. Wellhead/production deck
3. Cellar deck

The main function of the topside is to carry the load from the facilities and drilling equipment to the jacket structure. The percentage between the weights of the topside component and the total platform weight is about 30–45%.

The function of the jacket is to surround the piles and hold the pile extensions in position all the way from the mud line to the deck substructure. Moreover, the jacket provide support for boat landings, mooring bits, barges bumpers, corrosion protection system, and many other platform components. Example jacket drawings are presented in Figure 4.1. The plan view for highest jacket level are presented in Figure 4.2. Figure 4.3 presents the mud mat level, whereas Figure 4.4 presents the horizontal frames at different levels.

4.2 Guide for preliminary design

The topside structure is designed based on AISC WSD or LRFD specifications. Usually, the main supporting element is plate girder or tubular truss, but in most cases, it is preferred to select a tubular member, as it carries little wind load.

Marine Structural Design Calculations. DOI: http://dx.doi.org/10.1016/B978-0-08-099987-6.00004-0

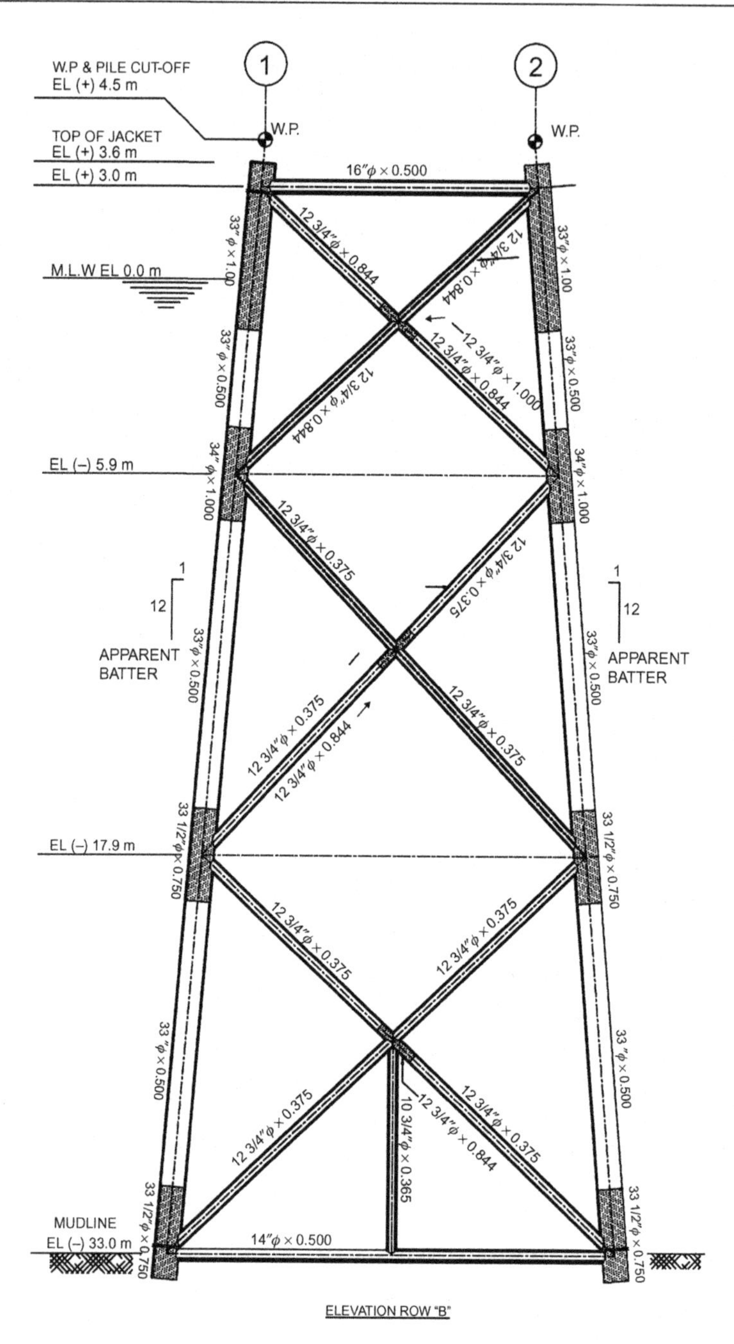

Figure 4.1 Jacket views.

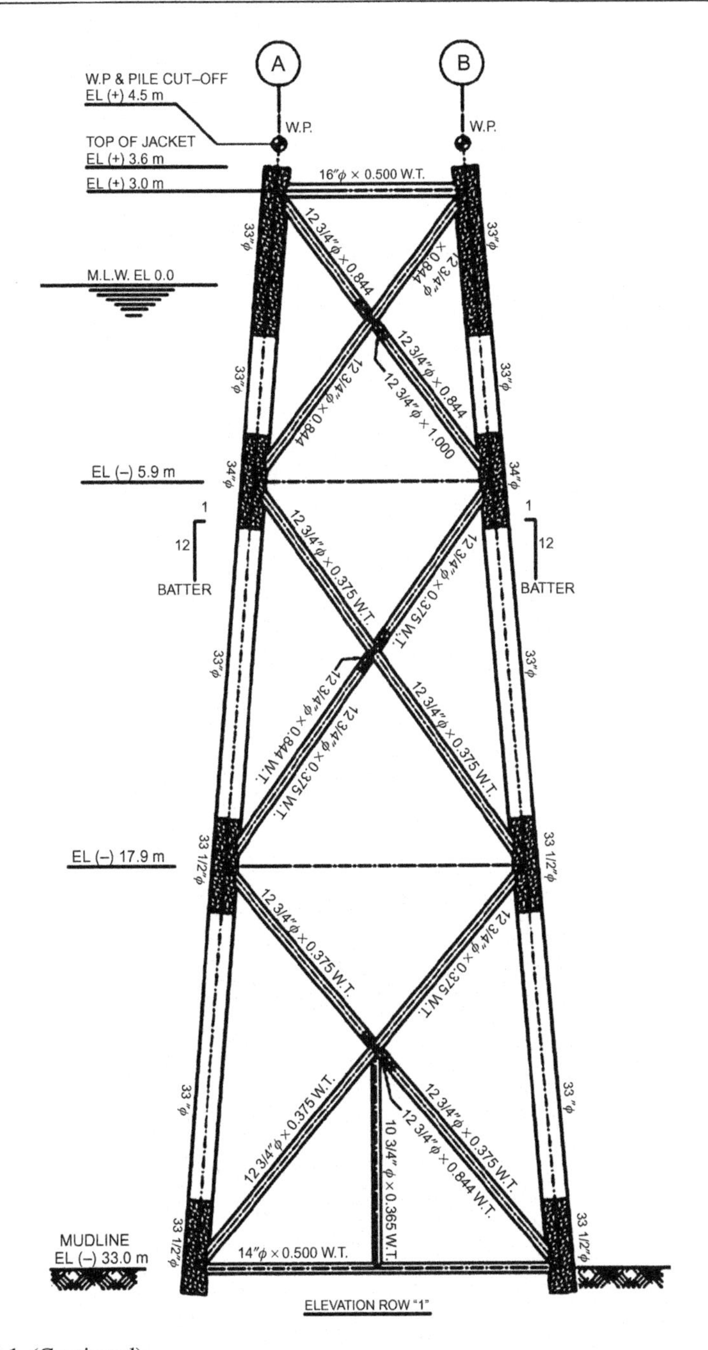

Figure 4.1 (Continued)

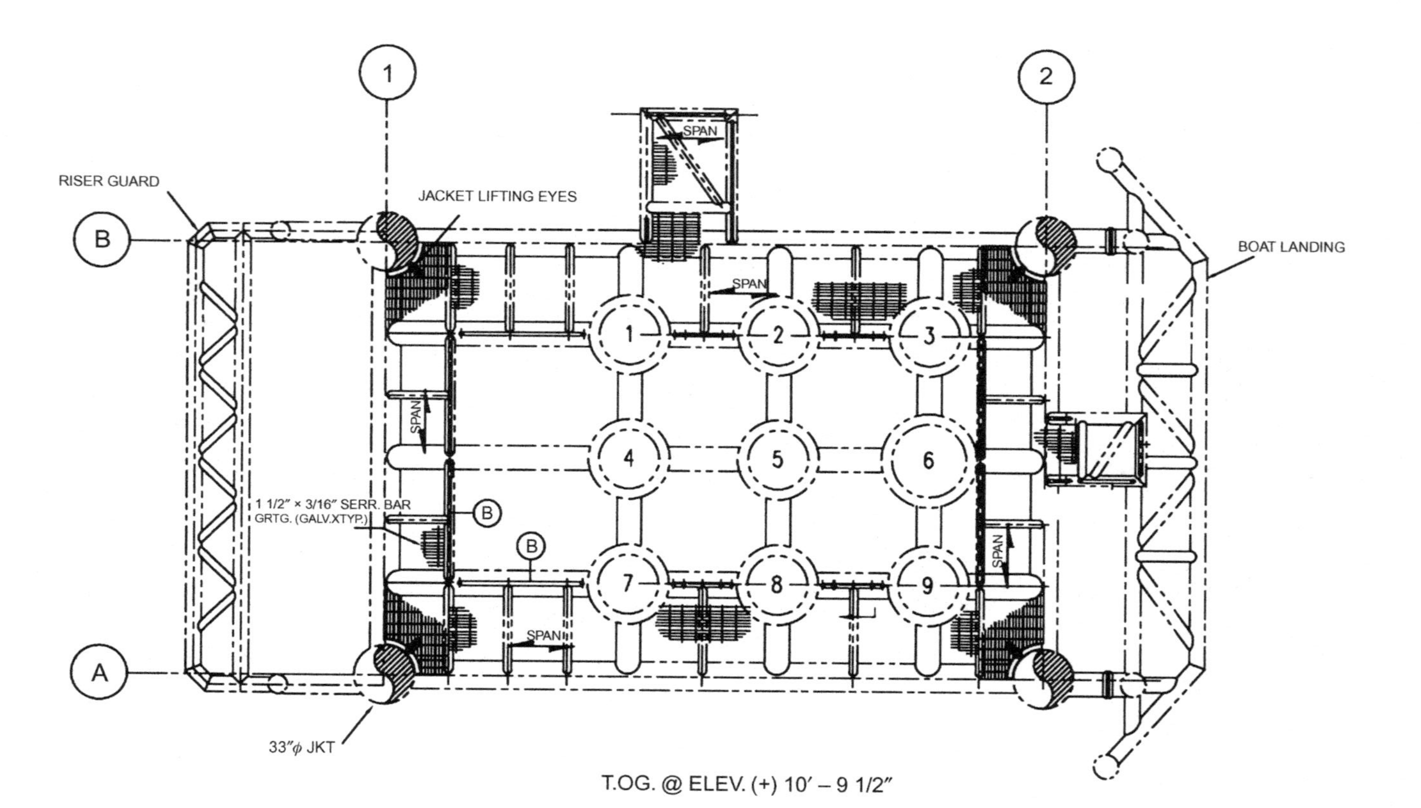

Figure 4.2 Plan view at highest jacket level.

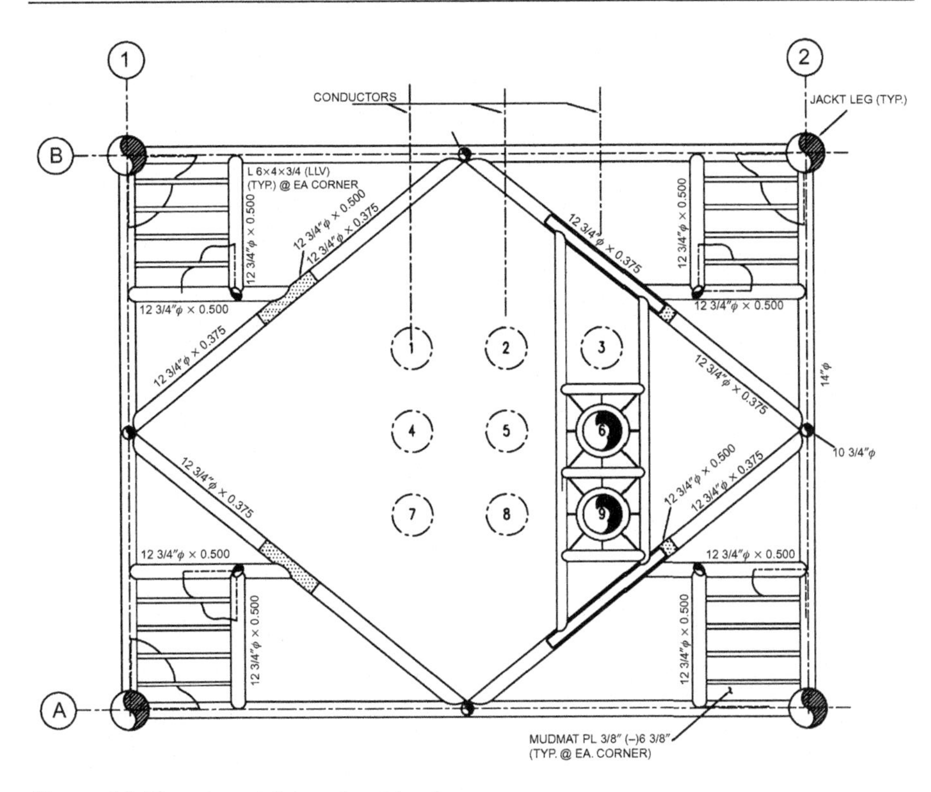

Figure 4.3 Plan view at the mud mat level.

The floor is covered with steel plates, usually about 38 mm (1½ in.) thick. The thickness of the deck framing depends on the spacing between floor beams and the anticipated load on the deck.

Figure 4.5 shows a platform elevation view with the main parts of the offshore platform and affected load.

It is worth mentioning that for the topside design requires no special software as there is no wave effect, so any software can perform the steel structure design, be used to do the analysis, and used for local check of the deck in case of adding small loads or a deck extension.

4.2.1 Approximate dimensions

- Large forces result when waves strike a platform's deck and equipment. Therefore, an air gap of at least 1.5 m (5 ft) is added to the crest height of the wave of the omnidirectional guideline wave heights with a normal return period of 100 years, as show in Figure 4.5.
- From a practical point of view, the sea deck level is usually stated at an elevation of 10–14 ft (3–4 m) above the mean water line (MWL).
- The jacket walkway is above the normal everyday waves that pass through the jacket.

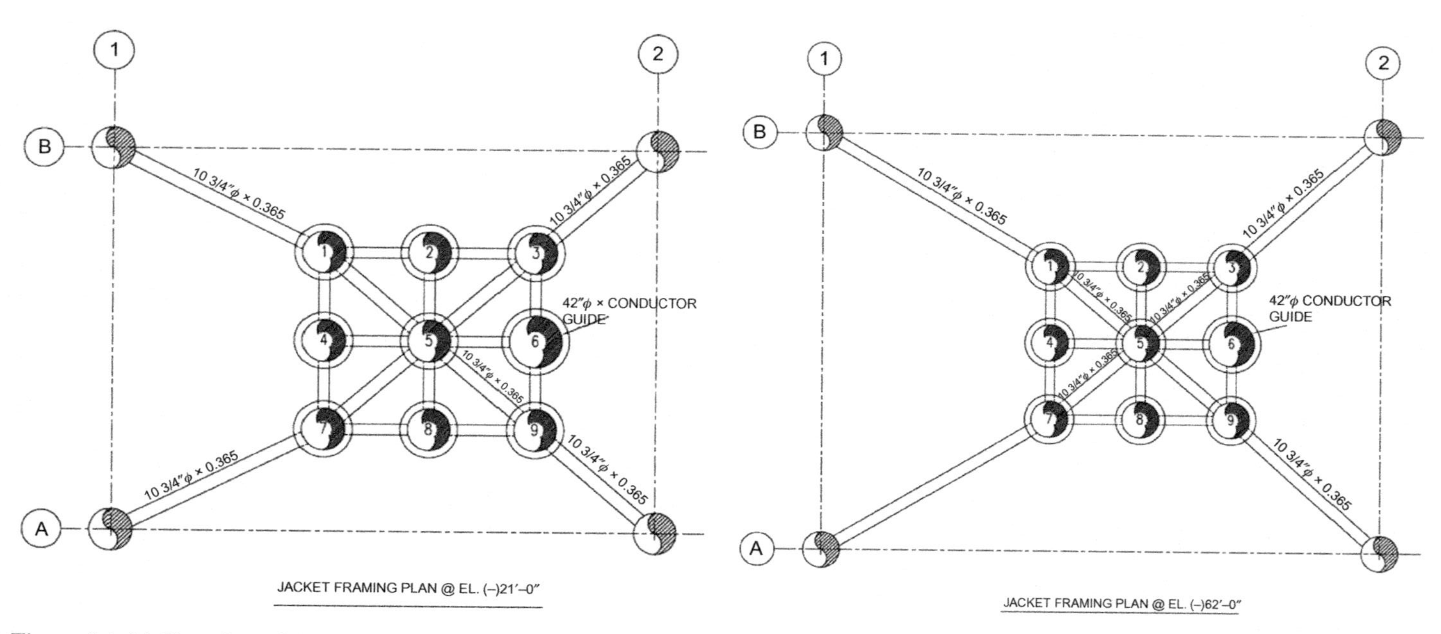

Figure 4.4 (a) Plan view of the jacket Elevation (−21′); (b) Plan view of the jacket at; at Elevation (−62′).

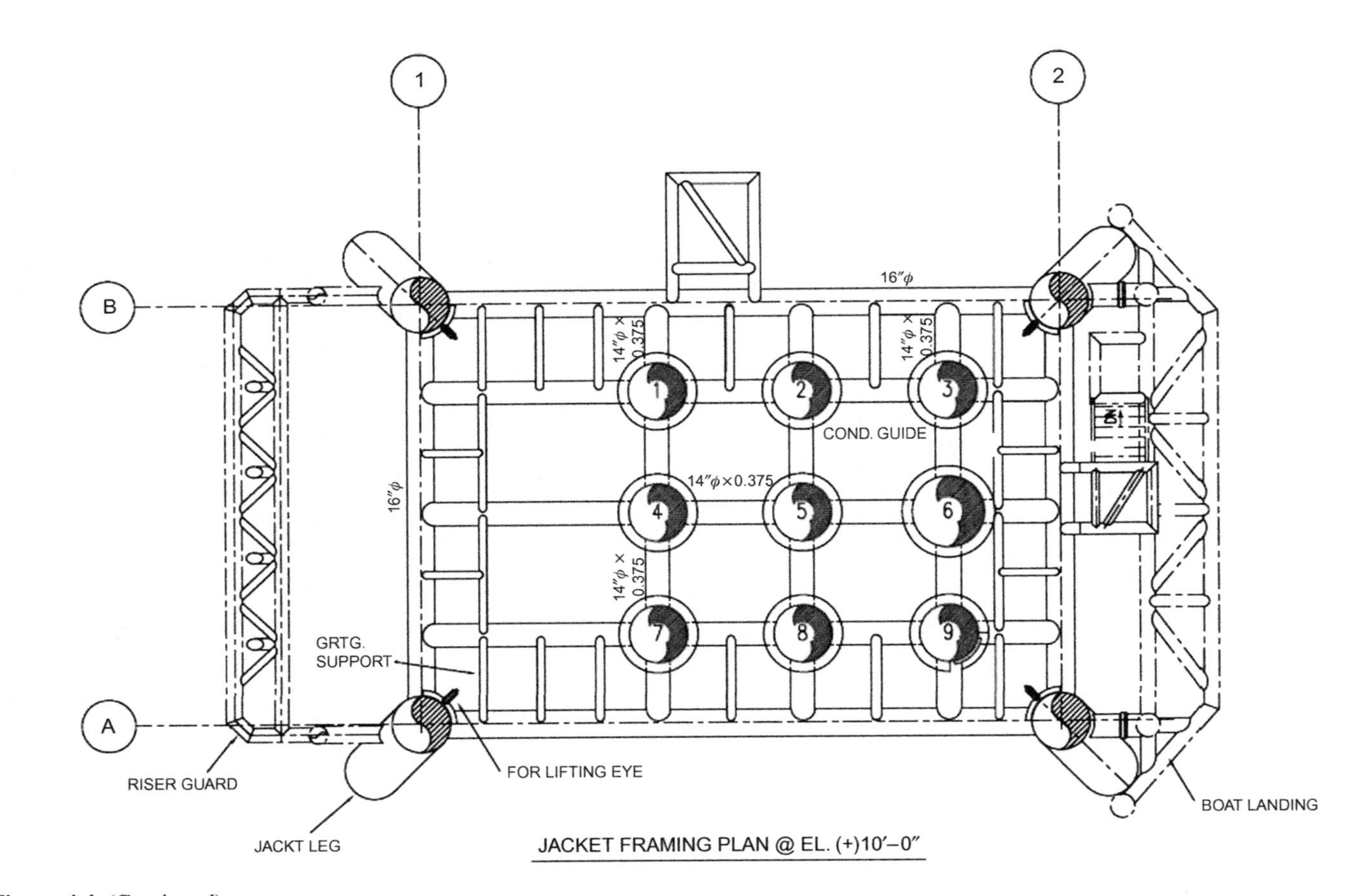

Figure 4.4 (Continued)

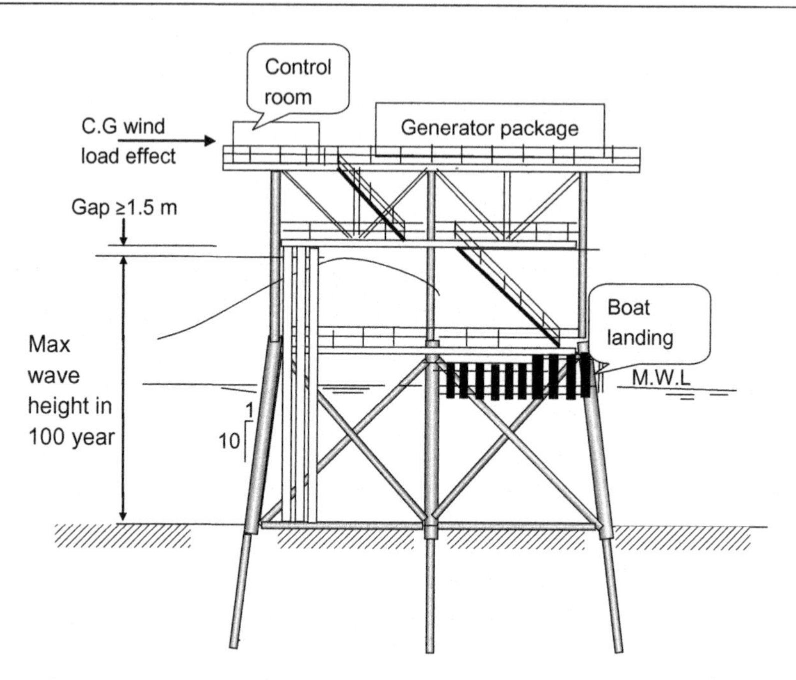

Figure 4.5 Platform elevation view.

- In case of the eight-legged Platform, the spacing between legs is about 12–18 m (40–60 ft), which is usually set by the availability of launch barges and the spacing of launch runners on these barges.
- In the transverse direction, the leg spacing is approx. 14 m (45 ft). This dimension is always constrained by the dimension of the drilling and production packages that will be placed on the deck.
- The length of the cantilever overhang is usually about (3.5–4.5 m) 12–15 ft.
- Allow 25 mm (1 in.) annular clearance between the pile and the inside of leg. In case of piles of 60″ and 48″ OD, the legs will have internal diameters of 62″ and 50″, respectively.
- Jacket legs are battened to provide a larger base for the jacket at the mudline and thus assist in resisting the environmentally induced overturning moments. The legs have battens for 1:8 or 1:7.
- Preliminary sizes are selected based on experience.
- Conductor and risers numbers and sizes are provided; the conductor is always (18″, 20″, 24″, or 30″), the risers are always (14–20″).

4.2.2 Bracing system

The bracing system consists of vertical, horizontal, and diagonal tubular members connected to jacket legs forming a stiff truss system. This system transfers the horizontal load acting on the platforms to the piles. There are variations in the platform bracing pattern. Every system has its advantages and disadvantages; for example, a K-brace pattern system has fewer members intersecting at joints so it reduces

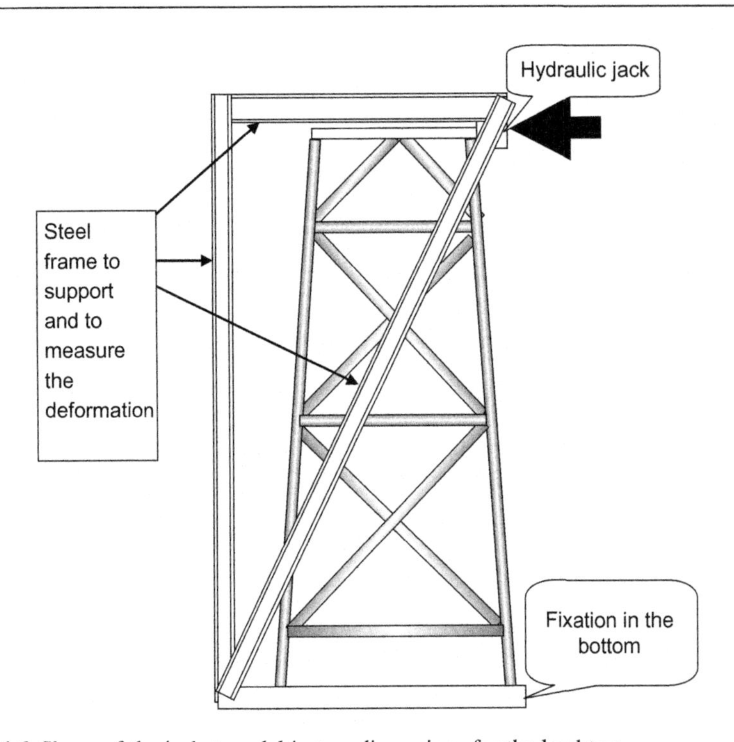

Figure 4.6 Shape of the jacket model in two dimensions for the load test.

welding and assembly costs, but it lacks the redundancy of X bracing. This was found in a study on joint industrial practice performed by different oil and gas companies.

Figure 4.6 presents a jacket frame under a lateral load in the test workshop. The load applied to the frame was increased gradually until complete failure. From this figure, one can see the buckling on the bracing member.

It is recommended that the designer choose a brace member diameter with a slenderness ratio (KL/*r*) in the range of 70 to 90. Limiting the ratio to a 70–90 range is an industry accepted practice. When the slenderness of a brace increases, its allowable axial stress (F_a) starts decreasing. At (KL/*r*) equal to 80, the allowable axial stress, F_a, for A36 steel is 71% of that allowable for a nonslender member, that is, KL/*r* equal to zero. In case of steel 50 KSi at (KL/*r*) equal to 80, F_a reduces to about 63% from the allowable stress for non slender member.

At a high KL/*r* ratio, the high-yield pipe is less efficient than at lesser values. Note that lower slenderness ratios also encourage higher *D*/*t* ratios for tubular members, which may compound local buckling problems.

For sizes up to and 450 mm (18 in.), use the wall thickness for a standard pipe as a starting point. For sizes up to and 700 mm (27 in.), try 12 mm. For 750 to 900 mm. start with 16 mm.

It is practical to keep the D/t ratio of the members between 19 and 60. Pipes with D/t less that 19 are difficult to buy or make. For A36 steel, a D/t higher than 60 can present local buckling problems.

From practical point of view, for water depth of h, in feet, begin to check for hydrostatic problems when D/t is higher than $250/(h)^{0.3333}$.

In general, the legs of the jacket are interconnected and rigidly held by the following kinds of bracings: diagonals in vertical planes, horizontal and diagonals in horizontal plans Often the plan of horizontal bracing spaced 12–16 m near the water surface has a span of approximately 12 m.

The benefits and general functions of a bracing system are as follows:

1. Transmission of the horizontal load to the foundation.
2. Providing structure integrity during fabrication and installation.
3. Resistance to the wrenching motion of the installed jacket/pile system.
4. Support for the corrosion anodes and well conductors.

4.2.3 Jacket design

Virtually all the decisions about the design depend on the jacket leg. The soil conditions and foundation requirements often tend to control the leg size.

A golden rule in jacket design is to minimize the projected area of members near the water surface (high-wave zone), minimize the load on the structure, and reduce the foundation requirements.

The results of the study presented in Figure 4.7. The relationship between the applied load and displacement in case of X bracing with horizontal bracing and without a joint can is presented in Figure 4.7(a). Figure 4.7(b) presents the relation between load and displacement in case of existing horizontal bracing and with a joint can. One can see that, in case of a joint can, the jacket can carry more load than that for the design based on API. In addition, the ductility is higher in case of a joint can.

If there is no horizontal bracing and without a joint can, the jacket, as in Figure 4.7(c), will carry less load than that in case of existing horizontal member.

The test was performed for K bracing with a different β value, which is the relation between the bracing diameter to the chord diameter (d/D) and with a different gap between the bracing and this gap, denoted by g.

Figure 4.8 presents the relation between the applied load and the frame displacement and compares the results with the design by API and DnV. It is found that, by decreasing, b values, the redundancy increases. By increasing the gap between the braces, the ductility increases, as per the comparison of cases C and D.

By comparing between x-bracing with horizontal bracing member to the k-bracing, it is found that the x-bracing can carry more load over the design load higher than in case of k-bracing. So in general the redundancy of x-bracing is higher than that in k-bracing.

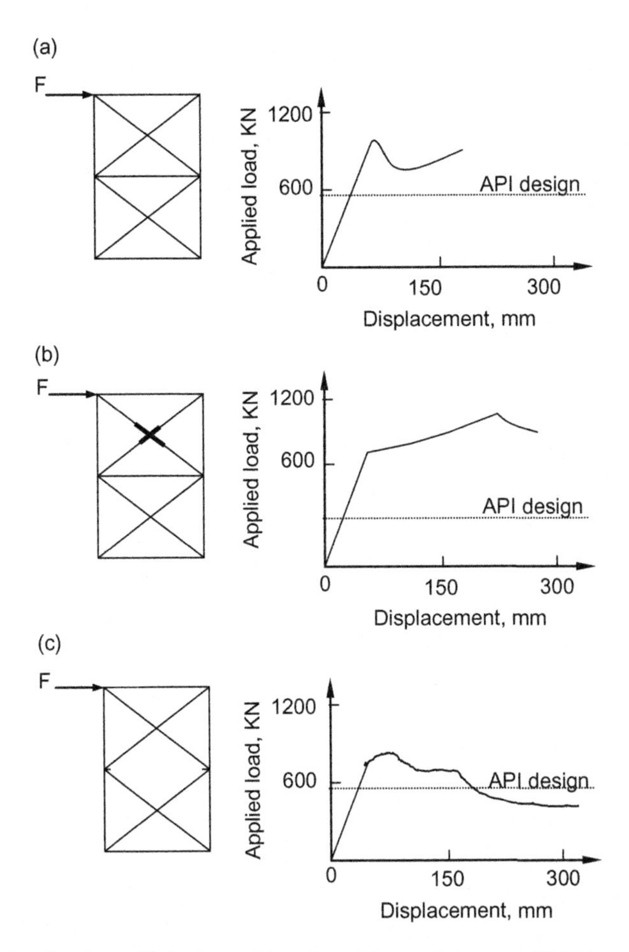

Figure 4.7 (a) X-bracing with horizontal bracing without a joint can; (b) X-bracing with horizontal bracing with a joint can; (c) X-bracing without horizontal bracing without joint can.

4.3 Structure analysis

For environmental and gravitational loads, all the necessary parameters required for the automatic load generation by the program should be input by the engineer. Most of software programs on the market use default values for many of these parameters. Usually, the engineer must override the defaults and enter project-specific values. The following should be defined in the software:

- Self weight.
- Buoyancy (flooded and unflooded members).
- Wind (direction, terrain category, gust duration, drag coefficients, etc.).

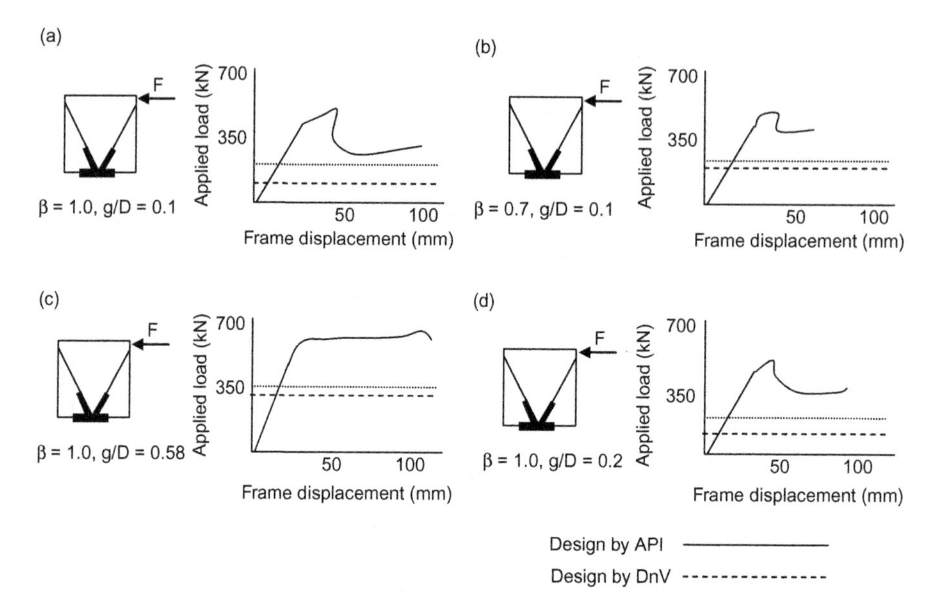

Figure 4.8 Relation between applied load and displacement for different K-bracing geometry.

- Waves (wave theory, direction, height, period, C_D, C_M, wave kinematics, etc.).
- Current (direction, speed variation with depth, current blockage factor, etc.).
- Marine growth (thickness variation with depth, roughness, etc.).

The software in most cases does not use the correct default coefficients, so it should be reviewed carefully.

The structure's foundations should be modeled sufficiently to reflect the actual stiffness of the foundation:

- Simple (fixed, pinned, sliding, etc.).
- Linear springs.
- Nonlinear stiffness method.

4.3.1 Global structure analysis

The steps for using software in design are as follows:

1. The designer must define the structure in terms of physical dimensions, member size, and material properties.
2. The designer must enter the soil conditions as interpreted by a soil specialist in some programs, which need a *P*-*y* curve.
3. All loads must be entered into the program
4. The design wave through the structure at several azimuth angles to determine the direction that produce highest reactions and the current load is entered as presented in Figure 4.9.

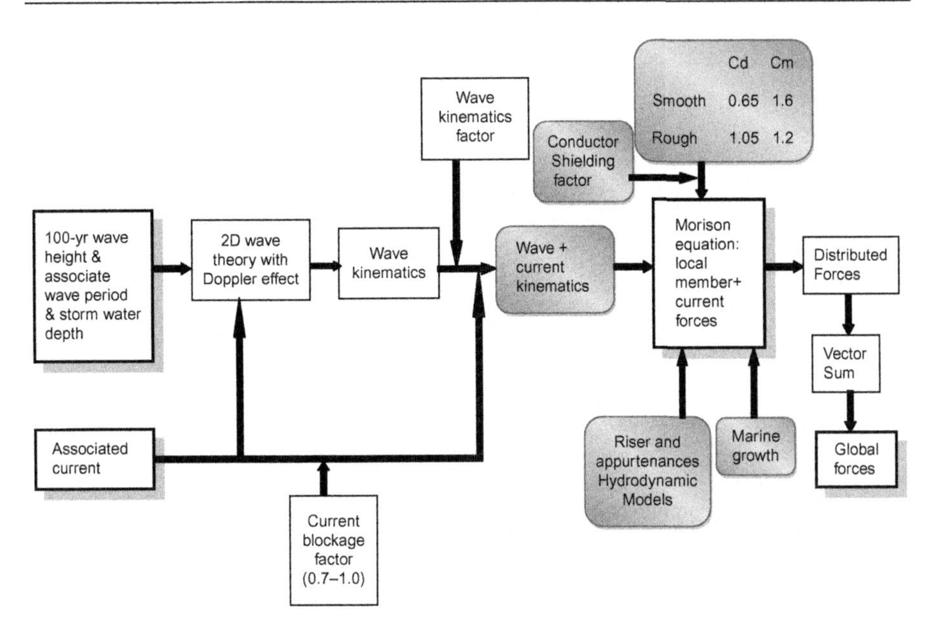

Figure 4.9 Applied of environmental load.

5. The computer advances the wave through the structure at specified increments, calculating total shear and overturning moment on the structure at the mudline.
6. For each load condition the computer analysis provides
   a. Total base shear and overturning moment.
   b. The member end forces and moments.
   c. Joint rotation and deflection.
   d. External support reaction.
7. After calculating the stresses, it compares them with the allowable stresses as defined by AISC.
8. The pile is replaced by lateral springs in two directions, an axial spring, and a moment spring.
9. The piling interacts with the surrounding soil in an inelastic manner, and the computer linearizes its response to generate the equivalent elastic spring, as shown in Figure 4.10.
10. In the case of sea ice abrasion, an allowance of 0.1 mm/annum is considered for all steel between the elevations of −1.3 m and 3.0 m and usually results in a minimum increase in thickness of 2.5 mm over the life of the platforms.

The process of applying the wave load and current to the offshore structure is illustrated in Figure 4.9 as described in API RP2A.

Bending stresses due to both horizontal and vertical forces should be investigated. However, the current velocity components should not be included in the wave kinematics when calculating wave loading. For X braces, members should be assumed to span the full length. Member lengthsare reduced to account for the jacket leg ratio.

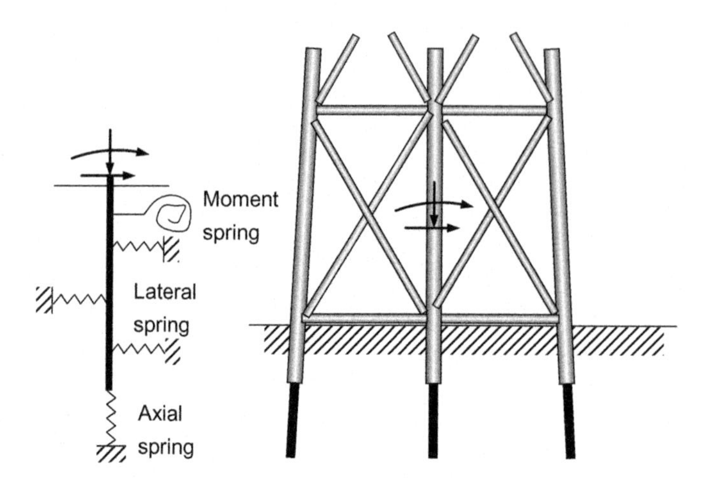

Figure 4.10 Foundation piling model.

The static structure analysis for the offshore structure is the same as in case of a normal structure as any software use the stiffness matrix to calculate the deflection then the internal forces and stresses to each member. But, in the case of offshore structure, the problem is the interaction between the structure and the pile, as the structure is elastic and the pile is inelastic. So the structure analysis steps are as follows:

1. Set up the geometry of the jacket with material specifications and preliminary member sections including the dimensions. The software calculates the stiffness matrix for the jacket excluding the piles in calculating the matrix.
2. Apply the loads on the structure jacket under different load cases but do not run the software, as there are no supports to apply to the structure system.
3. Ordering the nodes that reflect the degree of freedom in K, F so that the nodes p, which connect the piles, are together at the end of the stiffness matrix and follow the nodes j, which are slowly connected to jacket members. The stiffness matrix and force vector can be as follows:

$$\begin{bmatrix} j_1 & j_2 & p_1 & p_2 \\ & & & \\ k_{jj} & & k_{jp} & \\ k_{pj} & & k_{pp} & \end{bmatrix} \begin{Bmatrix} \delta_j \\ \\ \delta_p \end{Bmatrix} = \begin{Bmatrix} F_j \\ \\ F_p \end{Bmatrix} \tag{4.1}$$

4. To analyze the foundation behavior, the stiffness of the jacket and loading on the jacket, as felt by the piles, is required. The detailed behavior of the jacket is not required at this stage, as an equation of the form $K_s\delta_p = F_s$ is wanted, where K_s and F_s represent the stiffness and applied forces on the structure as seen at the pile connecting nodes P.
5. Form the model of the foundation by developing the stiffness matrix for the foundation at zero deflection

6. Assuming no load is applied directly to the piles, the only loading is applied through the structure. The foundation forcing vector contains the element from the forces on the structure but located in the appropriate positions for the same degree of freedom in the foundation stiffness matrix.
7. Add the foundation and the jacket substructure and solve, then recalculate the foundation stiffness, as the first nodal displacement along the pile and at the connection to the jacket was only a first estimate, based on step 5, the stiffness of *p-y* and *t-z* curves at zero deflection. Note that the pile deflections have been estimated, and a better estimate of the *p-y* and *t-z* stiffness can be made. The model of the foundation is shown in Figure 4.10. The stiffness may be represented by either a secant or tangent stiffness. Note that the secant stiffness method is generally slower but more stable than the tangent stiffness approach.
8. Repeat the sequence from step 5 until the stiffnesses converge, the nodal deflections can be used to determine the forces, shears, moment in the piles.
9. The deflections at the link to the jacket are also known now. These can be applied as the prescribed deflection to the pile nodes on the original jacket model for step 2.
10. Given the jacket deflections, the jacket member forces are calculated from separate member stiffness properties.

Figure 4.11 illustrates the global structure analysis procedure in a flowchart.

4.3.2 The loads on the piles

The load in the pile will be calculated by the software and can be calculated preliminary by hand calculation. The following method as shown in Figure 4.12 presents the calculation parameters.

$$N = \frac{(M - sh)\cos\alpha}{2(h + d_f)} \tag{4.2}$$

$$A_1 = \frac{1}{2}\left\{\frac{V}{\cos\alpha} - \left[s + \frac{(M - sh)\cos^2\alpha}{h + d_f}\right]\frac{1}{\sin\alpha}\right\} \tag{4.3}$$

$$A_2 = \frac{1}{2}\left\{\frac{V}{\cos\alpha} + \left[s + \frac{(M - sh)\cos^2\alpha}{h + d_f}\right]\frac{1}{\sin\alpha}\right\} \tag{4.4}$$

Example 4.1

A four-legged platform carries a vertical load of 500 tons with a shearing force of 200 tons. Estimate the load on each pile whereas the jacket leg battered is 1:8 and the data are as follows:

$V = 500$ ton
$S = 200$ ton
$M = 4000$ m ton
$h = 30$ m
$d_f = 3$ m

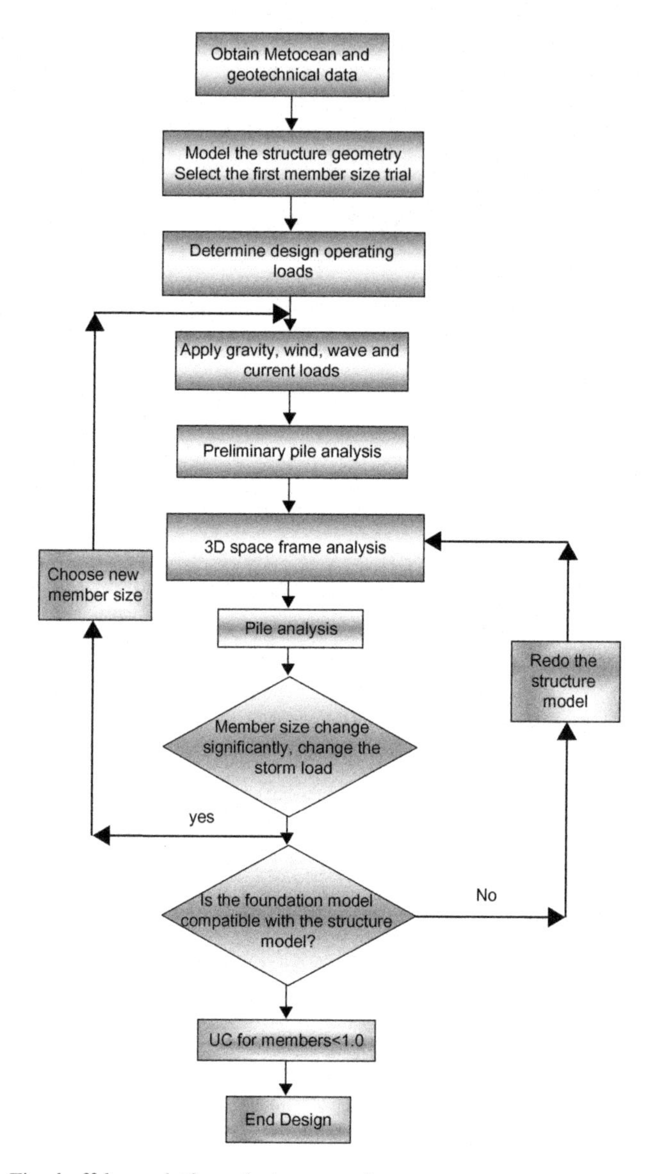

Figure 4.11 Fixed offshore platform design procedure.

Solution

$\alpha = 0.12435$ radians

$N = 10.9$ tons

$A1 = 326.963$ tons

$A2 = 587.574$ tons

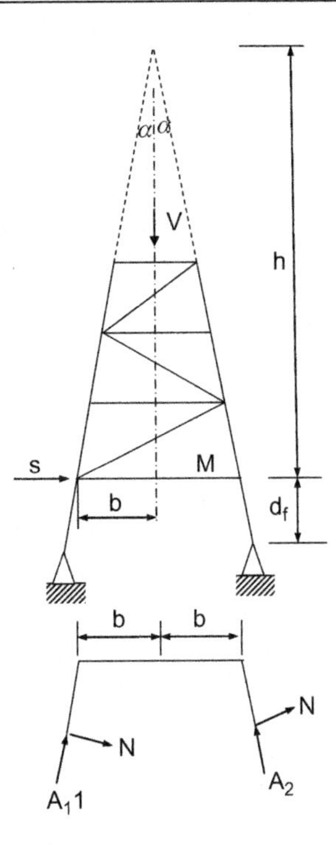

Figure 4.12 Calculation of loads on piles.

So, the load for a single pile is

$N = 5.45$ tons
$A1 = 165$ tons
$A2 = 294$ tons

The preceding is a rough estimate about the load on the pile for a preliminary calculation that can be done by hand.

There is another way to obtain the pile load by obtaining the overturning moment value at the mudline. The calculation of the reaction force at each pile is as shown Figure 4.13:

R = vertical pile reaction
M = total overturning moment
A = relative axial pile stiffness
$d_{x,y}$ = distance from neutral
θ = axis wave angle

The base is assumed to be rigid. The resultant force at each pile is calculated by assuming that M is constant with a varying Θ:

$$R = \frac{M_x d_y}{\sum A d_y^2} + \frac{M_y d_x}{\sum A d_x^2} \tag{4.5}$$

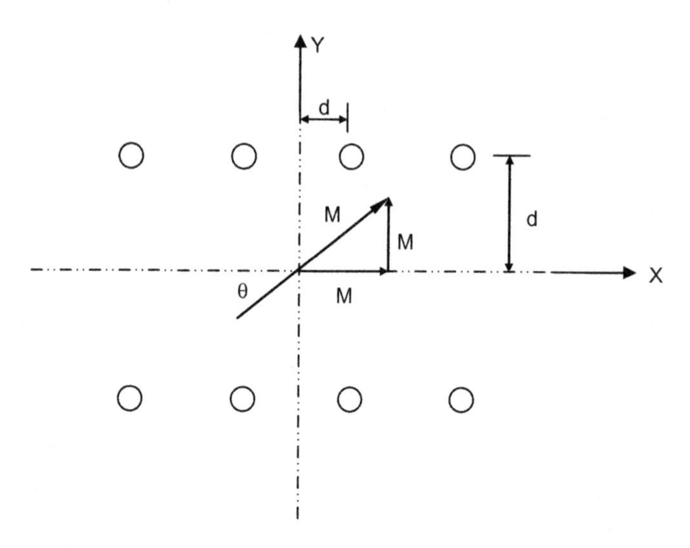

Figure 4.13 Jacket plan at the mudline.

4.3.3 Modeling techniques

The following are guidelines and recommendations for those using any software in modeling any steel structure and specific for offshore structure platforms.

As a guide, the global axis system should be orientated as noted here. The origin should be at the center of the platform or structure at chart datum, MSL, or mudline, as determined by the project.

- X axis pointing towards platform east
- Y pointing towards platform north
- Z vertically upwards

Joint numbers are assigned by the engineer. Allowing the program to automatically assign joint numbers is not permitted. It is important to follow a strict numbering system when creating or editing a model. This allows easier interpretation and use of the analysis results. The example modeling is as shown in Figure 4.14.

4.3.3.1 Joint coordinates

To facilitate the checking process when using software, the joint coordinates are always entered and presented using a single set of units (i.e., m or ft). Take care to avoid dual units as (m/cm or ft/in.), which is the regular mistake. As shown in Figure 4.15, the structure model consists of nodes and members; however, the pile model needs special data in the input file.

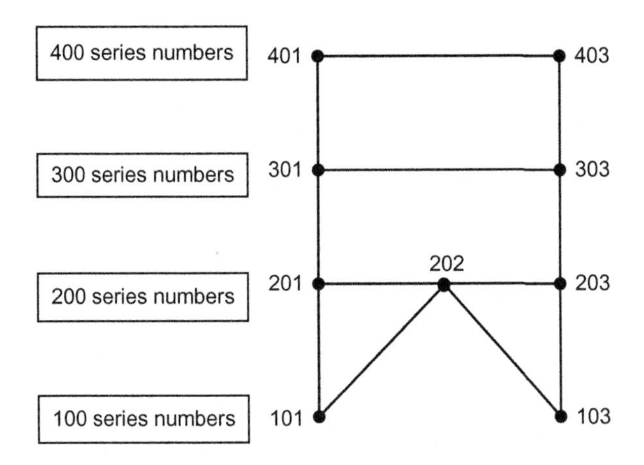

Figure 4.14 Nodes modeling consequence to enhance quality assurance.

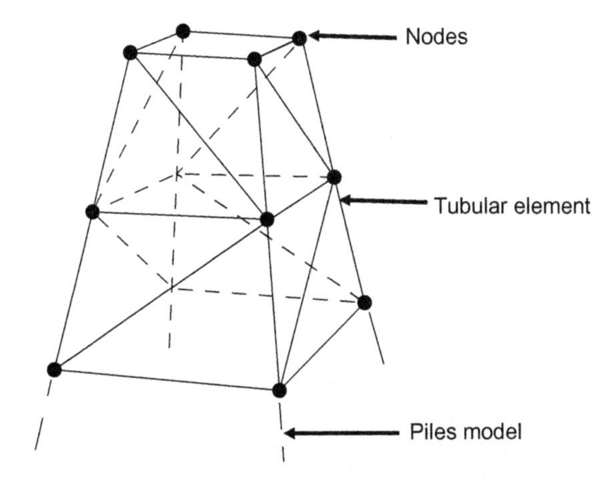

Figure 4.15 Modeling for the jacket structure.

Offshore structure fixed platforms usually have a sliding connection between the structure elements, which should be considered in the modeling. The most common two cases are

- Jacket piles are welded off at the top of the jacket and guided within the legs by spacers, as shown in Figure 4.16.
- Conductors are restrained horizontally but not vertically by conductor frames, as shown in Figure 4.17.

The boundary conditions should be clearly defined and reflect the actual support conditions for the structure.

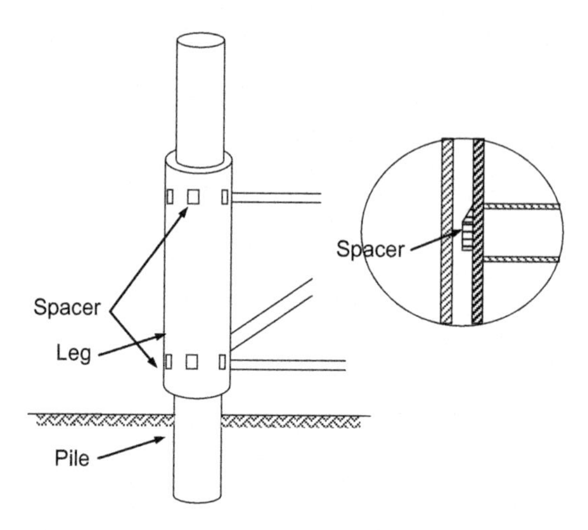

Figure 4.16 Spacer between legs and piles.

Figure 4.17 Conductor guides.

Any model should generally consist of all primary framing members. Secondary members do not need to be explicitly modeled unless they facilitate the input of loads or contribute to the structural action of a primary member.

Primary and secondary steelwork are defined as

- Topside primary steelwork includes all truss members, girders, and horizontal bracing.
- Jacket primary steelwork includes legs, diagonal bracing, horizontal bracing, and piles.
- Topside major secondary steelwork includes deck plate, grating, deck beams, walkways, stairs, and the crane pedestal.
- Jacket major secondary steelwork includes cathodic protection, boat landing, barge bumpers, walkways, appurtenance supports, and mud mats.

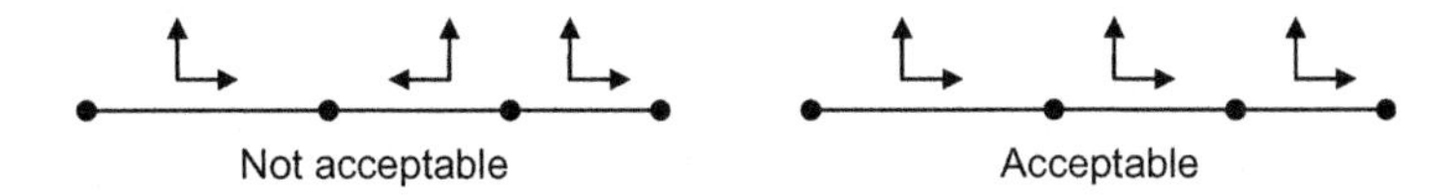

Figure 4.18 Local axis technique.

4.3.3.2 Local member axes

When constructing the model and running the software, the engineer should review and appreciate the program's default member axis system and adopt this system where possible. In addition to that, it should be taken into consideration with vertical members (especially I beams and channels), as the default local member axis system affects the orientation of the flanges.

The orientation of the members should follow a consistent format. That is, all like members should be orientated in the same direction sense, as shown in Figure 4.18.

Member end releases should be clearly defined and reflect the actual connection constraints for the member.

In most cases, the member end offsets may be used where there are large joint thicknesses. The offset should extend only to the face of the joint.

4.3.3.3 Member effective lengths

The effective length of a member under axial compression should reflect the relative joint stiffness at the end of the member. The appropriate effective length factor, K, should be selected from the recommended values in the design codes. Consideration should be given to the constraining effect provided by intermediate members along the length of the member. The effective length of a member buckling about its y-y axis is often different from the effective length about the z-z axis.

The compression flange (or critical flange) of a member may buckle under bending lateral torsion. The effective length of a member under bending should reflect the degree of torsional restraint offered by the end connections of the member and by intermediate members along the length of the member. The bending effective length of a member should be calculated using the appropriate factors given in the design codes.

4.3.3.4 Joint eccentricities

Eccentric joints in jacket structures should be modeled using member-end offsets. For topside-type structures, joint eccentricities should be modeled using discrete elements, thereby allowing easy extraction of joint forces from the output.

When required, the deck plate should be modeled as structural elements using it as a membrane plate. Note that the plate elements need not be offset.

Alternatively, pin ended axial brace members may be used in lieu of plate elements.

A problem always facing the structural engineer is how to model the pile inside the leg. Most software provide wishbone members that should be modeled at all horizontal bracing levels of the jacket to account for pile-to-jacket leg interaction, as shown in Figure 4.19. If the pile-to-jacket leg annulus is to be grouted, then a rigid connection between the pile and leg should be modeled.

Generally, appurtenances do not contribute to the structural stiffness of the primary structure. Appurtenances may be modeled to facilitate automatic load generation by the program and are sometimes referred to as *nonstructural members*.

Appurtenances may be modeled by assigning a small modulus of elasticity or small stiffness properties to these members. It is important to ensure any member end releases and the like accurately reflect the actual support conditions for the appurtenance and that no spurious forces enter the structure due to poor modeling techniques. When using this modeling method, the engineer should verify the analysis and ensure no compatibility problems are caused by the small stiffness.

4.4 Dynamic structure analysis

Dynamic analysis is becoming increasingly important for the following reasons:

1. For larger and more costly structures, dynamic analysis translate into huge construction cost savings.
2. For complex offshore structure, ordinary static analysis tends to result in higher risk.
3. The harsh environmental conditions at many deepwater sites cannot be adequately model by static analysis.
4. The software is available now, so it is easy to perform dynamic analysis.

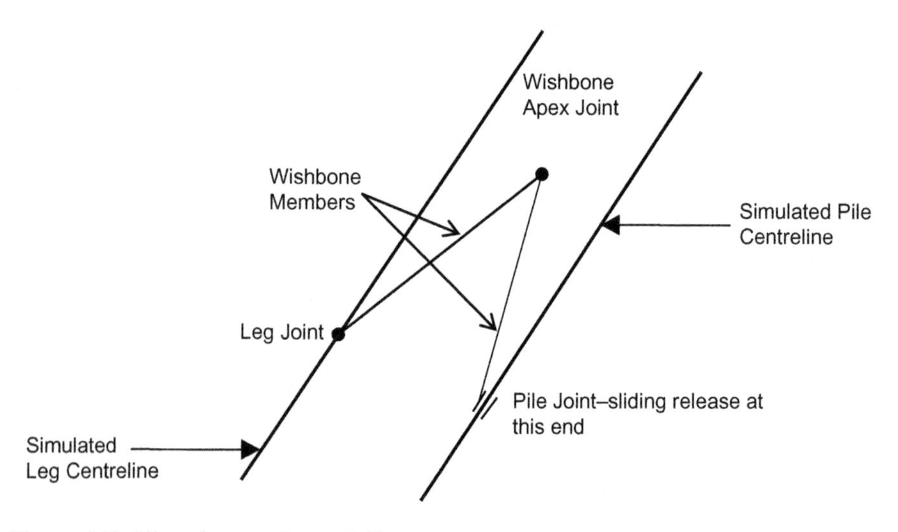

Figure 4.19 Pile to leg annulus modeling.

The equation of motion is as follows:

$$M\ddot{x} + C\dot{x} + kx = P(\dot{x}, \ddot{x}) \tag{4.6}$$

M = diagonal matrix of the virtual mass
C = matrix for structural and viscous damping
K = square linear structure stiffness matrix
$P(\dot{x}, \ddot{x})$ = the load vector, where
$\ddot{x}$ = structural acceleration
$\dot{x}$ = velocity
x = displacement

4.4.1 Natural frequency

Calculating the natural frequency of the structure is a first step in the dynamic analysis. In general there are two methods for calculating the natural frequency of linear, single degree of freedom structural models: the Rayleigh method, which is based on the energy principal, and the direct method, which is based on the equation of motion and is illustrated here.

The direct method of obtaining the natural frequency of a single degree of freedom structure is to use its equation of motion. The structure's natural frequency, ω_n, is defined as the frequency compatible with an undamped structure of constant mass, a restraint force that varies linearly with the displacement coordinate, and no external excitation force. Under these conditions, the governing equation of motion, written in terms of displacement coordinate, x, is

$$m\ddot{x} + kx = 0 \tag{4.7}$$

which is a special case of the general equation (4.7) if the applied force is zero and there is no damping, which is called *simple harmonic motion*:

$$x = x_o \sin \omega_n t \tag{4.8}$$

By substituting equation (4.8) and its second derivative into equation (4.7), the result is

$$(-m\omega_n^2 + k)x_o \sin \omega_n t = 0$$

However the term $\sin \omega_n t$ is not zero for all time t and therefore the term in parentheses must be zero. This leads to the following equation for the natural frequency of the structure:

$$\omega_n = \sqrt{\frac{k}{m}} \tag{4.9}$$

As ω_n is the natural frequency in rad/s.

The natural period T and the natural frequency, f, are calculated from the following equation:

$$T = 2\pi/\omega = 2\pi\sqrt{\frac{M}{k}}\ \text{s} \tag{4.10}$$

$$f = 1/T = 1/2\pi\sqrt{\frac{k}{M}}\ \text{cycle/s (Hz)} \tag{4.11}$$

in multiple degrees of freedom systems, as in the structure in general and the fixed offshore structure shown in the Figure 4.20. Every level of the structure has its mass own mass and its own stiffness. So the values of the mass, stiffness, displacement, and acceleration are resented in matrix form, as follows:

$$[M]\{\ddot{X}\} + [k]\{x\} = 0 \tag{4.12}$$

$$\{x\} = (A \sin \omega t + B \cos \omega t)\{\phi\} \tag{4.13}$$

where ϕ is the vibration shape.

$$\{\ddot{x}\} = -\omega^2(A \sin \omega t + B \cos \omega t)\{\phi\} \tag{4.14}$$

$$[k]\{x\} = \omega^2[M]\{x\} \tag{4.15}$$

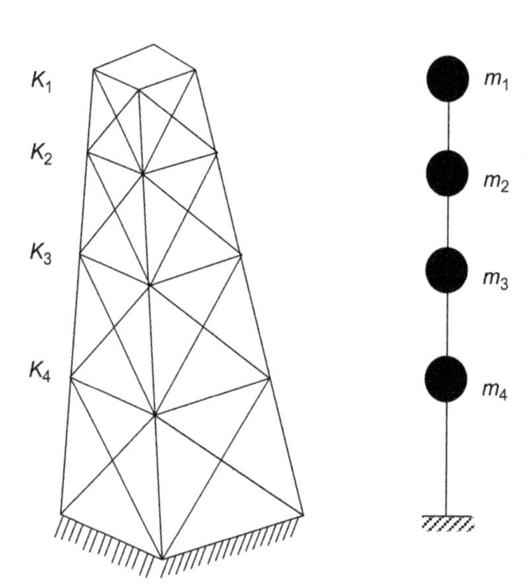

Figure 4.20 Mass distribution for the jacket.

Equation (4.12) is the standard eigenvalue problem, which may be solved by the natural frequency and natural modes.

The output results for the dynamic analysis are

1. Time history of member-end forces.
2. Time history of joint displacement.
3. Maximum values of joint displacement.
4. Time history of shear between stories.
5. Time history of the overturning moment.
6. Time history of the base shear.
7. Time history of axial pile loads.

$$\begin{bmatrix} m_1 & 0 & 0 & 0 \\ 0 & m_2 & 0 & 0 \\ 0 & 0 & m_3 & 0 \\ 0 & 0 & 0 & m_4 \end{bmatrix} \begin{Bmatrix} \ddot{x}_1 \\ \ddot{x}_2 \\ \ddot{x}_3 \\ \ddot{x}_4 \end{Bmatrix} + \begin{bmatrix} K_{11} & K_{12} & K_{13} & K_{14} \\ K_{21} & K_{22} & K_{23} & K_{24} \\ K_{31} & K_{32} & K_{33} & K_{34} \\ K_{41} & K_{42} & K_{43} & K_{44} \end{bmatrix} \begin{Bmatrix} x_1 \\ x_2 \\ x_3 \\ x_4 \end{Bmatrix} = \begin{Bmatrix} 0 \\ 0 \\ 0 \\ 0 \end{Bmatrix} \tag{4.16}$$

The modes are as shown above but in the space jacket the structure there are modes of direction as shown in Figures 4.21 and 4.22.

The analysis can be done by mode supposition as in this case the total response can be adding the individual modes as follow presented for plan structure for four levels of structure as shown in Figure 4.23:

$$\begin{Bmatrix} X_1 \\ X_2 \\ X_3 \\ X_4 \end{Bmatrix} = \{\phi_1\}x_1 + \{\phi_2\}x_2 + \{\phi_3\}x_3 \tag{4.17}$$

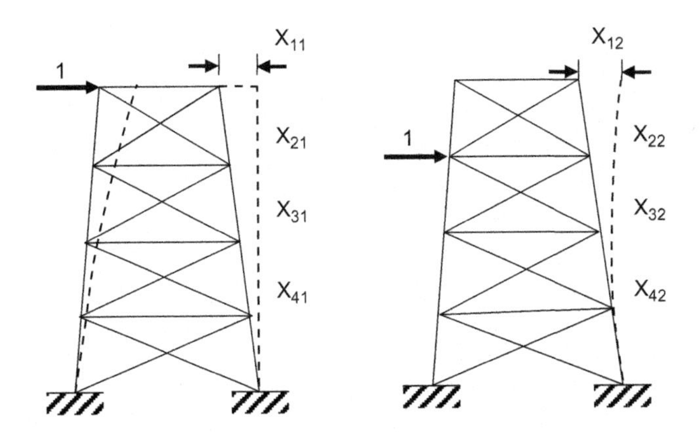

Figure 4.21 Relation between applied unit load at each level and deflection.

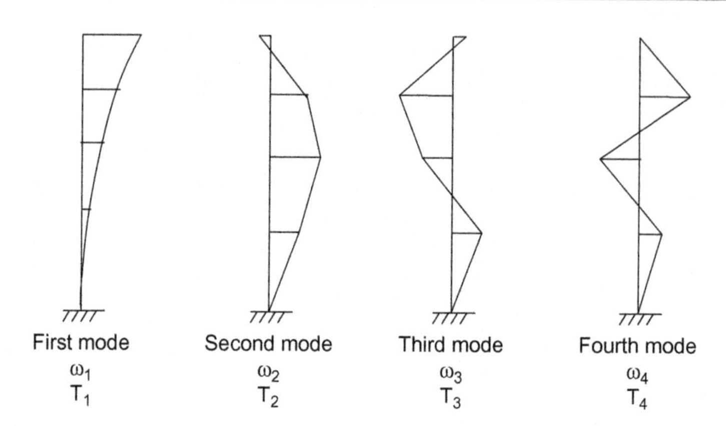

Figure 4.22 Modes of deformation.

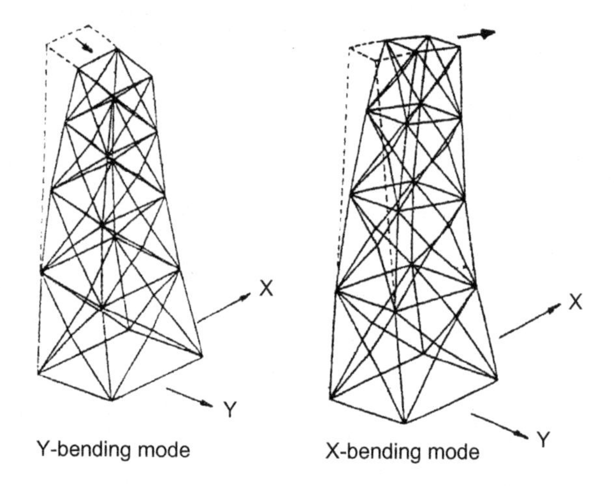

Figure 4.23 Mode of deflection.

The total procedure may be summarized as

1. Calculate the natural frequency (ω_n) and mode shapes (ϕ_n).
2. Calculate generalized masses for each mode:

$$M_m = \{\phi_m\}^T[M]\{\phi_m\} \tag{4.18}$$

3. Calculate generalized stiffness for each mode:

$$K_m = \{\phi_m\}^T[K]\{\phi_m\} \tag{4.19}$$

4. Calculate generalized force for each mode:

$$P_m = \{\phi_m\}^T\{P(t)\} \tag{4.20}$$

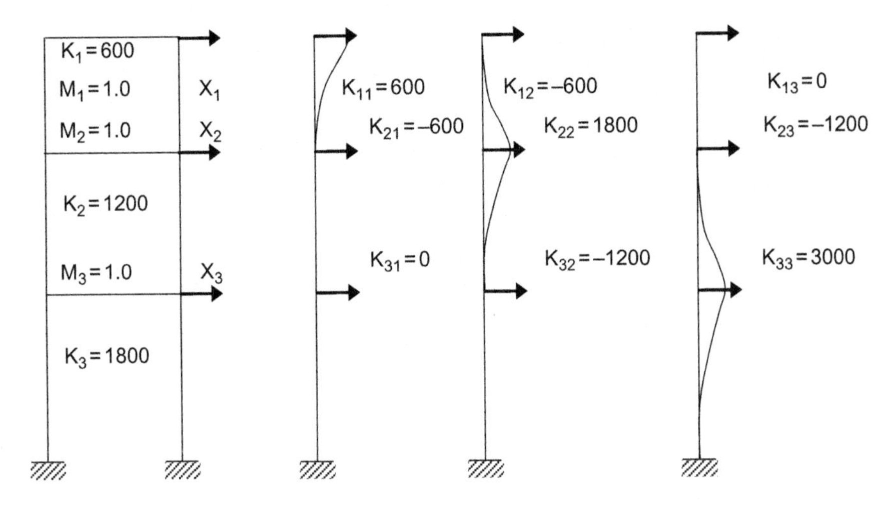

Figure 4.24 Example 4.2 applied load.

5. Calculate the response for each mode from the following equation:

$$M_m \ddot{X} + K_m = P_m(t) \tag{4.21}$$

6. The equations of step 5 are as follows:

$$\ddot{X}_1 + \omega_1^2 X_1 = P_m/M_m \qquad X_1(t) = \frac{P_m}{M_m\omega^2}(1 - \cos\omega_1 t) \tag{4.22}$$

$$\ddot{X}_2 + \omega_2^2 X_2 = P_m/M_m \qquad X_2(t) = \frac{P_m}{M_m\omega^2}(1 - \cos\omega_2 t) \tag{4.23}$$

$$\ddot{X}_3 + \omega_3^2 X_3 = P_m/M_m \qquad X_3(t) = \frac{P_m}{M_m\omega^2}(1 - \cos\omega_3 t) \tag{4.24}$$

$$\ddot{X}_4 + \omega_4^2 X_4 = P_m/M_m \qquad X_4(t) = \frac{P_m}{M_m\omega^2}(1 - \cos\omega_4 t) \tag{4.25}$$

$$\begin{Bmatrix} X_1 \\ X_2 \\ X_3 \\ X_4 \end{Bmatrix} = \{\phi_1\}X_1(t) + \{\phi_2\}X_2(t) + \{\phi_3\}X_3(t) + \{\phi_4\}X_4(t) \tag{4.26}$$

From these equations, one can define the values of drift for each floor with time.

Example 4.2

Calculate the natural frequency for the following structure as shown in Figure 4.24.

For free vibration to calculate the natural frequency, $F(t)=0$:

$$\begin{bmatrix} 1.0 & 0 & 0 \\ 0 & 1.5 & 0 \\ 0 & 0 & 2.0 \end{bmatrix} \begin{Bmatrix} \ddot{x}_1 \\ \ddot{x}_2 \\ \ddot{x}_3 \end{Bmatrix} + \begin{bmatrix} 600 & -600 & \\ -600 & 1800 & -1200 \\ & -1200 & 3000 \end{bmatrix} \begin{Bmatrix} x_1 \\ x_2 \\ x_3 \end{Bmatrix} = \begin{Bmatrix} 0 \\ 0 \\ 0 \end{Bmatrix}$$

$$\begin{bmatrix} 600-1.0\omega_n^2 & -600 & \\ -600 & 1800-1.5\omega_n^2 & -1200 \\ & -1200 & 3000-2.0\omega_n^2 \end{bmatrix} \begin{Bmatrix} x_1 \\ x_2 \\ x_3 \end{Bmatrix} = \begin{Bmatrix} 0 \\ 0 \\ 0 \end{Bmatrix}$$

By solving the three equations with different mode shapes as in Figure 4.25, the eigenvector and natural frequency will be as follows:

Case 1. $\omega_1 = 14.5$ $T_1 = 0.433$ s
Case 2. $\omega_2 = 14.5$ $T_2 = 0.433$ s
Case 3. $\omega_3 = 14.5$ $T_3 = 0.433$ s

By applying equations (4.10), (4.11), (4.12), and (4.13),

$$\phi_1 = \begin{Bmatrix} 1.00 \\ 0.65 \\ 0.30 \end{Bmatrix}, \quad \phi_2 = \begin{Bmatrix} 1.00 \\ -0.61 \\ -0.68 \end{Bmatrix}, \quad \phi_3 = \begin{Bmatrix} 1.00 \\ -2.54 \\ -2.44 \end{Bmatrix}$$

4.5 Cylinder member strength

The traditional member in the jacket and in some cases for the topside is a cylinder, so this section focuses on the design of tubular members based on ISO 19902, the

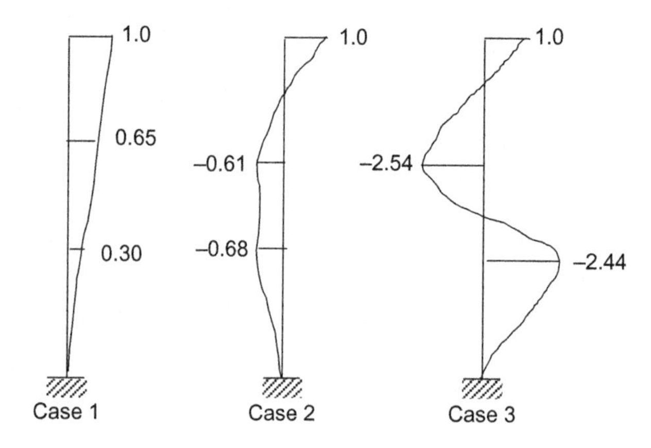

Figure 4.25 Eigenvector shapes.

principles of which emphasize load resistance factor design. In addition, calculation of the strength of the cylinder member also is presented based on API RP2A, which emphasizes the working stress design.

4.5.1 Cylinder member strength calculation by ISO19902

Based on ISO 19902, the tubular members subjected independently to axial tension, axial compression, bending, shear, or hydrostatic pressure are designed to satisfy the strength and stability requirements.

4.5.1.1 Axial tension

Tubular members subjected to axial tensile forces are designed to satisfy the following condition:

$$f_t \leq \frac{F_t}{\gamma_{R,t}} \tag{4.27}$$

where

f_t is the axial tensile stress due to forces from factored actions
F_t is the representative axial tensile strength, $F_t = F_y$
F_y is the representative yield strength, in stress units
$\gamma_{R,t}$ is the partial resistance factor for axial tensile strength, $\gamma_{R,t} = 1.05$

The member unity check, UC, under axial tension is calculated from equation (4.28):

$$\mathrm{UC} = \frac{f_t}{F_t/\gamma_{R,t}} \tag{4.28}$$

4.5.1.2 Axial compression

Tubular members subjected to axial compressive forces are designed to satisfy the following condition:

$$f_c \leq \frac{F_c}{\gamma_{R,c}} \tag{4.29}$$

where

f_c is the axial compressive stress due to forces from factored actions
F_c is the representative axial compressive strength, in stress units
$\gamma_{R,c}$ is the partial resistance factor for axial compressive strength $>\gamma_{R,c} = 1.18$

The member unity check, U_C, under axial compression shall be calculated from equation (4.30):

$$U_C = \frac{f_c}{F_C/\gamma_{R,c}} \tag{4.30}$$

4.5.1.3 Column buckling

In the absence of hydrostatic pressure, the representative axial compressive strength for tubular members should be the smaller of the in-plane and the out-of-plane buckling strengths determined from the following equations:

$$F_c = \left[1.0 - 0.278\lambda^2\right]F_{yc}, \quad \text{for } \lambda \le 1.34 \tag{4.31}$$

$$F_c = \frac{0.9}{\lambda^2}F_{yc}, \quad \text{for } \lambda > 1.34 \tag{4.32}$$

where

$$\lambda = \sqrt{\frac{F_{yc}}{F_e}} = \frac{KL}{\pi r}\sqrt{\frac{F_{yc}}{E}} \tag{4.33}$$

where

F_c is the representative axial compressive strength, in stress units
F_{yc} is the representative local buckling strength, in stress units
λ is the column slenderness parameter
F_e is the smaller of the Euler buckling strengths in the y- and z-directions, in stress units
E is the Young's modulus of elasticity
K is the effective length factor,
L is the unbraced length in the y- or z-direction
r is the radius of gyration

$$r = \sqrt{\frac{I}{A}}$$

I is the moment of inertia of the cross section
A is the cross-sectional area

4.5.1.4 Local buckling

The representative local buckling strength, F_{yc}, is calculated from the following equations:

$$F_{yc} = F_y, \quad \text{for } F_y/F_{xe} \le 0.170 \tag{4.34}$$

$$F_{yc} = \left[1.047 - 0.274\frac{F_y}{F_{xe}}\right]F_y, \quad \text{for } F_y/F_{xe} > 0.170 \tag{4.35}$$

$$F_{xe} = 2C_x\,Et/D \tag{4.36}$$

where

F_y is the representative yield strength, in stress units
F_{xe} is the representative elastic local buckling strength, in stress units
C_x is the critical elastic buckling coefficient, see below
E is Young's modulus of elasticity
D is the outside diameter of the member
t is the wall thickness of the member

The theoretical value of C_x for an ideal tubular member is 0.6. However, a reduced value of $C_x = 0.3$ should be used in equation (4.36) to account for the effect of initial geometric imperfections within the tolerance limits as presented in Chapter 5. A reduced value of $C_x = 0.3$ is also implicit in the limits for F_y/F_{xe} given in equations (4.34) and (4.35).

4.5.1.5 Bending

Tubular members subjected to bending moments are designed to satisfy the following condition:

$$f_b = \frac{M}{Z_e} \leq \frac{F_b}{\gamma_{R,b}} \tag{4.37}$$

where

f_b is the bending stress due to forces from factored actions, when $M > M_y$, f_b is considered an equivalent elastic bending stress, M/Z_e
F_b is the representative bending strength, in stress units, see later
$\gamma_{R,b}$ is the partial resistance factor for bending strength, $\gamma_{R,b} = 1.05$
M is the bending moment due to factored actions
M_y is the elastic yield moment
Z_e is the elastic section modulus

$$Z_e = \frac{\pi}{64}[D^4 - (D-2t)^4] \Big/ \left(\frac{D}{2}\right) \tag{4.38}$$

The utilization of a member as described by the ISO or unity check as used in most software, U_C, under bending moments is calculated from the following equation:

$$U_C = \frac{f_b}{F_b/\gamma_{R,b}} = \frac{M/Z_e}{F_b/\gamma_{R,b}} \tag{4.39}$$

The representative bending strength for tubular members are determined from:

$$F_b = \left(\frac{Z_p}{Z_e}\right)F_y, \text{ for } F_yD/Et \leq 0.0517 \tag{4.40}$$

$$F_b = \left[1.13 - 2.58\left(\frac{F_yD}{Et}\right)\right]\left(\frac{Z_p}{Z_e}\right)F_y, \text{ for } 0.0517 < F_yD/Et \leq 0.1034 \tag{4.41}$$

$$F_b = \left[0.94 - 0.76\left(\frac{F_yD}{Et}\right)\right]\left(\frac{Z_p}{Z_e}\right)F_y, \text{ for } 0.1034 < F_yD/Et \leq 120\, F_y/E \tag{4.42}$$

where, additionally,

F_y is the representative yield strength, in stress units
D is the outside diameter of the member
t is the wall thickness of the member
Z_p is the plastic section modulus, calculated from the following equation:

$$Z_p = \frac{1}{6}\left[D^3 - (D-2t)^3\right] \tag{4.43}$$

4.5.1.6 Shear

Tubular members subjected to beam shear forces are designed to satisfy the following condition:

$$f_{v,b} = \frac{2V}{A} \leq \frac{F_v}{\gamma_{R,v}}$$
$$F_v = F_y/\sqrt{3} \tag{4.44}$$

where

$f_{v,b}$ is the maximum beam shear stress due to forces from factored actions
F_v is the representative shear strength, in stress units,
$\gamma_{R,v}$ is the partial resistance factor for shear strength = 1.05
V is the beam shear due to factored actions, in force units
A is the cross-sectional area

The member unity check, U_c, under beam shear is calculated from the following equation:

$$U_C = \frac{f_{v,b}}{F_v/\gamma_{R,v}} = \frac{2V/A}{F_v/\gamma_{R,v}} \tag{4.45}$$

4.5.1.7 Torsional shear

Tubular members subjected to torsional shear forces are designed to satisfy the following condition:

$$f_{v,t} = \frac{M_{v,t}D}{2I_p} \leq \frac{F_v}{\gamma_{R,v}} \tag{4.46}$$

where

$f_{v,t}$ is the torsional shear stress due to forces from factored actions
$M_{v,t}$ is the torsional moment due to factored actions

$$I_p = \frac{\pi}{32}\left[D^4 - (D-2t)^4\right] \tag{4.47}$$

I_p is the polar moment of inertia

The partial resistance factor, $\gamma_{R,v}$, for shear is the same for both torsional shear and beam shear.

The member unity check, U_C, under torsional shear IS calculated from equation (4.48):

$$U_C = \frac{M_{v,t}D/2I_p}{F_v/\gamma_{R,v}} \tag{4.48}$$

4.5.1.8 Hydrostatic pressure

The effective depth at the location being checked is calculating taking into account the depth of the member below the still water level and the effect of passing waves. The factored water pressure, p, is calculated from the following equation:

$$p = \gamma_{,G1}\rho H_z \tag{4.49}$$

where

$\gamma_{,G1}$ is the partial action factor for permanent loads, as shown in Table 3.23 in chapter 3.
ρ is the density of the seawater, which may be taken as 1.025 kg/m^3
H_z is the effective hydrostatic head (m)

$$H_z = -z + \frac{H_w}{2}\frac{\cosh[k(d+z)]}{\cosh(kd)} \tag{4.50}$$

z is the depth of the member relative to the still water level (measured as positive upward),
d is the still water depth to the sea floor
H_w is the wave height
k is the wave number, $k = 2\pi/\lambda$, where λ is the wave length

For installation conditions, z is the maximum submergence during launch or the maximum differential head during upending. Also, the installation sequence plus an amount to allow for deviations from the planned sequence, and $\gamma_{f,G1}$ in equation (4.49) is replaced by $\gamma_{f,T}$, which is equal to 1.1 when permanent and variable actions predominate and equal 1.35 when the environmental load is predominant, as in case of transportation and installation calculation.

4.5.1.9 Hoop buckling

Tubular members subjected to external pressure aree designed to satisfy the following condition:

$$f_h = \frac{pD}{2t} \le \frac{F_h}{\gamma_{R,h}} \tag{4.51}$$

where

f_h is the hoop stress due to forces from factored hydrostatic pressure
p is the factored hydrostatic pressure, as calculated from equation (4.32)
D is the outside diameter of the member
t is the wall thickness of the member
F_h is the representative hoop buckling strength, in stress units, see the following
$\gamma_{R,h}$ is the partial resistance factor for hoop buckling strength, $\gamma_{R,h} = 1.25$

For tubular members satisfying out-of-roundness tolerances as presented in Chapter 5, F_h is determined from the following formulas:

$$F_h = F_y, \quad \text{for } F_{\text{he}} > 2.44F_y \tag{4.52}$$

$$F_h = 0.7(F_{\text{he}}/F_y)^{0.4}F_y \le F_y, \quad \text{for } 0.55F_y < F_{\text{he}} \le 2.44\,F_y \tag{4.53}$$

$$F_h = F_{\text{he}}, \quad \text{for } F_{\text{he}} \le 0.55\,F_y \tag{4.54}$$

where

F_y is the representative yield strength, in stress units
F_{he} is the elastic hoop buckling strength, in stress units

The elastic hoop buckling stress (F_{he}) is determined from the following equation:

$$F_{\text{he}} = 2C_h\,Et/D \tag{4.55}$$

where the critical elastic hoop buckling coefficient, C_h, is

$$\begin{aligned} &C_h = 0.44t/D, \text{ for } \mu \geq 1.6D/t \\ &C_h = 0.44t/D + 0.21(D/t)^3/\mu^4, \text{ for } 0.825D/t \leq \mu < 1.6D/t \\ &C_h = 0.737/(m - 0.579), \text{ for } 1.5 \leq \mu < 0.825D/t \\ &C_h = 0.80 \text{ for } \mu < 1.5 \end{aligned} \tag{4.56}$$

where

μ is a geometric parameter

$$\mu = \frac{L_r}{D}\sqrt{\frac{2D}{t}} \tag{4.57}$$

where Lr is the length of tubular member between stiffening rings, diaphragms, or end connections.

If a member violates the allowable tolerances and has ane out of roundness greater than 1% but less than 3%, the reduced value of F_{he} is as follows:

$$F'_{he} = F_{he}\left(1 - 0.2\sqrt{\frac{D_{max} - D_{min}}{0.01D_n}}\right)/0.8 \tag{4.58}$$

where D_{max} and D_{min} are the maximum and minimum values of any measured outside diameter at a cross section, and D_n is the nominal diameter.

The unit check of a member, U_C, under external pressure is calculated from the following equation;

$$U_C = \frac{pD/2t}{F_h/\gamma_{R,h}} \tag{4.59}$$

4.5.1.10 Tubular members subjected to combined forces without hydrostatic pressure

This subclause gives requirements for members subjected to combined forces, which gives rise to global and local interactions between axial forces and bending moments, without hydrostatic pressure. Generally, the secondary moments from factored global actions and the associated bending stresses (P-Δ effects) do not need to be considered. However, when the axial member force is substantial or when the component on which the axial force acts is very flexible, the secondary moments due to P-Δ effects from factored global actions should be taken into account.

4.5.1.10.1 Axial tension and bending

Tubular members subjected to combined axial tension and bending forces are designed to satisfy the following condition at all cross sections along their length:

$$\frac{\gamma_{R,t} f_t}{F_t} + \frac{\gamma_{R,b}\sqrt{f_{b,y}^2 + f_{b,z}^2}}{F_b} \leq 1.0 \tag{4.60}$$

where

$f_{b,y}$ is the bending stress about the member y-axis (in plane) due to forces from factored actions.
$f_{b,z}$ is the bending stress about the member z-axis (out of plane) due to forces from factored actions.

4.5.1.10.2 Axial compression and bending

Tubular members subjected to combined axial compression and bending forces are designed to satisfy the following conditions at all cross sections along their length:

$$\frac{\gamma_{R,c} f_c}{F_c} + \frac{\gamma_{R,b}}{F_b}\sqrt{\left[\left(\frac{C_{m,y} f_{b,y}}{1 - f_c/F_{e,y}}\right)^2 + \left(\frac{C_{m,z} f_{b,z}}{1 - f_c/F_{e,z}}\right)^2\right]} \leq 1.0 \tag{4.61}$$

and

$$\frac{\gamma_{R,c} f_c}{F_c} + \frac{\gamma_{R,b}\sqrt{f_{b,y}^2 + f_{b,z}^2}}{F_b} \leq 1.0 \tag{4.62}$$

where

$C_{m,y}$, $C_{m,z}$ are the moment reduction factors corresponding to member for y- and z-axes. respectively.
$F_{e,y}$, $F_{e,z}$ are the Euler buckling strengths corresponding to the member for y- and z-axes, respectively, in stress units.

$$F_{e,y} = \frac{\pi^2 E}{(K_y L_y / r_y)^2} \tag{4.63}$$

$$F_{e,z} = \frac{\pi^2 E}{(K_z L_z / r_z)^2} \tag{4.64}$$

where

K_y, K_z are the effective length factors in the y- and z- directions, respectively.
L_y, L_z are the unbraced lengths in the y- and z- directions, respectively.

4.5.1.11 Tubular members subjected to combined forces with hydrostatic pressure

A tubular member below the water line is subjected to hydrostatic pressure unless it has been flooded due to installation procedure requirements. Platform legs are normally flooded to assist in upending and placement and for pile installation. Even where members are flooded in the in-place condition, they can be subjected to hydrostatic pressures during launch and installation. The analysis of the structure can take the axial components of hydrostatic pressure on each member (capped-end actions) into account or these effects can be included subsequently.

The requirements are presented in terms of axial stresses, which include capped-end forces, f_{ac}. For analyses using factored actions that include capped-end actions, f_{ac} is the axial stress resulting from the analysis. For analyses using factored actions that do not include the capped-end actions,

$$f_{ac} = |f_a \pm \gamma_{f,G1} f_q| \tag{4.65}$$

where

f_a is the axial stress resulting from the analysis without capped-end actions.
f_q is the compressive axial stress due to the capped-end hydrostatic actions; f_q is added to f_a if f_a is compressive and subtracted from f_a if f_a is tensile.

Note that the condition for which f_a is tensile and $f_a < \gamma_{f,G1} f_q$ is one of axial compression

The capped-end stresses. f_q. may be approximated as half the hoop stress, due to forces from factored hydrostatic pressure; that is,

$$f_q = |0.5 f_h| \tag{4.66}$$

In reality, the magnitude of these stresses depends on the restraint on the member provided by the rest of the structure, and its value can be more or less than the value in equation (4.66). The approximation $|0,5\ f_h|$ may be replaced by a stress computed from a more rigorous analysis.

In all cases, equation (4.51) should be satisfied in addition to the requirements that follow.

4.5.1.11.1 Axial tension, bending, and hydrostatic pressure

Tubular members subjected to combined axial tension, bending, and hydrostatic pressure are designed to satisfy the following requirements at all cross sections along their length:

$$\frac{\gamma_{R,t} f_{ac}}{F_{t,h}} + \frac{\gamma_{R,b}\sqrt{f_{b,y}^2 + f_{b,z}^2}}{F_{b,h}} \leq 1.0 \tag{4.67}$$

where

f_{ac} is the tensile axial stress due to forces from factored actions that include capped-end actions. ($f_{ac} > 0$) $F_{t,h}$ is the representative axial tensile strength in the presence of external hydrostatic pressure, in stress units:

$$F_{t,h} = F_y\left[\sqrt{1 + 0.09\,B^2 - B^{2\eta}} - 0.3B\right] \tag{4.68}$$

$F_{b,h}$ is the representative bending strength in the presence of external hydrostatic pressure, in stress units:

$$B = \frac{\gamma_{R,h} f_h}{F_h},\ B \leq 1.0$$

$$\eta = 5 - 4\frac{F_h}{F_y}$$

4.5.1.11.2 Axial compression, bending, and hydrostatic pressure

Tubular members subjected to combined axial compression, bending, and hydrostatic pressure are designed to satisfy the following requirements at all cross sections along their length:

$$\frac{\gamma_{R,c} f_{ac}}{F_{yc}} + \frac{\gamma_{R,b}\sqrt{f_{b,y}^2 + f_{b,z}^2}}{F_{b,h}} \leq 1.0 \tag{4.69}$$

If $f_a < 0$; that is, the member is in compression regardless of the capped-end stresses, equation (4.70) should also be satisfied:

$$\frac{\gamma_{R,c} f_c}{F_{c,h}} + \frac{\gamma_{R,b}}{F_{b,h}}\sqrt{\left[\left(\frac{C_{m,y} f_{b,y}}{1 - f_c/F_{e,y}}\right)^2 + \left(\frac{C_{m,z} f_{b,z}}{1 - f_c/F_{e,z}}\right)^2\right]} \leq 1.0 \tag{4.70}$$

where, additionally, $F_{c,h}$ is the representative axial compressive strength in the presence of external hydrostatic pressure, in stress units:

$$F_{c,h} = 0.5F_{yc}\left[\left(1.0 - 0.278\lambda^2\right) - \frac{2f_q}{F_{yc}} + \sqrt{\left(1.0 - 0.278\lambda^2\right)^2 + 1.12\lambda^2\frac{f_q}{F_{yc}}}\right],$$

$$\text{for } \lambda \leq 1.34\sqrt{\left(1 - \frac{2f_q}{F_{yc}}\right)^{-1}} \tag{4.71}$$

$$F_{c,h} = \frac{0.9}{\lambda^2}F_{yc},\ \text{ for } \lambda > 1.34\sqrt{\left(1 - \frac{2f_q}{F_{yc}}\right)^{-1}} \tag{4.72}$$

If the maximum combined compressive stress $f_x = f_b + f_{ac}$ (if $f_{ac} \leq 0$) or $f_x = f_b - f_{ac}$ (if $f_{ac} < 0$) and the elastic local buckling strength F_{xe} exceed the limits given in equation (4.73), then equation (4.74) should also be satisfied:

$$f_x > 0.5\frac{F_{he}}{\gamma_{R,h}} \text{ and } \frac{F_{xe}}{\gamma_{R,c}} > 0.5\frac{F_{he}}{\gamma_{R,h}} \tag{4.73}$$

$$\frac{f_x - 0.5\frac{F_{he}}{\gamma_{R,h}}}{\frac{F_{xe}}{\gamma_{R,c}} - 0.5\frac{F_{he}}{\gamma_{R,h}}} + \left[\frac{\gamma_{R,h} f_h}{F_{he}}\right]^2 \leq 1.0 \tag{4.74}$$

where

F_{he} is the elastic hoop buckling strength.
F_{xe} is the representative elastic local buckling strength from equation (4.36).

4.5.1.12 Effective lengths and moment reduction factors

The effective lengths and moment reduction factors may be determined using a rational analysis that includes joint flexibility and side sway. In lieu of such a rational analysis, values of effective length factors (K) and moment reduction factors (Cm) may be taken from Table 4.1.

Values from this table it do not apply to cantilever members, and it is assumed that the both member ends are rotationally restrained in both planes of bending.

Lengths to which the effective length factors K are applied are normally measured from centertline to centerline of the end joints. However, for members framing legs, the following modified lengths may be used, provided no interaction between the buckling of members and legs affects the utilization of the legs:

- Face of leg to face-of-leg for main diagonal braces.
- Face of leg to centerline of end joint for K-braces.

Lower K factors than those of Table 4.1 may be used if these are supported by a more rigorous analysis. Where K can be obtained from the chart alignment in Figure 4.26.

In the figure, the subscripts A and R refer to the joints at the two ends of the column section being considered. G is defined as

$$G = \frac{\sum \frac{I_c}{L_c}}{\sum \frac{I_G}{L_G}} \tag{4.75}$$

in which $\sum$ indicates a summation of all members rigidly connected to that joint and lying in the plane in which buckling of the column is being considered. I_c is the moment of inertia, L_c is the unsupported length of the column section, I_G is the moment of inertia, and L_G the unsupported length of a girder or other restraining member. I_c and I_G are taken about axes perpendicular to the plane of buckling being considered.

Table 4.1 Effective length and moment reduction factors for member strength checking

Structural component	K	C_m [a]
Topsides legs		
Braced	1.0	0.85
Portal (unbraced)	K^b	0.85
Structure legs and piling		
Grouted composite section	1.0	$C_m = 1.0 - 0.4 \cdot (fc/Fe)$, or 0.85, whichever is less
Ungrouted jacket legs	1.0	$C_m = 1.0 - 0.4 \cdot (fc/Fe)$, or 0.85, whichever is less,
Ungrouted piling between shim points	1.0	$C_m = 0.6 - 0.4 \cdot M_1/M_2$
Structure brace members		$C_m = 0.6 - 0.4 \cdot M_1/M_2$ or $C_m = 1.0 - 0.4 \cdot (f_c/F_e)$ or 0.85, whichever is less
Primary diagonals and horizontals	0.7	
K- braces[c]	0.7	
X –braces	0.7	
Longer segment length[c]	0.8	$C_m = 1.0 - 0.4 \cdot (f_c/F_e)$ or 0.85, whichever is less
Full length[d]	0.7	
Secondary horizontals	0.7	

Notes:

[a] M_1/M_2 is the ratio of smaller to larger moments at the ends of the unfaced portion of the member in the plane of bending under consideration. M_1/M_2 is positive when the member is bent in reverse curvature, negative when bent in single curvature. $F_e = F_{ey}$ or F_{ez}, as appropriate.

[b] Use the effective length alignment chart as in Figure 4.26.

[c] For either in-plane or out-of-plane effective lengths, at least one pair of members framing into a K or X joint should be in tension, if the joint is not braced out of plane.

[d] When all members are in compression and the joint is not braced out of plane.

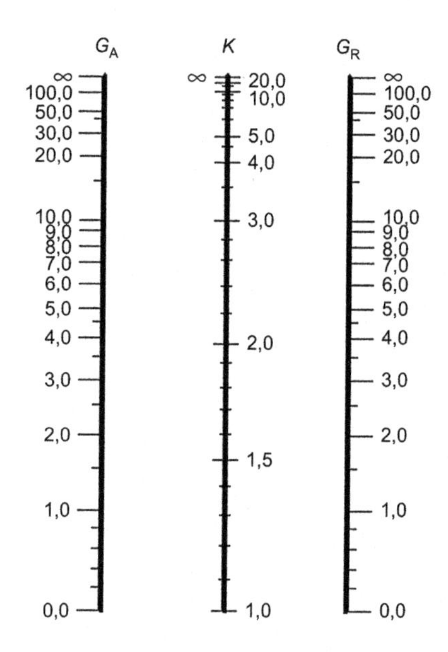

Figure 4.26 Chart alignment.

4.5.2 Cylinder member strength calculation by API RP2A

The design of cylinder member in API is based on AISC allowable stress design.

4.5.2.1 Axial tension

The allowable tensile stress, F_t, for cylindrical members subjected to axial tensile loads should be determined from

$$F_t = 0.6F_y \tag{4.76}$$

where

F_y = yield strength, ksi (MPa).

4.5.2.2 Axial compression

The allowable axial compressive stress, F_a, should be determined from the following AISC formulas for members with a *D/t* ratio equal to or less than 60:

$$F_a = \frac{\left[1 - \frac{(kl/r)^2}{2C_c^2}\right]F_y}{\frac{5}{3} + \frac{3(kl/r)}{8C_c} - \frac{(kl/r)^3}{8C_c^3}}, \quad \text{for } kl/r < C_c \tag{4.77}$$

$$F_a = \frac{12\pi^2 E}{23(kl/r)^2} \quad \text{for } kl/r \geq C_c \tag{4.78}$$

where

$$C_c = \sqrt{\frac{2\pi^2 E}{F_y}}$$

E = Young's modulus of elasticity, ksi (MPa)
K = effective length factor
l = unbraced length, m (in.)
r = radius of gyration, m (in.)

For members with a *D/t* ratio greater than 60, substitute the critical local buckling stress (F_{xe} or F_{xc}, whichever is smaller) for F_y in determining C_c and F_a.

4.5.2.3 Local buckling

Unstiffened cylindrical members fabricated from structural steels should be investigated for local buckling due to axial compression when the *D/t* ratio is greater than 60.

When the D/t ratio is greater than 60 and less than 300, with wall thickness $t > 0.25$ in. (6 mm), both the elastic (Fxe) and inelastic local buckling stress (Fxc) due to axial compression should be determined from Equation (4.90). Overall column buckling should be determined by substituting the critical local buckling stress (Fxe or Fxc, whichever is smaller) for Fy in Eq. (4.87) and in the equation for Cc.

1. **Elastic local buckling stress**. The elastic local buckling stress, F_{xe}, should be determined from equation (4.36):
2. **Inelastic local buckling stress.** The inelastic local buckling stress, F_{xc}, should be determined from

$$F_{xc} = F_y[1.64 - 0.23(D/t)^{0.25}] \leq F_{xe}$$
$$F_{xc} = F_y \text{ for } (D/t) \leq 60 \tag{4.79}$$

4.5.2.4 Bending

The allowable bending stress, F_b, should be determined from:

$F_b = 0.75F_y$, for $D/t \leq 10340/F_y$ (SI units)

$$F_b = \left[0.84 - 1.74\frac{F_y D}{Et}\right]F_y, \text{ for } 10340/F_y < D/t \leq 20680/F_y \text{ (SI units)} \tag{4.80}$$

$$F_b = \left[0.72 - 0.58\frac{F_y D}{Et}\right]F_y, \text{ for } 20680/F_y < D/t \leq 300 \text{ (SI units)} \tag{4.81}$$

For D/t ratios greater than 300, refer to API Bulletin 2U.

4.5.2.5 Shear

The maximum beam shear stress, f_v, for cylindrical members is

$$f_v = \frac{V}{0.5A} \tag{4.82}$$

where

f_v = the maximum shear stress, MPa (ksi)
V = the transverse shear force, MN (kips)
A = the cross sectional area, m^2 ($in.^2$)

The allowable beam shear stress, F_v, should be determined from

$$F_v = 0.4\, F_y \tag{4.83}$$

4.5.2.6 Torsional shear

The maximum torsional shear stress, F_v, on cylindrical members caused by torsion is

$$f_{vt} = \frac{M_t(D/2)}{I_p} \tag{4.84}$$

where

f_{vt} = maximum torsional shear stress, MPa
M_t = torsional moment, MN-m
I_p = polar moment of inertia, m^4

The allowable torsional shear stress, F_{vt}, should be determined from equation (4.83).

4.5.2.7 Pressure (stiffened and unstiffened cylinders)

For tubular platform members satisfying API Spec 2B out-of-roundness tolerances, the acting membrane stress, f_h, in ksi (MPa), should not exceed the critical hoop buckling stress, F_{hc}, divided by the appropriate safety factor:

$$f_h \leq F_{hc}/SF_h \tag{4.85}$$

$$f_h = pD/2t \tag{4.86}$$

where

f_h = hoop stress due to hydrostatic pressure, ksi (MPa)
p = hydrostatic pressure, ksi (MPa)
S_{Fh} = safety factor against hydrostatic collapse

4.5.2.8 Design hydrostatic head

The hydrostatic pressure calculation in API is the same as presented in equation (3.55) and the depth below the still water surface including tide, ft (m). The value of z is positive, measured downward from the still water surface. For installation, z should be the maximum submergence during the launch or differential head during the upending sequence, plus a reasonable increase in head to account for structural weight tolerances and deviations from the planned installation sequence. The seawater density is equal to 0.01005 MN/m^3 (64 lbs/ft^3).

4.5.2.9 Hoop buckling stress

The elastic hoop buckling stress, F_{he}, and the critical hoop buckling stress, F_{hc}, are determined from the following formulas.

1. **Elastic Hoop Buckling Stress.** Determination of the elastic hoop buckling stress is based on a linear stress-strain relationship from

$$F_{he} = 2C_h E\, t/D \tag{4.87}$$

where the critical hoop buckling coefficient, C_h, includes the effect of initial geometric imperfections within API Specification 2B tolerance limits.

$$\begin{aligned} C_h &= 0.44t/D, \text{ for } M \geq 1.6D/t \\ C_h &= 0.44t/D + 0.21(D/t)^3/M^4 0.825D/t \leq M < 1.6D/t \\ C_h &= 0.736/(M-0.636) 3.5 \leq M < 0.825D/t \\ C_h &= 0.736/(M-0.559)\ 1.5 \leq M < 3.5 \\ C_h &= 0.8\ M < 1.5 \end{aligned} \tag{4.88}$$

The geometric parameter, M, is defined as

$$M = \frac{L}{D}\sqrt{\frac{2D}{t}} \tag{4.89}$$

where

L = length of cylinder between stiffening rings, diaphragms, or end connections, in. (m)

Note: For $M > 1.6D/t$, the elastic buckling stress is approximately equal to that of a long unstiffened cylinder. Therefore, stiffening rings, if required, should be spaced such that $M < 1.6D/t$ in order to be beneficial.

2. **Critical Hoop Buckling Stress.** The material yield strength relative to the elastic hoop buckling stress determines whether elastic or inelastic hoop buckling occurs, and the critical hoop buckling stress, F_{hc}, in ksi (MPa) is defined by the appropriate formula. Fort elastic buckling,

$F_{hc} = F_{he}$, for $F_{he} \leq 0.55F_y$

For inelastic buckling,

$$\begin{aligned} F_{hc} &= 0.45F_y + 0.18 + F_{he}, \quad \text{for } 0.55F_y < F_{he} \leq 1.6F_y \\ F_{hc} &= \frac{1.31F_y}{1.15 + (F_y/F_{he})}, \quad \text{for } 1.6F_y < F_{he} < 6.2F_y \\ F_{hc} &= F_y F_{he} > 6.2F_y \end{aligned} \tag{4.90}$$

4.5.2.10 Combined stresses for cylindrical members

Discussion of the method of calculating the applied combined stress between bending with compressive and tensile stress in addition to the hydrostatic stress is based on AISC.

4.5.2.10.1 Combined axial compression and bending

Cylindrical members subjected to combined compression and flexure should be proportioned to satisfy both the following requirements at all points along their length:

$$\frac{f_a}{F_a} + \frac{C_m\sqrt{f_{bx}^2 + f_{by}^2}}{\left(1 - \frac{f_a}{F'_e}\right)F_b} \leq 1.0 \tag{4.91}$$

$$\frac{f_a}{0.6F_y} + \frac{\sqrt{f_{bx}^2 + f_{by}^2}}{F_b} \leq 1.0 \tag{4.92}$$

where the undefined terms used are as defined by the AISC *Specification for the Design, Fabrication, and Erection of Structural Steel for Buildings.*

When $f_a/F_a \leq 0.15$, the following formula may be used in lieu of the preceding two formulas:

$$\frac{f_a}{F_y} + \frac{\sqrt{f_{bx}^2 + f_{by}^2}}{F_b} \leq 1.0 \tag{4.93}$$

Equation (4.94) assumes that the same values of C_m and F'_e are appropriate for f_{bx} and f_{by}. If different values are applicable, the following formula or other rational analysis should be used instead of equation (4.94):

$$\frac{f_a}{F_a} + \frac{\sqrt{\left(\frac{C_{mx}f_{bx}}{1 - \frac{f_a}{F'_{ex}}}\right)^2 + \left(\frac{C_{my}f_{by}}{1 - \frac{f_a}{F'_{ey}}}\right)^2}}{F_b} \leq 1.0 \tag{4.94}$$

4.5.2.10.2 Member slenderness

Determination of the slenderness ratio *Kl/r* for cylindrical compression members should be in accordance with the AISC. A rational analysis for defining the effective length factors should consider joint fixity and joint movement. Moreover, a rational definition of the reduction factor should consider the character of the cross section and the loads acting on the member. In lieu of such an analysis, the values in Table 4.2 may be used.

4.5.2.11 Combined axial tension and bending

Cylindrical members subjected to combined tension and bending should be proportioned to satisfy equation (4.95) at all points along their length, where f_{bx} and f_{by} are the computed bending tensile stresses.

Table 4.2 **K and C_m values based on API RP2A**

Structure element	Effective length factor, K	Reduction factor C_m
Superstructure legs		
Braced	1.0	0.85
Portal (unbraced)	K**	0.85
Jacket legs and piling		
Grouted composite section	1.0	Min of $[1-0.4(f_a/F_e)]$ or 0.85
Ungrouted jacket legs	1.0	Min of $[1-0.4(f_a/F_e)]$ or 0.85
Ungrouted piling between shim points	1.0	$0.6-0.4(M_1/M_2)$, $0.4<C_m<0.85$
Deck truss web members		$0.6-0.4(M_1/M_2)$, $0.4<C_m<0.85$
In-plane action	0.8	$0.6-0.4(M_1/M_2)$, $0.4<C_m<0.85$ or 0.85
Out-of-plane action	1.0	0.85
Jacket braces		$0.6-0.4(M_1/M_2)$, $0.4<C_m<0.85$ or min of $[1-0.4(f_a/F_e)]$ or 0.85
Face-to-face length of main diagonals	0.8	
Face of leg to centerline of joint length of K braces*	0.8	Min of $[1-0.4(f_a/F_e)]$ or 0.85
Longer segment length of X braces*	0.9	Min of $[1-0.4(f_a/F_e)]$ or 0.85
Secondary horizontals	0.7	Min of $[1-0.4(f_a/F_e)]$ or 0.85
Deck truss chord members	1.0	0.85 or $0.6-0.4(M_1/M_2)$, $0.4<C_m<0.85$ or min of $[1-0.4(f_a/F_e)]$ or 0.85

*At least one pair of members framing into a joint must be in tension if the joint is not braced out of plane.
**As in Figure 4.27.

4.5.2.12 Axial tension and hydrostatic pressure

When member longitudinal tensile stresses and hoop compressive stresses occur simultaneously, the following interaction equation should be satisfied:

$$A^2 + B^2 + 2\nu AB \le 1.0 \tag{4.95}$$

where

$$A = \frac{f_a + f_b - (0.5f_h)}{F_y}(\mathrm{SF}_x) \tag{4.96}$$

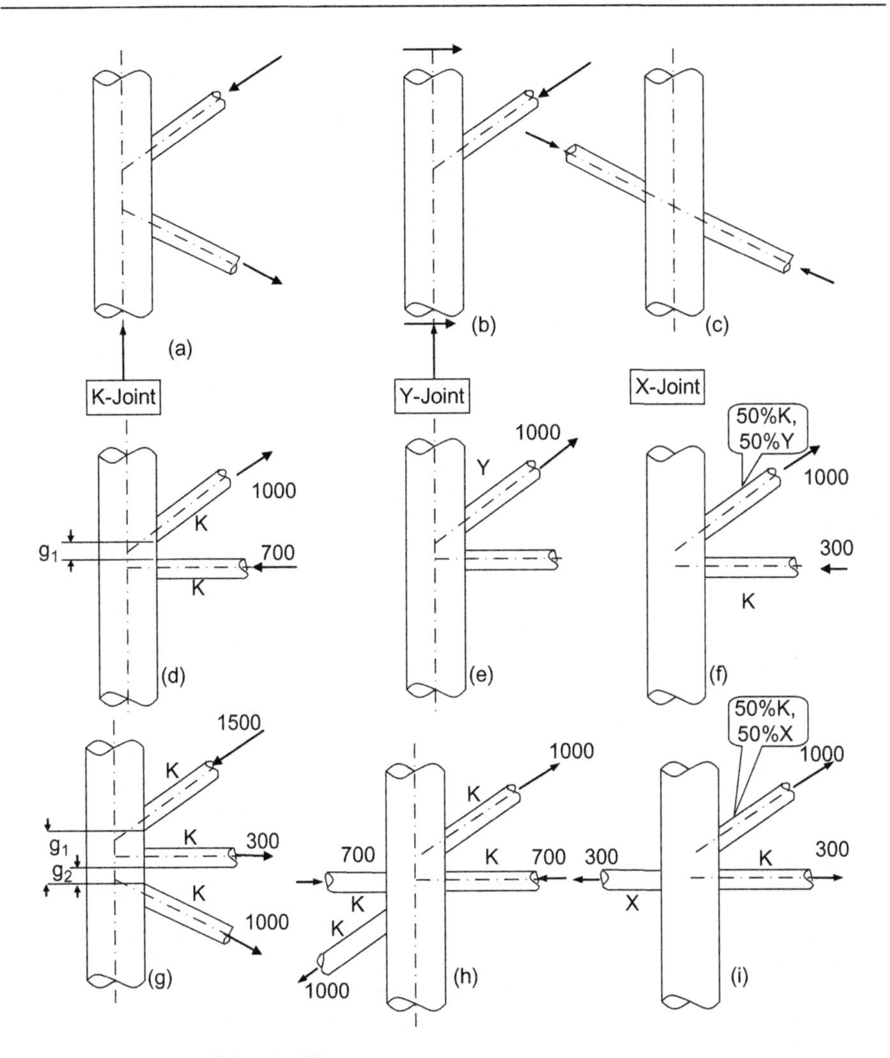

Figure 4.27 In-plane joint detailing.

The term A should reflect the maximum tensile stress combination.

$$B = f_h / F_{hc}(\mathrm{SF}_h) \tag{4.97}$$

ν = Poisson's ratio = 0.3
F_y = yield strength, ksi (MPa)
f_a = absolute value of acting axial stress, ksi (MPa)
f_b = absolute value of acting resultant bending stress, ksi (MPa) f_h = absolute value of hoop compression stress, ksi (MPa)
F_{hc} = critical hoop stress
SF_x = safety factor for axial tension
SF_h = safety factor for hoop compression

4.5.2.13 Axial compression and hydrostatic pressure

When longitudinal compressive stresses and hoop compressive stresses occur simultaneously, the following equations should be satisfied:

$$\frac{f_a + (0.5f_h)}{F_{xc}}(\mathrm{SF}_x) + \frac{f_b}{F_y}(\mathrm{SF}_b) \leq 1.0 \tag{4.98}$$

$$\mathrm{SF}_h \frac{f_h}{F_{hc}} \leq 1.0 \tag{4.99}$$

Equation (4.98) should reflect the maximum compressive stress combination. The following equation should also be satisfied when $f_x > 0.5\ F_h a$:

$$\frac{f_x - 0.5F_{ha}}{F_{aa} - 0.5F_{ha}} + \left(\frac{f_h}{F_{ha}}\right)^2 \leq 1.0 \tag{4.100}$$

where

$F_{aa} = F_{xe}/\mathrm{SF}_x$
$F_{ha} = F_{he}/\mathrm{SF}_h$
SF_x = safety factor for axial compression
SF_b = safety factor for bending
$f_x = f_a + f_b + (0.5f_h)$; f_x should reflect the maximum compressive stress combination

where F_{xe}, F_{xc}, F_{he}, and F_{hc} are given by Equations (4.36), (4.80), (4.89), and (4.91), respectively.

Note that, if $f_b > f_a + 0.5\ f_h$, both Equations (4.95) and (4.100) must be satisfied

4.5.2.14 Safety factors

To compute allowable stresses, the required safety factors should be used with the local buckling interaction equations presented in Table 4.3.

Example 4.3

A 240-ft wellhead jacket installation upends the leg A_1 from an elevation of −50 m to 170 m (−170 ft to 240 ft). The relevant data are as follows:

Outside diameter, D = 46 in. = 1150 mm
Wall thickness; t = 0.5 in. = 12.7 mm
Length of cylinder between stiffening rings, diaphragms, or end connection; L = 450 in. = 11.43 m
Safety factor against hydrostatic collapse, SF_h = 2.0
Depth below the still water surface, z = 85.74 ft = 26 m
Wave height, H_w = 4 ft = 1.2 m
Wave length, L_w = 6 ft = 1.8 m
Still water depth, d = 240 ft = 72 m
Yield strength, F_y = 36 ksi = 248 MPa

Table 4.3 Safety factor based on API RP2A

Design condition	Loading			
	Axial tension	Bending	Axial compression	Hoop compression
Where the basic allowable stresses are used, e.g., pressures that definitely are encountered during the installation or life of the structure	1.67	F_y/F_b	1.67 to 2.0	2.0
Where the one-third increase in allowable stresses is appropriate, e.g., when considering interaction with storm loads	1.25	$F_y/1.33\ F_b$	1.25 to 1.5	1.5

Young's modulus, $E = 29{,}000$ ksi = 2,000 MPa
Seawater density, $\gamma = 64.0$ Ibs/ft^3 = 1.02 t/m^3

Calculation Results Calculating the critical hoop buckling stress,

Geometric parameter, $M = 132.7$ from equation (4.89)
Critical buckling coefficient, $C_h = 0.005310$, from equation (4.88)
Elastic hoop buckling stress, $F_{he} = 3.348$ ksi = 23.1 MPa, from equation (4.90)
Critical hoop stress, $F_{hc} = 3.348$ ksi
$F_{hc}/\text{SF}_h = 3.348/2 = 1.674$

Calculating the acting membrane stress,

Wave number, $k = 1.0$
Design hydrostatic head = 85.74 ft
Hydrostatic pressure, $p = 0.0381$ ksi
Hoop stress duet to hydrostatic pressure, $f_h = 1.753$ ksi, from equation (4.86)
$\text{UC} = (F_{hc}/\text{SF}_h)/f_h = 1.047 > 1.0$ unsafe; you should reduce the distance between rings or increase thickness

Example 4.4

Calculate the utilization factor without and with rings. The relevant data are as follows:

Depth below the still water surface, $z = 230.0$ ft = 70 m
Outside diameter, $D = 52.0$ in. = 1320 mm
Wall thickness, $t = 0.5$ in. = 12.7 mm
Member yield strength, $F_y = 50.0$ ksi = 345 MPa
Safety factor, $\text{SF}_h = 2.0$

The ring type is external.

Calculations For stress calculation without rings,

Geometric parameter; $M = 231.96$

$D/t = 104$
Critical hoop buckling coefficient; $C_h = 0.004$

For stress calculation with rings,

Geometric parameter, $M = 38.274$
Critical hoop buckling coefficient, $C_h = 0.020$
Assumed flange width, $b_f = 5.609$ in. = 142.5 mm
Centroid of ring, $y = 4.57$ in. = 116.1 mm
Moment of inertia of ring, $I_{xx} = 24.368$ in.^{4} = 10,142,727.4 mm^4

The calculation results for members without rings are

Length of unstiffened member, L_{uns}. = 836.4 in. = 21244.6 mm
Hoop stress due to hydrostatic pressure, $f_h = 5.316$ ksi = 36.7 MPa
Elastic hoop buckling stress, $F_{be} = 2.359$ ksi = 16.3 MPa
Critical hoop buckling stress, $F_{bc} = 2.359$ ksi = 16.3 MPa
U.F = 4.506

The calculation results for members with rings are

Web thickness, $t_w = 0.5$ in. = 12.7 mm
Web depth, $d_w = 6.0$ in. = 152.4 mm
Length of member between stiffening rings, $L_{stiff} = 138.0$ in. = 3.51 m
Elastic hoop buckling stress, $F_{he} = 10.905$ ksi = 75.3 MPa
Critical hoop buckling stress, $F_{hc} = 10.905$ ksi = 75.3 MPa
Required moment of inertia for ring composite section, $I_c = 12.124$ in.^{4} = 5046389.8 mm^4
Member unity check, U.F. = 0.975
Ring U.F. = 0.498

Example 4.5

As in example 4.1, designing the ring is required. the relevant data are as follows:

The member outside diameter, $D = 56$ in. = 1422.2 mm
Wall thickness, $t = 0.75$ = 19 mm
Length between stiffening rings or end connection, $L = 20.0$ ft = 6.0 m
Ring thickness, $t_w = 0.50$ in. = 12.7 mm
Ring web depth, where $h/t_w < (E/F_y)^{1/2}$, $h = 6.0$ in. = 152.4 mm
Minimum yield stress, $F_y = 50$ ksi = 348 MPa
Young's modulus of elasticity, $E = 2{,}900$ ksi = 20,000 MPa

Hydrostatic data

Depth below the still water surface including tides, $z = 240$ ft = 73.2 m
Wave height, $H_w = 0$ ft
Wave length, $L_w = 0$ ft
Still water depth, = 240 ft
Seawater density; $\gamma = 64.2$ Ibs/ft^3

Section properties

D/t ratio = 74.7
Shell effective width, $1(Dt)^{1/2} = 7.129$ in.
Composite ring section N.A. (from shell out) n.a. = 1.588 in. = 40.3 mm
Ring composite section moment of inertia, I' = 31.14 in.^{4} = 12961446.6 mm^4

Hydrostatic properties

$2\pi/L_{w;K}$, no wave
Design hydrostatic head, $h_z = 240.0$ ft $= 73.2$ m
Hydrostatic pressure, $p = 0.107$ ksi $= 0.737$ MPa

Acting stress

Acting membrane hoop stress, $f_h = 3.99$ ksi $= 27.51$ MPa, from equation (4.86)

Allowable stress

Elastic hoop buckling stress geometric parameter, $M = 52.4$, from equation (4.89)
$3.5 \leq M \leq 0.825\ D/t$, from equation (4.88)
Critical hoop buckling coefficient; $C_h = 0.014$, from equation (4.86)
Elastic hoop buckling stress, $F_{he} = 11.05$ MPa, from equation (4.87)
$F_{he} \leq 0.55\ F_y$
Critical hoop buckling stress; $F_{hc} = 11.05$ MPa, from equation (4.90)
Safety factor for hoop compression, $SF_h = 2.00$

Code check

Hydrostatic pressure check, $f_h SF_h / F_{hc} = 0.723$

Ring design

Ring composite sect required moment of inertia,. $I_c = 26.89$ in.^{4} $= 11{,}192{,}463.03$ mm^4
Ring moment of inertia unity check, $I_c/I' = 0.863$

4.6 Tubular joint design

In an old API RP2A, the punching shear governs the design of the tubular joints. It is appropriate to summarize the historical development of the API RP 2 A-WSD provisions and the background to the most recent major updates, as incorporated into this supplement to the 21st edition. In the third edition of API RP 2AWSD, issued in 1972, according to Marshall and Toprac (1974), some simple recommendations were introduced based on punching shear principles. In the fourth edition, factors were introduced to allow for the presence of load in the chord and the brace-to-chord diameter ratio (β). In the ninth edition, issued in 1977, differentiation was introduced in the allowable stress formulations for the joint and loading configuration, that is, *T/Y*, *X*, and *K*.

Much work was done over the period 1977 to 1983, including large-scale load tests to failure, to improve the understanding and prediction of joint behavior. This work cumulated in the issue of the 14th edition of API RP 2AWSD, in which the punching shear stress formulations were considerably modified and included a more realistic expression to account for the effect of chord loads as well as providing an interaction equation for the combined effect of brace axial and bending stresses. Also introduced in the 14th edition was the alternative nominal load approach, which gives equivalent results to the punching shear method. The guidance then essentially remained unchanged for all editions up to the 21st, although further recommendations were added on load transfer through the chord in the 20th edition (1993).

Regardless of API RP 2A-WSD stability, much further knowledge, including both experimental data and numerical studies, has been gained on the behavior of joints since the 14th edition was issued. Over the period 1994 to 1996, MSL Engineering, under the auspices of a joint industry project, undertook an update to the tubular joint database and guidance as presented by MSL report (1996) and Dier and Lalani (1995). This work and more recent studies, notably by API/EWI and the University of Illinois, have formed the basis of the tubular joint strength provisions of ISO19902 (ISO, 2004). The ISO drafting committee took, as a starting point, the relevant provisions from API RP 2A-LRFD first edition (similar to API RP 2A-WSD 20th edition) because ISO is in LRFD format. However, the API RP 2A-WSD provisions were greatly modified during the drafting process to take account of the greater knowledge.

For the purposes of this supplement to the 21st edition of API RP 2A-WSD, the draft ISO provisions, in turn, have been used as a starting basis. Additional studies, not available at the time of the preparation of the draft ISO guidance, have been incorporated into this supplement to the 21st edition of API RP 2A-WSD. The major updates between the 21st edition and this supplement to the 21st edition are detailed in the following subsections but, in summary, involve a relaxation of the two-thirds limit on tensile strength, additional guidance on detailing practice, removal of the punching shear approach, new Q_u and Q_f formulations, and a change in the form of the brace load interaction equation. But, here, we present the old method of calculating the tubular joint capacity based on the punching shear that was removed from the new API RP2A (2007).

4.6.1 Simple joint calculation from API RP2A (2007)

4.6.1.1 Joint classification and detailing

The first step is to define the joint classification under axial load, of which K, X, and Y are the components of loading corresponding to the three joint types for which capacity equations exist. Such subdivision normally considers all the members *in one plane* at a joint.

Based on API RP2A 2007, the brace planes within $\pm 15°$ of each other may be considered in a common plane. Each brace in the plane can have a unique classification that could vary with load condition. Noting that, the classification can be a mixture of the main three joint types, which are X, T, and K.

Some simple examples of joint classification is presented in Figure 4.27. For a brace to be considered in the K joint classification, the axial load in the brace should be balanced to within 10% by loads in other braces on the same plane and on the same side of the joint. For Y joint classification, the axial load in the brace is reacted as beam shear in the chord. For X joint classification, the axial load in the brace is transferred through the chord to the opposite side (e.g., to braces, padeyes, launch rails).

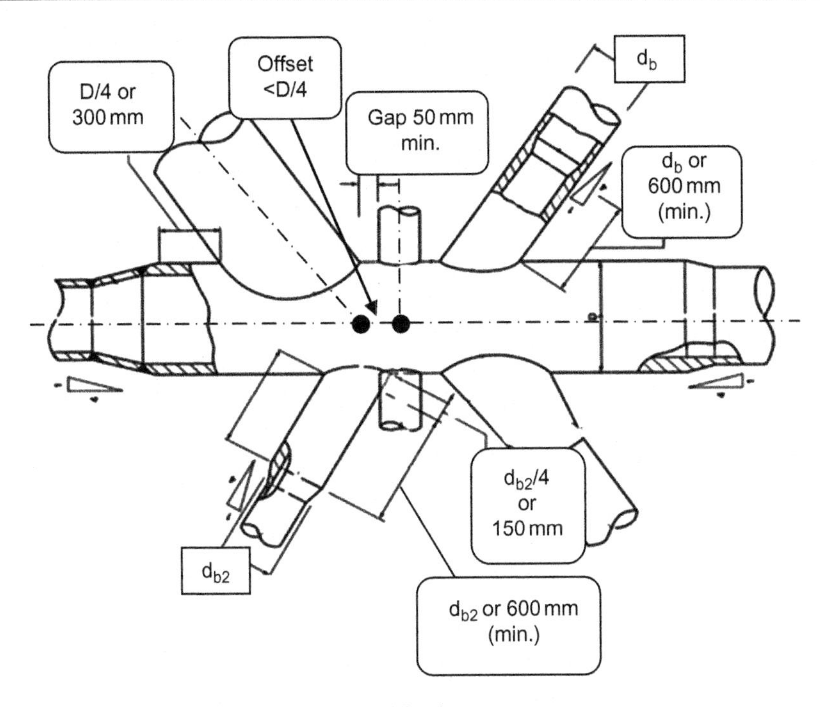

Figure 4.28 Different shapes of joints classification.

Case (h) in Figure 4.27 is a good example of the loading and classification hierarchy that should be adopted in the classification of joints. Replacement of brace load by a combination of tension and compression load to give the same net load is not permitted.

For example, replacing the load in the horizontal brace on the left-hand side of the joint by a compression load of 1000 and tension load of 500 is not permitted, as this may result in an inappropriate X classification for this horizontal brace and a K classification for the diagonal brace.

Special consideration should be given to establishing the proper gap if a portion of the load is related to K joint behavior. The most obvious case is in Figure 4.28 (a), for which the appropriate gap is between adjacent braces. However, if an intermediate brace exists, as in case (d), the appropriate gap is between the outer loaded braces. In this case, since the gap is often large, the K joint capacity could revert to that of a Y joint. Case (e) is instructive in that the appropriate gap for the middle brace is gap 1, whereas for the top brace it is gap 2. Although the bottom brace is treated as a 100% K classification, a weighted average in capacity is required, depending on how much of the acting axial load in this brace is balanced by the middle brace (gap 1) and how much is balanced by the top brace (gap 2).

In some instances, the joint behavior is more difficult to define or is apparently worse than predicted by the preceding approach to classification. Two of the more common cases in the latter category are launch truss loading and in-situ loading of skirt pile sleeves.

Joint detailing is an essential element of joint design. For unreinforced joints, the recommended detailing nomenclature and dimensioning is shown are Figures 4.28 and 4.29. These figures indicate that, if an increased chord wall thickness is required, it should extend past the outside edge of incoming bracing a minimum of one quarter of the chord diameter or 12 inches (300 mm), whichever is greater. Even greater lengths of increased wall thickness or special steel may be needed to avoid downgrading the joint capacity. If an increased wall thickness of brace or special steel is required, it should extend a minimum of one brace diameter or 24 inches (600 mm), whichever is greater. Neither the cited chord can nor brace stub dimension includes the length over which the 1:4 thicknesses taper occurs. In situations where fatigue has a major effect, tapering on the inside may have the undesirable consequence of fatigue cracking originating on the inside surface and be difficult to inspect.

The minimum nominal gap between adjacent braces, whether in or out of plane, is 2 inches (50 mm). Care should be taken to ensure that an overlap of welds at the toes of the joint is avoided.

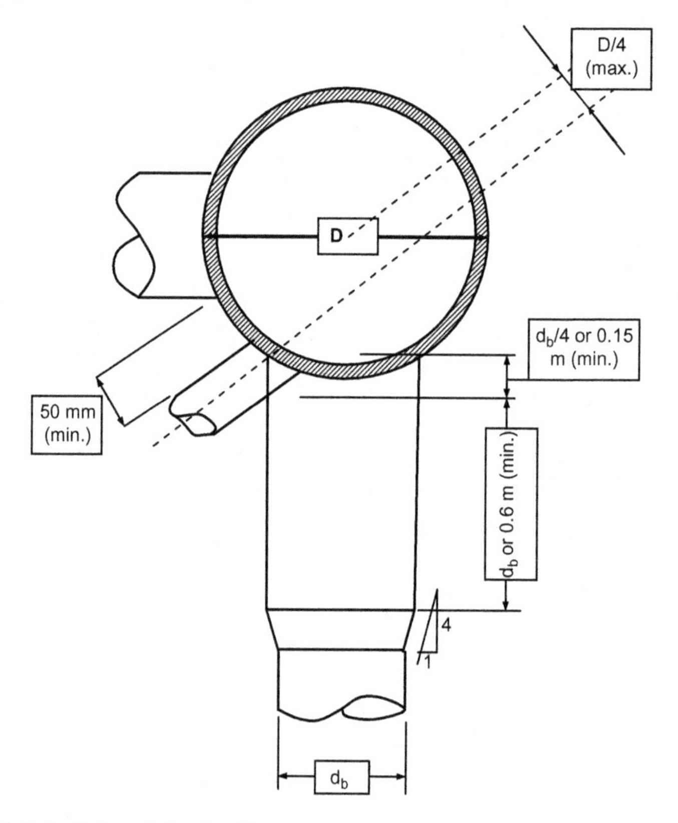

Figure 4.29 Out-of-plane joint detailing.

When overlapping braces occur, the amount of overlap should be at least $d/4$ (where d is the diameter of the through brace) or 6 inches (150 mm), whichever is greater. This dimension is measured along the axis of the through member.

Where an overlapping of braces is necessary or preferred and the braces differ in nominal thickness by more than 10%, the brace with the larger wall thickness should be the through brace and fully welded to the chord. Further, where a substantial overlap occurs, the larger diameter brace should be specified as the through member. This brace may require an end stub to ensure that the thickness is at least equal to that of the overlapping brace.

Longitudinal seam welds and girth welds should be located to minimize or eliminate their impact on joint performance. The longitudinal seam weld of the chord should be separated from incoming braces by at least 12 inches (300 mm), as shown in Figure 4.29. The longitudinal seam weld of a brace should be located near the crown heel of the joint. Longer chord cans may require a girth weld. This weld should be positioned at a lightly loaded brace intersection, between the saddle and crown locations, as shown in Figure 4.28.

4.6.1.2 Simple tubular joint calculation

The simple joints calculation is valid based on the following criteria:

0.2 ≤ .2 in1.0
10 ≤ 00 in50
30° 0 0 0 90°
$Fy \leq$ y72 ksi (500 MPa)
$g/D > -0.6$ (for K joints)

Tubular joints without the overlap of principal braces and having no gussets, diaphragms, grout, or stiffeners should be designed using the following guidelines:

$$P_a = Q_u Q_f \frac{F_{yc} T^2}{\mathrm{SF} \sin\theta} \tag{4.101}$$

$$M_a = Q_u Q_f \frac{F_{yc} T^2 d}{\mathrm{SF} \sin\theta} \tag{4.102}$$

(plus a one-third increase in both cases where applicable) where

P_a = allowable capacity for brace axial load
M_a = allowable capacity for brace bending moment
F_{yc} = the yield stress of the chord member at the joint (or 0.8 of the tensile strength, if less), ksi (MPa)
SF = safety factor = 1.60

Table 4.4 Values for Q_u

Joint classification	Brace load			
	Axial tension	Axial compression	In-plane bending	Out-of-plane bending
K	$(16 + 1.2\gamma)\beta^{1.2}Q_g$ but $\leq 40\beta^{1.2}Q_g$			
T/Y	30β	$2.8+(20+0.8\gamma)\beta^{1.6}$ but $\leq 2.8+36\beta^{1.6}$	$(5 + 0.7\gamma)\beta^{1.2}$	$2.5 + (4.5 + 0.2\gamma)\beta^{2.6}$
X	23β for $\beta \leq 0.9$ $20.7 + (\beta-0.9)(17\gamma - 220)$ for $\beta > 90$	$[2.8 + (12 + 0.1\gamma)\beta]Q_\beta$		

For joints with thickened cans, Pa should not exceed the capacity limits.

For axially loaded braces with a classification that is a mixture of K, Y, and X joints, take a weighted average of P_a based on the portion of each in the total load.

4.6.1.2.1 Strength factor Q_u

The strength factor Q_u varies with the joint and load type, as given in Table 4.4. Where the working points of members at a gap connection are separated by more than $D/4$ along the chord centerline or where a connection has simultaneously loaded branch members in more than one plane, the connection may be classified as a general or multiplanar connection.

Q_β is a geometric factor defined by

$$Q_\beta = \frac{0.3}{\beta(1-0.833\beta)}, \text{ for } \beta > 0.6$$
$$Q_\beta = 1.0, \text{ for } \beta \leq 0.6 \tag{4.103}$$

Q_g is the gap factor defined by

$Q_g = 1 + 0.2\,[1-2.8\, g/D]^3$, for $g/D \geq 0.05$ but ≥ 1.0
$Q_g = 0.13 + 0.65\,\phi\,\gamma^{0.5}$, for $g/D \leq -0.05$

where

$\phi = t\,F_{yb}/(TF_y)$

The overlap should not be less than $0.25\beta D$. Linear interpolation between the limiting values of the preceding two Q_g expressions may be used for $-0.05 < g/D < 0.05$ when this is otherwise permissible or unavoidable.

F_{yb} = yield stress of brace or brace stub if present (or 0.8 times the tensile strength if less), ksi (MPa)

The Q_u term for tension loading is based on limiting the capacity to the first crack. The Q_u associated with full ultimate capacity of tension loaded Y and X joints is given in the commentary.

The X joint, axial tension, Q_u term for $\beta > 0.9$ applies to coaxial braces (i.e., $e/D \leq /0.2$ where e is the eccentricity of the two braces). If the braces are not coaxial ($e/D > 0.2$), then 23β should be used over the full range of β.

4.6.1.2.2 Chord load factor Q_f

The chord load factor, Q_f, is a factor to account for the presence of nominal loads in the chord and is calculated from the following equation:

$$Q_f = \left[1 + C_1\left(\frac{SFP_c}{P_y}\right) - C_2\left(\frac{SFM_{ipb}}{M_p}\right) - C_3A^2\right] \tag{4.104}$$

The parameter A is defined as follows:

$$A = \sqrt{\left[\left(\frac{SFP_c}{P_y}\right)^2 + \left(\frac{SFM_c}{M_p}\right)^2\right]} \tag{4.105}$$

where a one-third increase applicable, SF = 1.20 in equations (4.104) and (4.105).

Where P_c and M_c are the nominal axial load and bending results in the chord, P_y is the yield axial capacity of the chord, M_p is the plastic moment capacity of the chord. Table 4.5 presents the values of the coefficients C_1, C_2, and C_3, which are dependent on joint and load type.

The average of the chord loads and bending moments on either side of the brace intersection should be used in Equations (4.104) and (4.105). Chord axial load is positive in tension, chord in-plane bending moment is positive when it produces compression on the joint footprint. The chord thickness at the joint should be used in the above calculations.

Table 4.5 Values C_1, C_2, and C_3

Joint type	C_1	C_2	C_3
K joints under brace axial loading	0.2	0.2	0.3
T/Y joints under brace axial loading	0.3	0	0.8
X joints under brace axial loading*			
$\beta \leq 0.9$	0.2	0	0.5
$\beta = 1.0$	−0.2	0	0.2
All joints under brace moment loading	0.2	0	0.4

*Linearly interpolated values between $\beta = 0.9$ and $\beta = 1.0$ for X joints under brace axial loading.

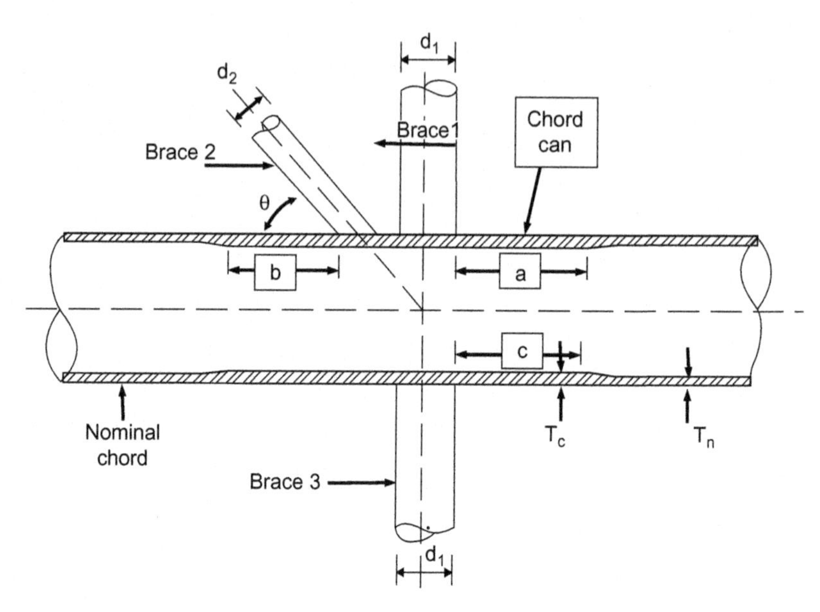

Figure 4.30 Chord length, L_c.

4.6.1.2.3 Joints with thickened cans

For simple, axially loaded Y and X joints where a thickened joint can is specified, the joint allowable capacity may be calculated as follows:

$$P_a = [r + (1 - r)(T_n/T_c)^2](P_a)_c \tag{4.106}$$

where

$(P_a)_c = P_a$ from equation (4.101) based on chord can geometric and material properties, including Q_f calculated with respect to chord can
T_n = nominal chord member thickness
T_c = chord can thickness
$r = L_c/(2.5\ D)$ for joints with $\beta \leq$ f0.9
$= (4\beta - 3)\ L_c/(1.5\ D)$ for joints with $\beta > 0.9$
L_c = effective total length

Figure 4.30 gives examples for calculation of L_c. Note that r has no value greater than unity.

Alternatively, an approximate closed ring analysis may be employed, including plastic analysis with appropriate safety factors, using an effective chord length up to 1.25D on either side of the line of action of the branch loads at the chord face but no more than the actual distance to the end of the can. Special consideration is required for more complex joints. For multiple branches on the same plane, dominantly loaded in the same sense, the relevant crushing load is $\Sigma_t P_i \sin\ \theta_t$. Any reinforcement within this dimension, such as diaphragms, rings, gussets, or the stiffening effect of out of plane members, may be considered in the analysis, although its effectiveness decreases with distance from the branch footprint.

4.6.1.2.4 Strength check

The joint interaction ratio, IR, for axial loads and bending moments in the brace should be calculated using the following expression:

$$\mathrm{IR} = \left|\frac{P}{P_a}\right| + \left(\frac{M}{M_a}\right)^2_{ipb} + \left|\frac{M}{M_a}\right|_{opb} \leq 1.0 \tag{4.107}$$

4.6.1.2.5 Overlapping joints

Braces that overlap in or out of plane at the chord member form overlapping joints.

Joints that have an in-plane overlap involving two or more braces on a single plane (e.g., K and KT joints), may be designed using the simple joint equation, using the negative gap in Q_g, with the following exceptions and additions:

1. Shear parallel to the chord face is a potential failure mode and should be checked.
2. Joints within the joint can equation do not apply to overlapping joints with balanced loads.
3. If axial forces in the overlapping and through braces have the same sign, the combined axial force representing that in the through brace plus a portion of the overlapping brace forces should be used to check the through brace intersection capacity. The portion of the overlapping brace force can be calculated as the ratio of the cross-sectional area of the brace that bears onto the through brace to the full area.
4. For either in-plane or out-of-plane bending moments, the combined moment of the overlapping and through braces should be used to check the through brace intersection capacity. This combined moment should account for the sign of the moments. Where combined nominal axial and bending stresses in the overlapping brace peak in the overlap region, the overlapping brace should also be checked on the basis of its chord being the through brace, using $Q_g = 1.0$. That is, the through brace capacity should be checked for combined axial and moment loading in the overlapping brace. In this instance, the Q_f associated with the through brace should be used.

Joints having out-of-plane overlap may be assessed on the same general basis as in-plane overlapping joints, with the exception that axial load capacity may be calculated as for multiplanar joints.

4.6.1.2.6 Grouted joints

Two varieties of grouted joints commonly occur in practice. The first relates to a fully grouted chord. The second is the double-skin type, where grout is placed in the annulus between a chord member and an internal member. In both cases, the grout is unreinforced and, as far as joint behavior is concerned, benefit shear keys is neglected.

The value of Q_u for grouted joints is calculated from the following equation. In case of axial tension,

$$Q_u = 2.5\gamma\beta\, K_a \tag{4.108}$$

where

$K_a = 0.5\ (1 + 1/\sin \Theta)$

In the case of bending,

$$Q_u = 1.5\gamma\beta \tag{4.109}$$

Note that no term is provided for axial compression, since most grouted joints cannot fail under compression; compression capacity is limited by that of the brace.

For grouted joints that are otherwise simple in configuration, the simple joint provisions may be used with the following modifications and limitations.

For fully grouted and double-skin joints, the Q_u values may be replaced with the values pertinent to grouted joints given in the preceding equations (4.108) and (4.109) in case of axial tension and bending, respectively. The classification and joint can derating may be disregarded. The adopted Q_u values should not be less than those for simple joints.

For double-skin joints, failure may also occur by chord ovalization. The ovalization capacity can be estimated by substituting the following effective thickness into the simple joint equations:

$$T_e = \sqrt{\left(T^2 + T_p^2\right)} \tag{4.110}$$

Figure 4.30 presents an example of calculating the chord length (L_c). For different braces, the chord length is calculated as follows:

Brace1. $L_c = 2a + d_1$
Brace2. $L_c = 2b + d_2/\sin \Theta$
Brace3. $L_c = 2c + d_3$

where a, b, and c are obtained as shown in Figure 4.23.

T_e = effective thickness, in. (mm)
T = wall thickness of chord, in. (mm)
T_p = wall thickness of inner member, in. (mm)

T_e should be used in place of T in the simple joint equations, including the γ term.

The Q_f calculation for both fully grouted and double skinned joints should be based on T; it is presumed that calculation of Q_f has already accounted for load sharing between the chord and inner member, such that further consideration of the effect of grout on that term is unnecessary. However, for fully grouted joints, Q_f may normally be set to unity, except in the instance of high β (≥ 0.9) X joints with brace tension/OPB and chord compression/OPB.

4.6.2 Joint calculation from API RP2A (2000)

Figure 4.31 shows the geometry and parameters for a tubular joint:

t is the brace thickness.
g is gap between brace.
T is the chord thickness.
d is the brace diameter.
D is the chord diameter.
θ is the brace angle measured from the chord.

$\iota = t/T$
$\beta = d/D$
$\gamma = D/2T$

4.6.2.1 Punching shear

$$V_p = \tau f \sin\theta \tag{4.111}$$

where

f = nominal axial, in-plane bending, or out-of-plane bending stressing the brace

The allowable punching shear stress in the chord wall is the lesser of AISC shear allowable or

$$V_{pa} = Q_q Q_f (F_{yc}/0.6g) \tag{4.112}$$

$$Q_f = 1.0 - \lambda\gamma A^2 \tag{4.113}$$

where

$\lambda = 0.030$ for brace axial stress
$= 0.045$ for brace in-plane bending stress
$= 0.021$ for brace out-of-plane bending stress
$A = [(f_{AX})^2 + (f_{IPB})^2 + (f_{OPB})^2]^{0.5}/0.6F_{ye}$
$Q_f = 1.0$ when all extreme fiber stresses in the chord are tensile

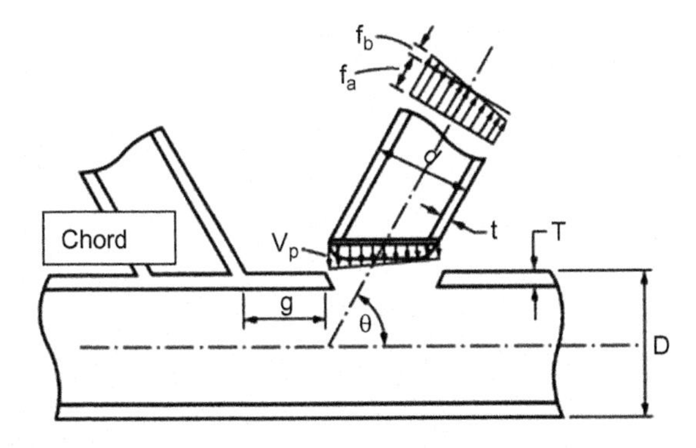

Figure 4.31 Geometry and parameters for a tubular joint.

Table 4.6 **Values of Q_q**

	Axial compression	Axial tension	In-plane bending	Out-of-plane bending
K (gap)	$(1.10 + 0.2/\beta)Q_g$			
T and Y	$(1.10 + 0.2/\beta)$			
Cross without diaphragm	$(1.10 + 0.20/\beta)$	$(0.75 + 0.20/\beta)$ Q_β	$(3.27 + 0.67/\beta)$	$(1.37 + 0.67/\beta)$ Q_β
Cross with diaphragm	$(1.10 + 0.20/\beta)$			

The value of Q_q can be obtaied from Table 4.6.

$$Q_\beta = 0.3/[\beta(1 - 0.833\beta)], \text{ for } \beta > 0.6$$
$$Q_\beta = 1.0, \text{ for } \beta \leq 0.6 \tag{4.114}$$

$$Q_g = 1.8 - 0.1(g/T), \text{ for } \gamma \leq 20$$
$$Q_g = 1.8 - 4(g/D) \text{ for } \gamma > 20 \tag{4.115}$$

In any case, Q_g should be higher than or equal to 1.0.

The joint classification as K, T, Y or cross should apply to individual braces according to their load pattern for each load case. To be considered a K joint, the punching load in a brace should be essentially balanced by loads on other braces on the same plane on the same side of the joint. In T and Y joints, the punching load is reached as beam shear in the chord. In cross joints, the punching load is carried through the chord to braces on the opposite side.

4.6.2.2 Allowable joint capacity

The allowable joint capacity is calculated as follows:

$$P_a = Q_u Q_f F_{yc} T^2/1.7 \sin\theta \tag{4.116}$$

$$M_a = Q_u Q_f F_{yc} T^2/1.7 \sin\theta(0.8d) \tag{4.117}$$

P_a and M_a are the allowable capacity for brace axial load and bending moment, respectively. The values of Q_u are as shown in Table 4.7.

Example 4.6

In a tubular joint design, the chord and brace properties are as follows:

Chord diameter, $D = 56$ in. = 142.24 mm
Chord thickness, $T = 1.25$ in. = 3.175 mm

Table 4.7 **Values of Q_u**

	Axial compression	**Axial tension**	**In-plane bending**	**Out-of-plane bending**
K (gap)	$(3.4+19\beta)Q_g$			
T and Y	$(3.4+19\beta)$			
Cross without diaphragm	$(3.4+19\beta)$	$(3.4+13\beta)Q_\beta$	$(3.4+19\beta)$	$(3.4+7\beta)Q_\beta$
Cross with diaphragm	$(3.4+19\beta)$			

Brace diameter, $d = 16$ in. = 40.64 mm
Brace thickness, $t = 0.5$ in. = 1.27 mm
Brace angle from chord, $\theta = 86.71°$
Gap, $g = 0$

Define the joint type as 9 K, K overlap, T and Y, X, or X with diaphragm.

Yield strength of chord at joint, $F_{yc} = 50$ ksi = 344.7379 MPa
Tensile strength of chord, $F_{uc} = 65$ ksi = 448.16 MPa
Yield strength of brace, $F_{yb} = 50$ ksi = 344.7379 MPa

The internal chord and brace forces are as follows:

Chord axial force, $P^C_{AX} = -366.1$ kips = -1628.494 kN
Chord in-plane bending moment, $M^C_{IPB} = 11{,}826.4$ in. kips = 1,336,203.782 kN mm
Chord out-of-plane bending moment, $M^C_{OPB} = 5896.3$ in. kips = 666192.45 kN mm
Brace axial force, $P^B_{AX} = 322.2$ kips = 143.217 kN
Brace in-plane bending moment, $M^B_{IPB} = 258.9$ in. kips = 1151.645 kN mm
Brace out-of-plane bending moment, $M^B_{OPB} = 234.2$ in. kips = 1041.774 kN mm

The section properties are as follows:

Chord section area, $A^C = 215$ in.^{2} = 138,711.1724 mm^2
Chord section modulus, $Z^C = 2{,}878.7$ in^3 = 47,172,805.72 mm^3
Brace section area = $A^B = 24.3$ in.^{2} = 15,707.93185 mm^2
Brace section modulus, $Z^B = 91.5$ in.^{3} = 1,499,297.552 mm^3
Thickness ratio (t/T), $\tau = 0.40$
Diameter ratio (d/D), $\beta = 0.29$
$D/2T$ ratio, $\gamma = 22.40$

The acting stresses are as follows:

Nominal chord axial stress, $f^C_{AX} = -1.70$ kips = -11.74 MPa
Nominal chord in-plane bending stress, $f^C_{IPB} = 4.11$ ksi = 28.33 MPa
Nominal chord out-of-plane bending stress, $f^C_{OPB} = 2.05$ ksi = 14.12 MPa
Nominal brace axial stress, $f^B_{AX} = 13.23$ ksi = 91.24 MPa
Brace in-plane bending moment, $f^B_{IPB} = 2.83$ ksi = 19.51 MPa
Brace out-of-plane bending moment, $f^B_{OPB} = 2.56$ ksi = 17.65 MPa

The punching shear and nominal loads are as follows:

Minimum of chord yield or 2/3 of tensile strength, $F_{yc} = 43.3$ ksi = 298.77 MPa
Acting axial punching shear stress, $v_{pAX} = 5.28$ ksi = 36.44 MPa
Acting IP bending punching shear stress, $v_{pIPB} = 1.13$ ksi = 7.79 MPa, from equation, (4.111)
Acting OP bending punching shear stress, $v_{pOPB} = 1.02$ ksi = 7.05 MPa, from equation (4.111)
$\beta \leq 0.6$
$Q_{\beta} = 1.0$, from Table 4.6
$\gamma > 20$, from Table 4.6 $Q_g = 1.80$, from Table 4.6
Loading and geometry factor – axial tension, $Q_{qAXT} = 1.8$, from Table 4.6
Loading and geometry factor – axial compression; $Q_{qAXC} = 1.8$, from Table 4.6
Loading and geometry factor – IP bending, $Q_{qIPB} = 6.07$
Loading and geometry factor – OP bending, $Q_{qOPB} = 3.72$
$A = 0.19$
Chord longitudinal stress factor – axial, $Q_{fAX} = 0.98$
Chord longitudinal stress factor – IP bending, $Q_{fIPB} = 0.96$
Chord longitudinal stress factor – OP bending, $Q_{fOPB} = 0.98$
Allowable axial punching shear stress; $v_{paAX} = 5.67$ ksi = 39.06 MPa, from Table 4.7
Allowable IP bending punching shear stress, $v_{paIPB} = 18.86$ ksi = 129.72 MPa
Allowable OP bending punching shear stress, $v_{paOPB} = 11.78$ ksi = 81.28 MPa
Ultimate strength factor - axial tension, $Q_{uAXT} = 8.83$, from Table 4.7
Ultimate strength factor - axial compression, $Q_{uAXC} = 8.83$, from Table 4.7
Ultimate strength factor - in-plane bending, $Q_{uIPB} = 8.83$, from Table 4.7
Ultimate strength factor - out-of-plane bending; $Q_{uOPB} = 5.40$, from Table 4.7
Allowable capacity for brace axial load, $P_a = 343.82$ kips = 2372.36 MPa
Allowable capacity for brace IP bending moment, $M_{aIPB} = 4{,}347.11$ in. kips = 29,995 MPa
Allowable capacity for brace OP bending moment, $M_{aOPB} = 2{,}711.49$ in. kips = 18,709.3 MPa

The utilizations are as follows:

Strength 50% check = 0.635
Punching shear check = 1.0
Nominal loads check = 1.004

4.6.2.3 Tubularjoint punching failure

The potential for a punch-through of a compression K joint member into chord due to plastic deformations around intersection under cyclic loads are as shown in the following figures, where punching shear is obvious in Figure 4.32. and the buckling of the tubular joint is illustrated in Figure 4.33. Limited deformation occurs at $\beta = 1.0$ compression X joint as the maximum load is attached and repeatable under cyclic loads. Figure 4.34 presents a tear in the tubular joint due to direct tensile force. This causes a flattening of $\beta = 1.0$ tension X joint as the chord yields, splitting the paint but with no cracks in the steel.

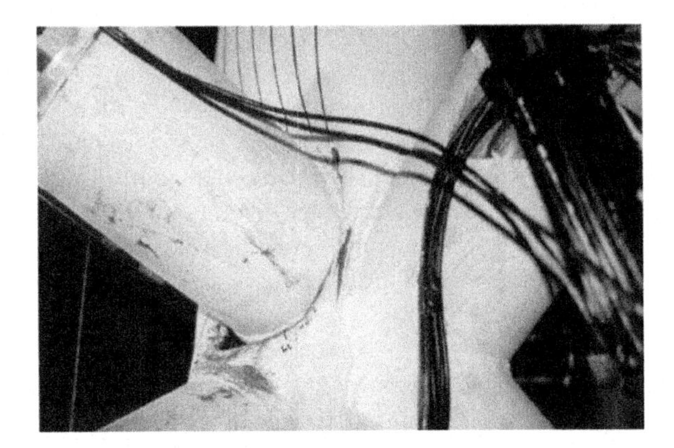

Figure 4.32 Punch-through shear of a tubular joint.

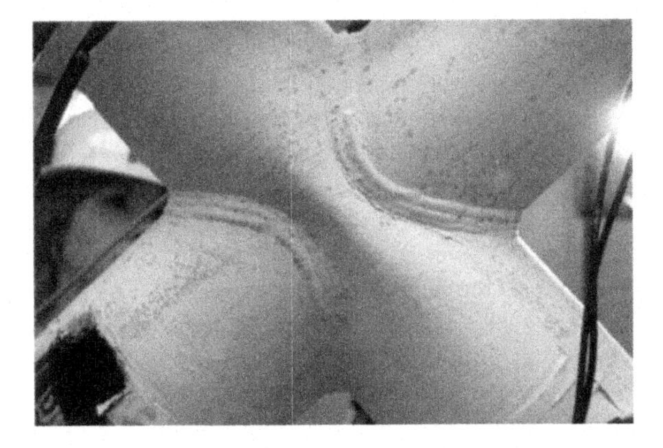

Figure 4.33 Buckling in the tubular joint.

Figure 4.34 Tearing in the tubular joint.

4.7 Fatigue analysis

Experience over the last 60 years and many laboratories tests have proven that a metal may fracture after relatively low stress if that stress is applied a great number of times. An offshore structure, particularly its tubular joints, must resist progressive damage due to fatigue that results from continuous wave action during the 20- to 30-year design life of the structure. Over the years, platforms are subjected to a wide variety of sea states, and within each sea state, the structure experiences many cycles of stress at various levels.

The purpose of fatigue analyses is to account for the fact that the number of cycles of stress that a structural component can withstand varies with the magnitude of the cyclic stress.

Dynamic analysis is used in the fatigue evaluation to predict the number of cycles and magnitude of stresses that occur in various sea states. As in extreme wave analysis, dynamic effects become increasingly important for deepwater structures carrying a heavy deck load.

Fatigue cracks grow because of tensile stress; corrosion of a metal is accelerated if the metal is subjected to tensile stress. Therefore, the effects of corrosion and fatigue are combined in the case of an offshore platform.

Kuang, Potvin, and Leick (1975) describe the design of tubular joints for offshore structures as an iterative procedure. The process begins with the sizing of the jacket piles according to the requirements of the specific soil and foundation of the platform. Sizing determines the diameter of the jacket legs and allows clearance for the piles to go through them. Once the trusswork geometry is selected, the column buckling characteristics determine the diameters of the various jacket braces. The initial wall thicknesses of chords and braces are determined by structural analysis. The next cycle of calculation involves increasing the chord wall thicknesses with heavy joint cans to ensure sufficient static strength to meet code or specification requirements. The next iteration involves calculating the fatigue strength of the joint to determine if it is compatible with the service life requirement of the platform. Depending on the method of fatigue analysis used, allowable stress concentration factors must be either specified for each joint or built into the method of analysis.

For each location around each member intersection of interest in the structure, the stress response for each sea state should be computed, giving adequate consideration to both global and local stress effects.

The stress responses should be combined into the long-term stress distribution, which should then be used to calculate the cumulative fatigue damage ratio, D, where

$$D = \sum (n/N) \quad (4.118)$$

where

n = number of cycles applied at a given stress range

N = number of cycles for which the given stress range would be allowed by the appropriate S-N curve

In most cases, the damage ratio is calculated for each sea state and combined to obtain the cumulative damage ratio.

In general, the designed fatigue life of each joint and member should not be less than the intended service life of the structure multiplied by a safety factor. The design fatigue life, D, should not exceed unity.

For in-situ conditions, the safety factor for the fatigue of steel components should depend on the failure consequence and its impact on cost, the environment and others, and in-service inspectability.

Critical elements are those whose sole failure could be catastrophic. In lieu of a more detailed safety assessment of Category L-1 structures for staffed and nonevacuated sites, a safety factor of 2.0 is recommended for inspectable, non-failure-critical, connections. For failure-critical and noninspectable connections, increased safety factors are recommended, as shown in Table 4.8. A reduced safety factor is recommended for Category L-2 and L-3, which are for staffed evacuated or unstaffed structures, respectively, for conventional steel jacket structures on the basis of in-service performance data:

SF = 1.0 for redundant diver or Remote Operating Vehicle (ROV) inspectable framing, with safety factors for other cases being half those in the table.

When fatigue damage can occur due to other cyclic loadings, such as transportation, the following equation should be satisfied:

$$\sum_i \mathrm{SF}_i D_i < 1.0 \tag{4.119}$$

where

D_i = the fatigue damage ratio for each type of loading
SF_i = the associated safety factor

For transportation where long-term wave distributions are used to predict short-term damage, a larger safety factor should be considered.

4.7.1 Stress concentration factors

The stress concentration factor is different from one joint geometry to another. It is known that the applied loads on tubular joints cause stresses at certain points along the intersection weld to be many times the nominal stress acting in the members. The stress concentration factor (SCF) is as multiplier applied to the nominal stress to reach the peak or maximum stress at the hot spot location.

Table 4.8 Safety factor for fatigue life

Failure criticality	Inspectable	Not inspectable
No	2	5
Yes	5	10

The hot spot is the location in the tubular joint where the maximum applied tensile stress occurs. To do a fatigue analysis of a certain selected tubular joints in an offshore structure, the stress history of the hot spots in those joints must first be determined; three basic stress types contribute to the development of hot spots, which are as follows:

Primary stresses are caused by axial forces and moments resulting from the combined truss and frame action of the jacket. As shown in the in-plan tubular joint hot spot in Figure 4.35, the location of hotspots in locations 1, 2, 3, 4, and 6 are most effected by axial forces and in-plane bending moments where regions 2 and 5 are most affected by the axial forces and circumferential moments in braces.

The secondary stresses have two causes. Theys may be due to the structure detail of the connection. such as poor joint geometry, poor fit-up, local variation within the joint due to rigid reinforcement. or the braces are restrained by a circumferential weld; these secondary stresses amplify the primary stresses. The second cause of secondary stresses are by the metallurgical factors that results from faulty welding practice, insufficient weld penetration, heavy beading, weld porosity, or varying cooling rates. These stresses mainly affect location, and their effect is significant at locations 3, 4, and 6.

As the metallurgical effect is essential in fatigue stresses on the tubular joint so the quality control for constructing this connection should receive more attention, as is describe in Chapter 5.

The most fatigue sensitive areas in offshore platforms are the welds at tubular joints because of high local stress concentrations. Shortened life due to fatigue at these locations is estimated by evaluating the hot spot stress range (HSSR) and using it as input into the appropriate S-N curve.

SCFs may be derived from finite element (FE) analyses, model tests, or empirical equations based on such methods. When deriving SCFs using FE analysis, it is recommended to use volume (brick and thick shell) elements to represent the weld region and adjoining shell, as opposed to thin shell elements. In such models the SCFs may be derived by extrapolating stress components to the relevant weld toes and combining these to obtain the maximum principal stress and, hence, the SCF. The extrapolation direction should be normal to the weld toes.

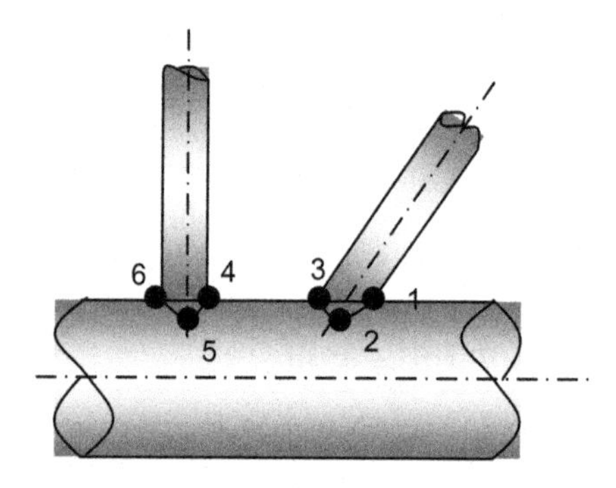

Figure 4.35 Hot spots of an in-plane tubular joint.

According to Healy and Buitrago (1994) and Neimi et al. (1995), if thin shell elements are used, the results should be interpreted carefully, since no single method is guaranteed to provide consistently accurate stresses.

Extrapolation to the midsurface intersection generally overpredicts SCFs but not consistently, whereas truncation at the notional weld toes generally underpredicts SCFs. In place of extrapolation, it is possible to directly use the nodal average stresses at the midsurface intersection. This generally overpredict stresses, especially on the brace side. This last method is expected to be more sensitive to the local mesh size than the extrapolation methods.

For each tubular joint configuration and each type of brace loading, the SCF is defined as

SCF = HSSR/nominal brace stress range

The nominal brace stress range should be based on the section properties of the brace-end under consideration, taking due account of the brace stub or a flared member end, if present. Likewise, the SCF evaluation should be based on the same section dimensions.

Nominal cyclic stress in the chord may also influence the HSSR and should be considered. The SCF should include all stress-raising effects associated with the joint geometry and type of loading, except the local (microscopic) weld notch effect, which is included in the S-N curve. SCFs may be derived from finite element analyses, model tests, or empirical equations based on such methods. In general. the SCFs depend on the type of brace cyclic loading (i.e., brace axial load, in-plane bending, out-of plane bending), the joint type, and details of the geometry. The SCF varies around the joint, even for a single type of brace loading.

When combining the contributions from the various loading modes, phase differences between them should be accounted for, with the design HSSR at each location being the range of hot-spot stress resulting from the point-in-time contribution of all loading components.

In general, for all welded tubular joints under all three types of loading, a minimum value of SCF equal to 1.5 should be used.

For unstiffened welded tubular joints, SCFs should be evaluated using the Efthymiou equations, as will be discussed later in thickness effect on the SCF.

The linearly extrapolated hot-spot stress from Efthymiou may be adjusted to account for the actual weld toe position, where this systematically differs from the assumed AWS basic profiles.

The SCF also applies to internally ring-stiffened joints, including the stresses in the stiffeners and the stiffener-to-chord weld. Noting that special consideration should be given to these locations, SCFs for internally ring-stiffened joints can be determined by applying the Lloyd's reduction factors based on the Lloyd's Report (1988) to the SCFs for the equivalent unstiffened joint. For ring-stiffened joints analyzed by such means, the minimum SCF for the brace side under axial or OPB loading should be taken as 2.0.

4.7.1.1 SCFs in grouted joints

The grouting joints are usually used in repairing or strengthen the platform. Grouting tends to reduce the SCF of the joint, since the grout reduces the chord deformations. In general, the larger is the ungrouted SCF, the greater the reduction in SCF with grouting. Hence, the reductions are typically greater for X and T joints than for Y and K joints. More discussion about the effect of grouting in the strengthen method is presented in Chapter 7.

4.7.1.2 SCFs in cast nodes

For cast joints, the SCF is derived from the maximum principal stress at any point on the surface of the casting (including the inside surface) divided by the nominal brace stress outside the casting. The SCFs for castings are not extrapolated values but are based on directly measured or calculated values at any given point, using an analysis that is sufficiently detailed to pick up the local notch effects of fillet radii and so forth. Consideration should also be given to the brace-to casting girth weld, which can be the most critical location for fatigue.

4.7.2 S-N curves for all members and connections, except tubular connections

Nontubular members and connections in deck structures, appurtenances, and equipment and tubular members and attachments to them, including ring stiffeners, may be subject to variations of stress due to environmental loads or operational loads. Operational loads include those associated with machine vibration, crane usage, and filling and emptying of tanks. Where variations of stress are applied to conventional weld details, identified in ANSI/AWS D1.1- 2002, Table 2.4, the associated S-N curves provided in AWS Figure 2.11 should be used, dependent on degree of redundancy.

Where such variations of stress are applied to tubular nominal stress situations identified in ANSI/AWS D1.1-2002. The associated S-N curves provided in AWS Figure 2.13 should be used. Stress categories DT, ET, FT, Kl, and K2, refer to tubular connections where the SCF is not known.

Where the hot-spot stress concentration factor can be determined, practice take precedence for service conditions where details may be exposed to random variable loads, seawater corrosion, or submerged service with effective cathodic protection.

The referenced S-N curves in ANSI/AWS D1.1.-2002, Figure 2.11, are class curves. For such curves, the nominal stress range in the vicinity of the detail should be used. Due to load attraction, shell bending, and the like not present in the class type test specimens, the appropriate stress may be larger than the nominal stress in the gross member. Geometrical stress concentration and notch effects associated with the detail itself are included in the curves.

Reference may alternatively be made to the S-N criteria similar to the other joint (OJ) curves contained within ISO DIS. The ISO 19902:2004 code proposal uses a weld detail classification system whereby the OJ curves include an allowance for notch stress and modest geometric stress concentration.

4.7.3 S-N curves for tubular connections

Design S-N curves follow for welded tubular and cast joints. The basic design S-N curve is of the form

$$\log_{10}(N) = \log_{10}(k_1) - m\log_{10}(S) \tag{4.120}$$

where

N = the predicted number of cycles to failure under stress range S
k_1 = a constant
m = the inverse slope of the S-N curve

Table 4.9 presents the basic welded joint (WJ) and cast joint (CJ) curves. These S-N curves are based on steels with yield strength less than 500 MPa (72 ksi). For welded tubular joints exposed to random variations of stress due to environmental or operational loads, the WJ curve should be used. The brace-to-chord tubular intersection for ring-stiffened joints should be designed using the WJ curve. For cast joints, the CJ curve should be used. For other details, including plated joints and, for ring-stiffened joints, the ring stiffener-to-chord connection and the ring inner edge are used.

The basic allowable cyclic stress should be corrected empirically for seawater effects, the apparent thickness effect, with the exponent dependentg on the profile, and the weld improvement factor on S. An example of S-N curve construction is given in Figure 4.36.

The basic design S-N curves given in Table 4.9 are applicable for joints in air and submerged coated joints. For welded joints in seawater with adequate cathodic protection, the $m = 3$ branch of the S-N curve should be reduced by a factor of 2.0 on life, with the $m = 5$ branch remaining unchanged and the position of the slope change adjusted accordingly.

Fabrication of welded joints should be in accordance with the quality control procedure. The curve for cast joints is applicable only to castings having an adequate fabrication inspection plan.

Table 4.9 Basic design S-N curves

Curve	Log_{10} (k_1) S, in ksi	Log_{10} (k_1) S, in MPa	m
Welding joints	9.95	12.48	3 for $N < 10^7$
	11.92	16.13	5 for $N > 10^7$
Cast joints	11.80	15.17	4 for $N < 10^7$
	13.00	17.21	5 for $N > 10^7$

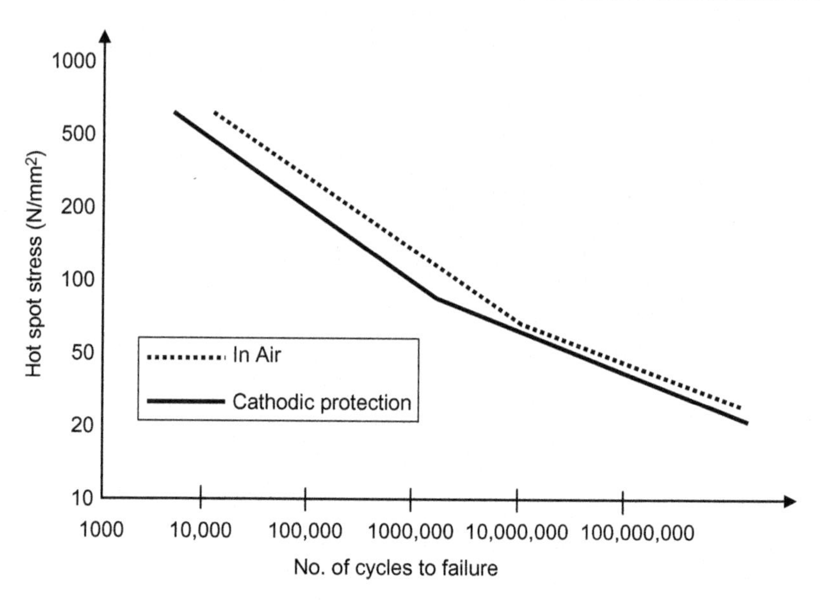

Figure 4.36 Tubular joint S-N curve for T = 16 mm.

4.7.3.1 *Thickness effect*

The welding joint curve is based on ⅝-in. (16-mm) reference thickness. For material thickness above the reference thickness, the following thickness effect should be applied or as welded joints:

$$S = S_o(t_{\mathrm{ref}}/t)^{0.25} \tag{4.121}$$

where

t_{ref} = the reference thickness, $^5/_8$-inch (16 mm)
S = allowable stress range
S_o = the allowable stress range from the S-N curve, as in Figure 4.36
t = member thickness for which the fatigue life is predicted

If the weld has profile control, the exponent in this equation may be taken as 0.20. If the weld toe has been ground or peened, the exponent in the equation may be taken as 0.15.

The material thickness effect for castings is given by

$$S = S_o(t_{\mathrm{ref}}/t)^{0.15} \tag{4.122}$$

where the reference thickness t_{ref} is 1.5 in. (38 mm).

No effect shall be applied to material thickness less than the reference thickness.

For any type of connection analyzed on a chord hot-spot basis, the thickness for the chord side of tubular joint should be used in the foregoing equations. For the brace-side hot spot, the brace thickness may be used.

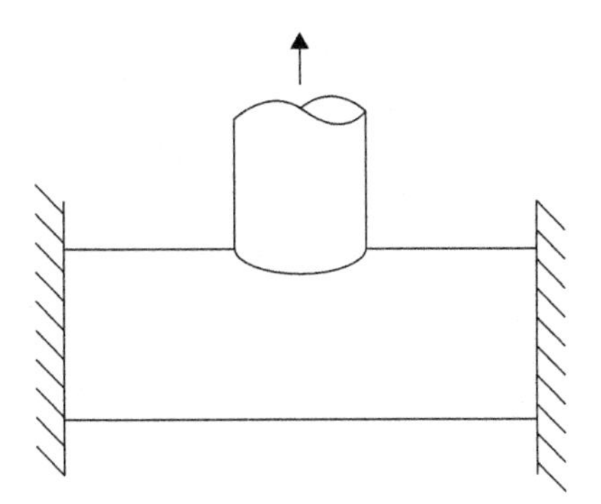

Figure 4.37 Axial load, chord ends fixed.

Use of the Efthymiou SCF equations is recommended because this set of equations is considered to offer the best option for all joint types and load types and is the only widely vetted set that covers overlapped K and KT joints.

"Mix and match" between different sets of equations is not recommended. The Efthymiou equations are also recommended in the *Proposed Revisions for Fatigue Design of Welded Connections* for adoption by International Institute of Welding, Eurocode 3, and ISO DIS 14347.

The Efthymiou equations are presented. The validity ranges for the Efthymiou parametric SCF equations are as follows:

β from 0.2 to 1.0.
τ from 0.2 to 1.0.
γ from 8 to 32.
α (length) from 4 to 40.
θ from 20 to 90 degrees.
ζ (gap) from $-0.6\ \beta/\sin\theta$ to 1.0.

For cases where one or more parameters fall outside this range, the following procedure may be adopted:

1. Evaluate SCFs using the actual values of geometric parameters/
2. Evaluate SCFs using the limit values of geometric parameters/
3. Use the greater of 1 or 2 in the fatigue analysis.

4.7.3.1.1 Axial load, chord ends fixed

See Figure 4.37. The chord saddle is determined by

$$SCF = \gamma\tau^{1.1}[1.11 - 3(\beta - 0.52)^2]\sin^{1.6}\theta \tag{4.123}$$

For a short chord, there is a correction factor, $F1$:

Short chord correction factors ($a < 12$)

$$F1 = 1 - (0.83\beta - 0.56\beta^2 - 0.02)\gamma^{0.23}\exp[-0.21\gamma^{-1.16}\alpha^{2.5}] \tag{4.124}$$

where $\exp(x) = e^x$

The chord crown is calculated as follows:

$$SCF = \gamma^{0.2}\tau[2.65 + 5(\beta - 0.65)^2] + \tau\beta(0.25\alpha - 3)\sin\theta \quad (4.125)$$

The *brace saddle*

$$SCF = 1.3 + \gamma\tau^{0.52}\alpha 0.1[0.187 - 1.25\beta 1.1(\beta - 0.96)]\sin(2.7 - 0.01\alpha)\theta \quad (4.126)$$

For a short chord, there is a correction factor, $F1$.
The *brace crown*

$$SCF = 3 + \gamma^{1.2}[0.12\exp(-4\beta) + 0.011\beta^2 - 0.045] + \beta\tau(0.1\alpha - 1.2) \quad (4.127)$$

4.7.3.1.2 Axial load, general fixity conditions

See Figure 4.38. The chord saddle

$$SCF = [\text{Equation (4.38)}]C_1(0.8\alpha - 6)\tau\beta^2(1 - \beta^2)^{0.5} + \sin^2 2\theta \quad (4.128)$$

For a short chord, there is a correction factor $F2$, where $F2$

$$F2 = 1 - (1.43\beta - 0.97\beta^2 - 0.03)\gamma^{0.04}\exp[-0.71\gamma^{-1.38}\alpha^{2.5}] \quad (4.129)$$

The chord crown

$$SCF = \gamma^{0.2}\tau[2.65 + 5(\beta - 0.65)^2] + \tau\beta(C_2\alpha - 3)\sin\theta \quad (4.130)$$

The brace saddle: is calculated using equation (4.126).
The brace crown is calculated as follows:

$$SCF = 3 + \gamma^{1.2}[0.12\exp(-4\beta) + 0.011\beta^2 - 0.045] + \beta\tau(C_3\alpha - 1.2) \quad (4.131)$$

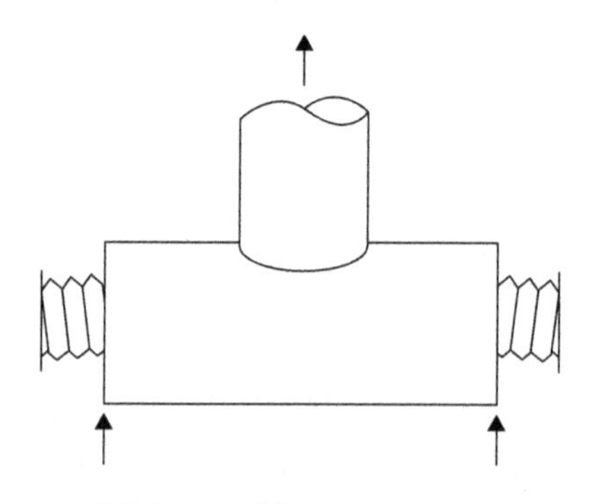

Figure 4.38 Axial load, general fixity conditions.

Note that the chord-end fixity parameter C is as follows:

$0.5 \leq C \leq 1.0$, typically $C = 0.7$
$C_1 = 2(C - 0.5)$
$C_2 = C/2$
$C_3 = C/5$

4.7.3.1.3 In-plane bending

See Figure 4.39. The *chord crown*

$$\text{SCF} = 1.45\beta\tau^{0.85}\gamma(1 - 0.68\beta)\sin^{0.7}\theta \tag{4.132}$$

For a short chord, there is a correction factor $F3$, where $F3$ is

$$F3 = 1 - 0.55\beta^{1.8}\gamma 0.16 \exp[-0.49\gamma^{-0.89}\alpha^{1.8}] \tag{4.133}$$

The *brace crown*

$$\text{SCF} = 1 + 0.65\beta\tau^{0.4}\gamma(1.09 - 0.77\beta) + \sin^{(0.06\gamma - 1.16)}\theta \tag{4.134}$$

4.7.3.1.4 Out-of-plane bending

See Figure 4.40. The *chord saddle*

$$\gamma\tau\beta(1.7 - 1.05\beta^3)\sin^{1.6}\theta \tag{4.135}$$

The brace saddle is calculated as follows:

$$\tau^{-0.54}\gamma^{-0.05}(0.99 - 0.47\beta + 0.08\beta^4) \times [\text{equation (4.135)}] \tag{4.136}$$

Chord-end fixity parameter C

4.7.3.1.5 Axial load, balanced

See Figure 4.41. The chord saddle is calculated as follows:

$$3.87\gamma\tau\beta(1.10 - \beta^{1.8})\sin^{1.7}\theta \tag{4.137}$$

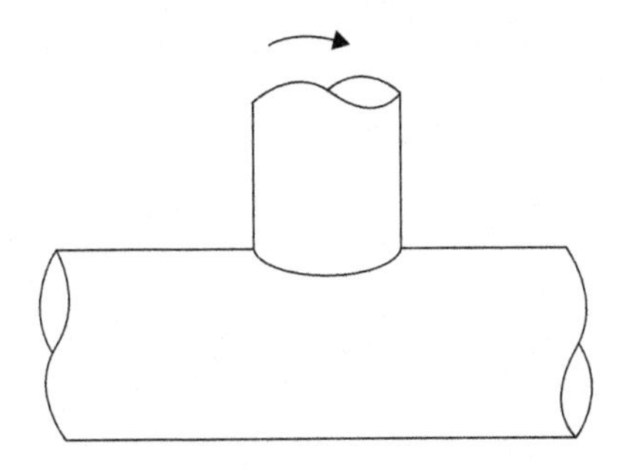

Figure 4.39 In-plane bending.

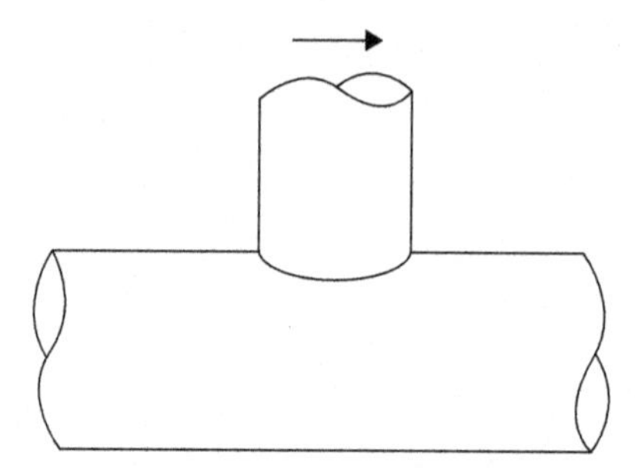

Figure 4.40 Out-of-plane bending.

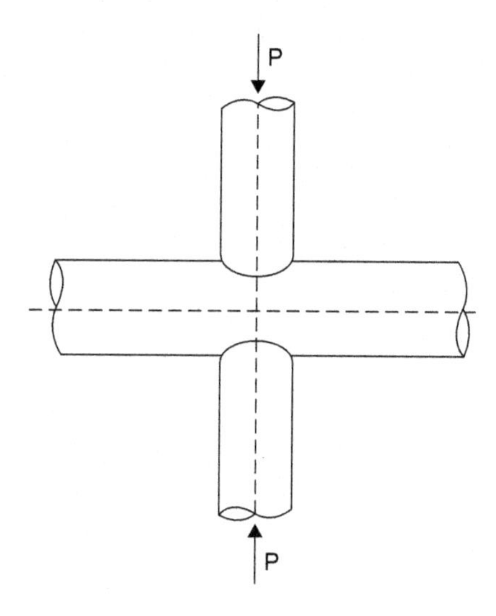

Figure 4.41 Axial load, balanced.

The chord crown is calculated as follows:

$$\gamma^{0.2}\tau[2.65 + 5(\beta - 0.65)^2] + 3\tau\beta \sin \theta \tag{4.138}$$

The brace saddle is calculated as follows:

$$1 + 1.9\gamma\tau^{0.5}\beta^{0.9}(1.09 - \beta^{1.7})\sin^{2.5}\theta \tag{4.139}$$

The brace crown is calculated as follows:

$$3 + \gamma^{1.2}[0.12\exp(-4\beta) + 0.011\beta^2 - 0.045] \tag{4.140}$$

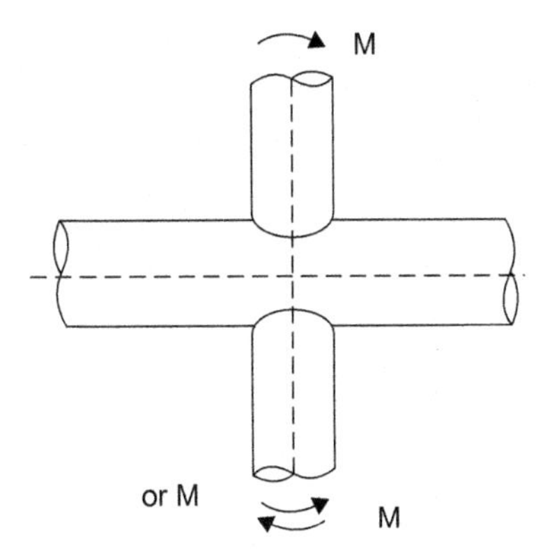

Figure 4.42 In-plane bending.

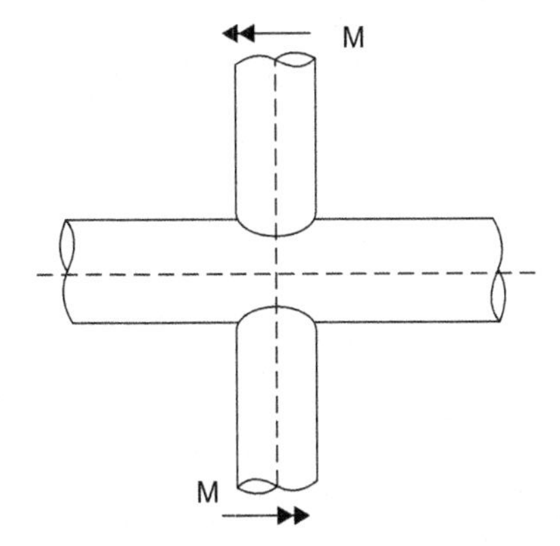

Figure 4.43 Out-of-plane bending, balanced.

In joints with short chords ($\alpha < 12$) and closed ends, the saddle SCFs can be reduced by the short chord factors $F1$ or $F2$, where

$$F1 = 1 - (0.83\beta - 0.56\beta^2 - 0.02)\gamma^{0.23}\exp[-0.21\gamma^{-1.16}\alpha^{2.5}] \tag{4.141}$$

$$F2 = 1 - (1.43\beta - 0.97\beta^2 - 0.03)\gamma^{0.04}\exp[-0.71\gamma^{-1.38}\alpha^{2.5}] \tag{4.142}$$

4.7.3.1.6 In-plane bending

See Figure 4.42. To calculate the chord crown, use equation (4.132). To calculate the brace crown, use equation (4.134).

4.7.3.1.7 Out-of-plane bending, balanced

See Figure 4.43. The chord saddle is calculated as follows:

$$\gamma\tau\beta(1.56 - 1.34\beta^4)\sin^{1.6}\theta \tag{4.143}$$

The brace saddle is calculated as follows:

$$\tau^{-0.54}\gamma^{-0.05}(0.99 - 0.47\beta + 0.08\beta^4) \times [\text{equation (4.132)}] \quad (4.144)$$

In joints with short chords ($\alpha < 12$) and closed ends, equations (4.143) and (4.144) can be reduced by the short chord factor $F3$, where

$$F3 = 1 - 0.55\beta^{1.8}\gamma^{0.16}\exp[-0.49\gamma^{-0.89}\alpha^{1.8}] \quad (4.145)$$

4.7.3.1.8 Balanced axial load

See Figure 4.44. The *chord SCF* is calculated as follows:

$$\tau^{0.9}\gamma^{0.5}(0.67 - \beta^2 + 1.16\beta)\sin\theta\left[\frac{\sin\theta_{max}}{\sin\theta_{min}}\right]^{0.30}\left[\frac{\beta_{max}}{\beta_{min}}\right]^{0.30}\left[1.64 + 0.29\beta^{-0.38}ATAN(8\zeta)\right] \quad (4.146)$$

The *brace SCF* is calculated as follows:

$$1 + [\text{equation 4.135}](1.97 - 1.57\beta^{0.25})\tau^{-0.14}\sin^{0.7}\theta + C\beta^{1.5}\gamma^{0.5}\tau^{-1.22}\sin^{1.8}(\theta_{max} + \theta_{min}) \times [0.131 - 0.084\,\text{atan}(14\zeta + 4.2\beta)] \quad (4.147)$$

where

$C = 0$ for gap joints
$C = 1$ for the through brace
$C = 0.5$ for the overlapping brace

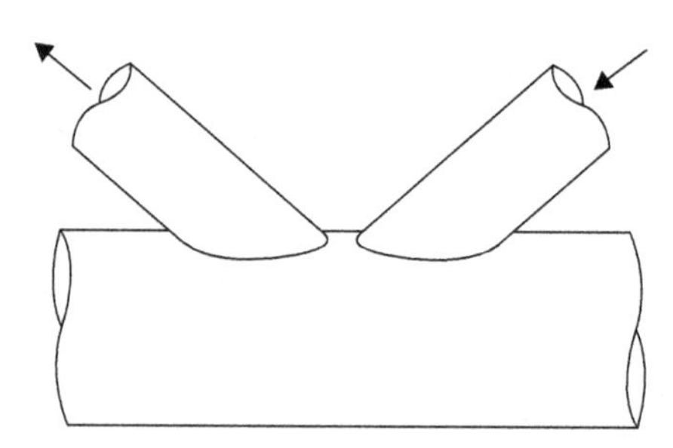

Figure 4.44 Balanced axial load.

Note that τ, β, θ, and the nominal stress relate to the brace under consideration; atan is arc tangent evaluated in radians.

4.7.3.1.9 Unbalanced in-plane bending

See Figure 4.45. To calculate the chord crown SCF use Eqn. T8. (For overlaps exceeding 30% of contact length, use $1.2 \times$ [equation. (4.132)). To calculate the

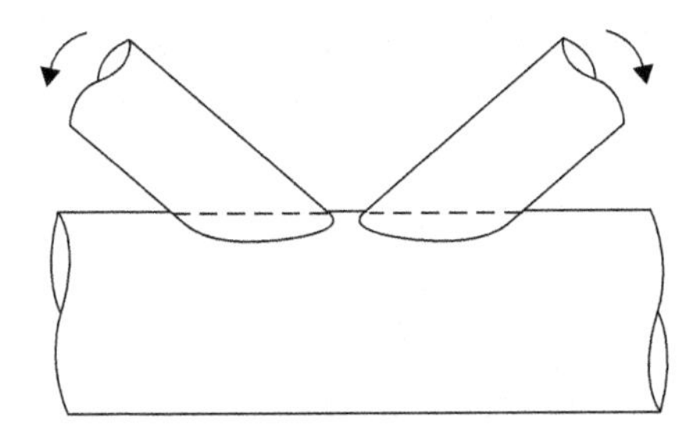

Figure 4.45 Unbalanced IPB.

gap joint–brace crown SCF, use equation. (4.134). Calculate the overlap joint–brace crown SCF: as follows: [equation. (4.134)] × (0.9 + 0.4).

4.7.3.1.10 Unbalanced out-of plane bending OPB

See Figure 4.46. The chord saddle SCF adjacent to brace A is calculated as follows:

$$[\text{equation (4.135)}]_A[1-0.08(\beta_{B\gamma})^{0.5}\exp(-0.8x)]+[\text{equation (3.135)}]_B[1-0.08(\beta_{A\gamma})^{0.5}\exp(-0.8x)][2.05\,(\beta_{max})^{0.5}\exp(-1.3x)] \quad (4.148)$$

where

$$x=1+\frac{\zeta\sin\theta_A}{\beta_A}$$

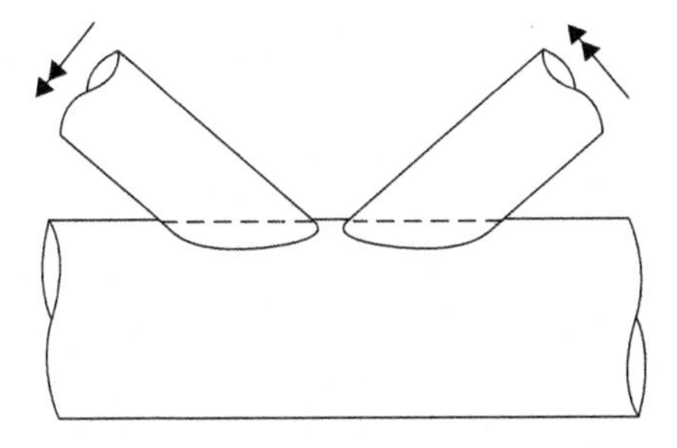

Figure 4.46 Unbalanced OPB.

The brace A saddle SCF is calculated as follows:

$$\tau^{-0.54}\gamma^{-0.05}(0.99-0.47\beta+0.08\beta^4)\times[\text{equation (4.148)}] \quad (4.149)$$

$$F4=1-1.07\beta^{1.88}\exp[-0.16\gamma^{-1.06}\alpha^{2.4}] \quad (4.150)$$

Note that [equation (4.135)]$_A$ is the chord SCF adjacent to brace A as estimated from equation (4.135). The designation of braces A and B does not depend on the geometry. It is nominated by the user.

4.7.3.1.11 Balanced axial load for three braces

See Figure 4.47. To calculate the chord SCF use equation (4.146). To calculate the brace SCF use equation (4.147). For the diagonal braces, A and C, use $\zeta = \zeta_{AB} + \zeta_{BC} + \beta_B$. For the central brace, B, use ζ = maximum of ζ_{AB} or ζ_{BC}.

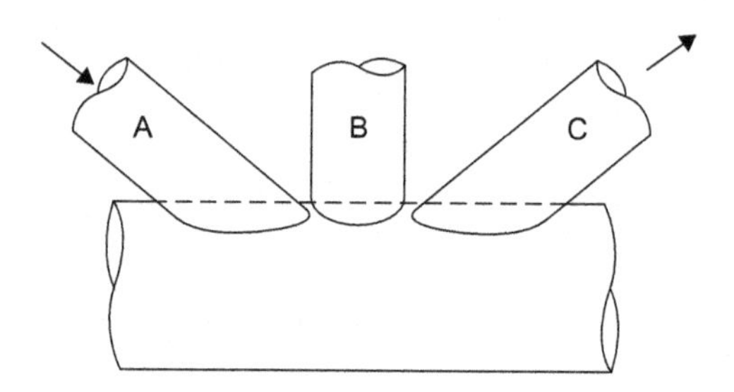

Figure 4.47 Balanced axial load for three braces.

4.7.3.1.12 In-plane bending for three braces

To calculate the chord crown SCF use equation (4.132). To calculate the brace crown SCF use equation (4.134).

4.7.3.1.13 Unbalanced out-of-plane bending for three braces

See Figure 4.48. The chord saddle SCF adjacent to diagonal brace A is calculated as follows:

$$\begin{aligned}&[\text{equation (4.135)}]_A[1 - 0.08(\beta_{B\gamma})^{0.5}\exp(-0.8x_{AB})] \cdot [1 - 0.08(\beta_{C\gamma})0.5\exp(-0.8x_{AC})]\\&\quad + [\text{equation (4.135)}]_B[1 - 0.08(\beta_{A\gamma})^{0.5}\exp(-0.8x_{AB})][2.05(\beta_{\max})^{0.5}\exp(-1.3x_{AB})]\\&\quad + [\text{equation (4.135)}]_C[1 - 0.08(\beta_{A\gamma})^{0.5}\exp(-0.8x_{AC})][2.05(\beta_{\max})^{0.5}\exp(-1.3x_{AC})]\end{aligned} \tag{4.151}$$

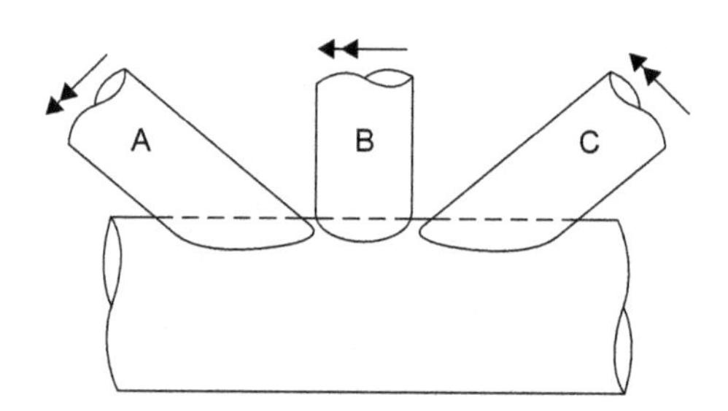

Figure 4.48 Unbalanced out-of-plane bending for three braces.

where

$$x_{AB} = 1 + \frac{\zeta_{AB}\sin\theta_A}{\beta_A} \tag{4.152}$$

$$x_{AC} = 1 + \frac{(\zeta_{AB} + \zeta_{BC} + \beta_B)\sin\theta_A}{\beta_A} \tag{4.153}$$

The chord saddle SCF adjacent to central brace A is calculated as follows:

$$\begin{aligned}&[\text{equation (4.135)}]_B[1-0.08(\beta_{A\gamma})^{0.5}\exp(-0.8x_{AB})]^{m}\cdot[1-0.08(\beta C_{\gamma})^{0.5}\exp(-0.8x_{BC})]^{m}\\&+[\text{equation (4.135)}]_A\cdot[1-0.08(\beta_{B\gamma})^{0.5}\exp(-0.8x_{AB})]\cdot[2.05(\beta_{max})^{0.5}\exp(-1.3x_{AB})]\\&+[\text{equation (4.135)}]_C\cdot[1-0.08(\beta_{B\gamma})^{0.5}\exp(-0.8x_{BC})]\cdot[2.05(\beta_{max})^{0.5}\exp(-1.3x_{BC})]\end{aligned} \tag{4.154}$$

where

$$m = (\beta_A/\beta_B)^2$$

$$x_{AB} = 1 + \frac{\zeta_{AB}\sin\theta_A}{\beta_B} \quad \text{and} \quad x_{BC} = 1 + \frac{\zeta_{BC}\sin\theta_B}{\beta_B}$$

The brace saddle SCFs under OPB are obtained from the adjacent chord SCFs using

$$\tau^{-0.54}\gamma^{-0.05}(0.99 - 0.47\beta + 0.08\beta^4) \times \text{SCF chord} \tag{4.155}$$

where

SCF chord = equation (4.140)KT1 or equation (4.154).

In joints with short chords ($\alpha < 12$) equations. (4.151), (4.154), and (4.155) can be reduced by the short chord factor $F4$, where $F4 = 1.-1.07\beta^{1.88}\exp[-0.16\gamma^{-1.06}\alpha^{2.4}]$.

4.7.3.2 Effect of weld toe position

Ideally, the SCF should be invariant, given the tubular connection's geometry (γ, τ, β, θ, and ζ). This is how Efthymiou and all the other SCF equations are formulated. Hot-spot stress is calculated from the linear trend of notch-free stress extrapolated to the toe of the basic standard weld profile, with nominal weld toe position as defined in AWS D1.1, Figure 3.8. When this is done, size and profile effects must be accounted for in the S-N curve, regardless of the underlying cause. This is how the previous API rules were set up.

Influenced by deBack and others, international thinking tends to suggest that weld profile effects (mainly the variable position of the actual weld toe) should be reflected in the SCF, rather than in the S-N curve. This is consistent with how experimental hot-spot stresses were measured to define the basic international S-N curve for hot-spot fatigue in 16 mm thick tubular joints. One tentative method for correcting the analytical SCF for weld toe position was presented in the seminal volume for deBack's retirement, Marshal (1989). Based on Marshall, Bucknell, and Mohr (2005), a more robust formulation is now proposed:

$$SCF_{corr} = 1 - (L_a - L)/L_{mp} \tag{4.156}$$

where:

SCF_{corr} = the correction factor applied to the Efthymiou SCF
L_a = the actual weld toe position for typical of yard practice
L = the nominal weld toe position
L_{mp} = the moment persistence length (distance from nominal toe to reversal of shell bending stress)

Various expressions for L_{mp} are shown in Table 4.10 as a function of joint type, load type, and hot-spot orientation.

In the table, R and T are radius and thickness, respectively, of the joint can. Consistency in format with the rules for strain gauge placement at crown and saddle position may be noted.

Attempts to produce an improved as-welded profile often result in overwelding. As such, high estimates of L_{mp} (low estimates of local stress gradient) produce conservative corrections. This approach assumes that the weld is not so massive as to change the overall load distribution in the joint can nor so finely tapered that positions other than the weld toe become critical and that local hotspot stresses are dominated by shell bending stress.

Failure is expressed as damage or fatigue life damage, so the fatigue life damage is the number of cycles of a particular stress range divided by the allowable number of cycles for that range from the S-N curve.

Table 4.10 Expressions for L_{mp}

Circumferential stress at saddle	
All loading modes	$L_{mp} = (0.42 - 0.28\beta)R$ Angle = $(24 - 16\beta)$ degrees
Longitudinal stress at crown	
Axis symmetric	$L_{mp} = 0.6(RT)^{0.5}$
Gap (g) of K joint	L_{mp} = lesser of $0.6\ (RT)^{0.5}$ or $g/2$
Outer heel/toe, axial	$L_{mp} = 1.5(RT)^{0.5}$
In-plane bending	$L_{mp} = 0.9(RT)^{0.5}$

Example 4.7

Calculate SCF calculation for a K joint using the input data in Table 4.11. It is assumed that both braces are identical, that is, of the same diameter, thickness, and angle to chord. It is also assumed that the braces do not overlap.

Table 4.12 contains the stress concentration factors for loading on one brace only.

Table 4.11 Input data for Example 4.7

Title	**Example only**
D = chord diameter	= 16 inches
T = chord thickness	= 0.625 inches
L = chord length	= 50 inches
C = chord-end fixity parameter	= 0.7
$0.5 < C < 1.0$	Typically $C = 0.7$
d = brace diameter	= 12.75 inches
t = brace thickness	= 0.375 inches
θ = brace angle (from chord)	= 60 degrees
g = gap between braces	= 2 inches
Geometrical parameters	
$\beta = d/D$	= 0.80
$\gamma = D/2T$	= 12.80
$\tau = t/T$	= 0.60
$\alpha = 2L/D$	= 6.25
$\zeta = g/D$	= 0.13
Validity range	
$0.2 < \beta < 1.0$	Satisfied
$0.2 < \tau < 1.0$	Satisfied
$8 < \gamma < 32$	Satisfied
$4 < \alpha < 40$	Satisfied
$20 < \theta < 90$	Satisfied
Chord-end fixity parameter	
$C1 = 2\,(C - 0.5)$	0.4
$C2 = C/2$	0.35
$C3 = C/5$	0.14
Short chord correction factors ($\alpha < 12$)	
$F1 = 1 - (0.83\beta - 0.56\beta^2 - 0.02)\gamma^{0.23} \exp[-0.21\gamma^{-1.16.16}\alpha^{2.5}]$*	= 0.82
$F2 = 1 - (1.43\beta - 0.97\beta^2 - 0.03)\gamma^{0.04} \exp[-0.71\gamma^{-1.38}\alpha^{2.5}]$	= 0.93
$F3 = 1 - 0.55\beta^{1.8}\gamma^{0.16} \exp[-0.49\gamma^{-0.89}\alpha^{1.8}]$	= 0.86
$F4 = 1 - 0.07\beta^{1.88} \exp[-0.16\gamma^{-1.06}\alpha^{2.4})]$	= 0.71

*Where $\exp[x]$ = ex.

Table 4.12 **Stress concentration factors for Example 4.7**

Load type and fixity conditions	Chord		Brace	
	Saddle	Crown	Saddle	Crown
Axial load on one brace only	4.67	2.42	3.37	2.14
IPB on one brace only		2.18		2.50
OPB on one brace only	4.39		4.25	

For a joint with loading on both braces.

Balanced axial loading (chord SCF) = 3.54
Balanced axial loading (brace SCF) = 2.68
Unbalanced IPB (chord crown SCF) = 2.18
Unbalanced IPB (brace crown SCF) = 2.50
Unbalanced OPB (chord saddle SCF) = 5.12
Unbalanced OPB (brace saddle SCF) = 3.85

Example 4.8

Calculate the SCF for T/Y joint using the input data in Table 4.13.

The stress concentration factors are shown in Table 4.14.

Example 4.9

Calculate the SCF for a T joint using the input data of Table 4.15.

The stress concentration factors for joint SCFs are shown in Table 4.16.

Example 4.10

Table 4.17 is an example of a case study of fatigue analysis. We have the stress range, the corresponding number of cycles of stress occurrence, and the allowable number of cycles based on the S-N curve and assume a point is subject to five cyclic stress ranges (due to waves).

$D_5 = 10 \times 10^6/50 \times 10^6 = 0.2$
$D_{10} = 0.1$
$D_{20} = 0.2$
$D_{50} = 0.05$
$D_{90} = 0.025$
Total damage = 0.575

If these waves occur over 10 years, then

Life = 10/0.575 = 17.4 years

4.7.4 Jacket fatigue design

Dynamic analysis is carried out to predict the fundamental periods of the platforms in order to confirm the sensitivity of the structure to wave-induced excitation.

Table 4.13 Input data for Example 4.8

Title	Example only
D = chord diameter T = chord thickness L = chord length C = chord-end fixity parameter $0.5 < C < 1.0$ d = brace diameter t = brace thickness θ = brace angle (from chord)	= 16 inches = 0.625 inches = 376.8 inches = 0.7 Typically C = 0.7 = 12.75 inches = 0.375 inches = 45 degrees
Geometrical parameters	
$\beta = d/D$ $\gamma = D/2T$ $\tau = t/T$ $\alpha = 2L/D$	= 0.80 = 12.80 = 0.60 = 47.10
Validity range	
$0.2 < \beta < 1.0$ $0.2 < \tau < 1.0$ $8 < \gamma < 32$ $4 < \alpha < 40$ $20 < \theta < 90$	Satisfied Satisfied Satisfied *Not* satisfied Satisfied
Chord-end fixity parameter	
$C1 = 2(C - 0.5)$ $C2 = C/2$ $C3 = C5$	= 0.4 = 0.35 = 0.14
Short chord correction factors ($a < 12$)	
$F1 = 1 - (0.83\beta - 0.56\beta^2 - 0.02)\gamma^{0.23}\exp[-0.21\gamma^{-1.16}\alpha^{2.5}]$ $F2 = 1 - (1.43\beta - 0.97\beta^2 - 0.03)\gamma^{0.04}\exp[-0.71\gamma^{-1.38}\alpha^{2.5}]$* $F3 = 1 - 0.55\beta^{1.8}\gamma^{0.16}\exp[-0.49\gamma^{-0.89}\alpha^{1.8}]$	= 1.00 = 1.00 = 1.00

*Where $\exp[x] = ex$.

Table 4.14 Stress concentration factors for Example 4.8

Load type and fixity conditions	Chord		Brace	
	Saddle	Crown	Saddle	Crown
Axial load, chord ends fixed	3.69	5.72	3.60	3.97
Axial load, chord ends fixed	6.61	7.31	3.60	4.87
In-plane bending		1.89		2.63
Out-of-plane bending	4.11		3.09	

Table 4.15 **Input data for Example 4.9**

Title	**Example only**
Chord	
Chord diameter = D Chord thickness = T Chord length = L Chord-end fixity value = $C(0.5 \leq C \leq 1.0)$	= 24 inches = 0.500 inches = 70.0 inches = 0.7
Brace	
Brace diameter = d Brace thickness = t Brace to chord inclination (loaded brace) = $\emptyset a$	= 16 inches = 0.375 inches = 50 degrees
Geometric parameters	
β γ τ α	= 0.67 = 24.00 = 0.75 = 5.83
Validity range	
$0.2 < \beta < 1.0$ $0.2 < \tau < 1.0$ $8 < \gamma < 32$ $4 < \alpha < 40$ $20 < \theta < 90$	Satisfied Satisfied Satisfied Satisfied Satisfied
Chord-end fixity parameter	**Short chord correction factors**
$C1 = 0.4$ $C2 = 0.35$ $C3 = 0.14$	F1 = 0.617 F2 = 0.730 F3 = 0.779

Notes

Validity range of equations:

$0.2 \leq b \leq 1.0$

$0.2 \leq t \leq 1.0$

$4 \leq a \leq 40$

$8 \leq g \leq 32$

$0 \leq q \leq 90$

Chord end fixity parameters:

C = 0.5 Chord ends are fully fixed.

C = 1.0 Chord ends are pinned.

C = 0.7 is considered typical.

Table 4.16 **Stress concentration factors for joint SCFs**

Load type	Chord		Brace	
	Saddle	Crown	Saddle	Crown
Axial	11.94	3.16	6.44	1.25
Short chord corrected		7.36		3.97
Axial, general fixity	11.81	3.39	6.44	1.37
Short chord corrected		8.62		4.70
In-plane bending		3.57		3.24
Out-of-plane bending	10.88		7.51	
Short chord corrected		8.48		5.85

Table 4.17 **Fatigue analysis**

Stress range	Occurrence	Allowable
5	10×10^6	50×10^6
10	4×10^5	4×10^6
20	6×10^4	3×10^5
50	5×10^2	1×10^4
90	25	1×10^3

The fundamental sway periods are used to derive the dynamic amplification for the in-place analysis loading conditions.

Fatigue analysis is performed for the jacket structures using methods appropriate to the sensitivity to dynamic loading. A spectral approach is deemed adequate for platforms with fundamental periods of less than 3 seconds.

From practical point of view, the in-service fatigue design life of the joints should be at least two times the service life of the platform (i.e., 50 years).

Fatigue analysis using software includes the following:

1. The environmental loads parameters.
2. A sufficient number of wave directions. For each direction, a minimum of four wave heights should be used to compute stress range versus wave height. The directions, wave heights, and exceedances selected should be those closest to the directions indicated in metocean data. A sufficient number of wave crest positions should also be considered.
3. Dynamic amplification effects. These are taken into account in calculation of the cycle loading should the natural period of the structure exceed 3 seconds.
4. Where significant cyclic stresses may be induced by the action of wind, waves, changes in member buoyancy, and the like. Such stresses should be combined with those due to wave action to obtain the total effective stress spectrum for a particular member or joint. Fatigue damage accumulated during fabrication and transportation should also be considered.

After that, the analysis procedure is through the following steps.

1. For each joint and type of failure under consideration, the stress range spectra are computed at a minimum of eight positions around the joint periphery.
2. Two types of failure are considered using the appropriate stress concentration factors (i.e., brace to weld failure and chord to weld failure).
3. For joints other than those between tubular members, individual detailed consideration is given with due regard being paid to published, reliable experimental data.

Stress concentration factors (SCF) are determined. For tubular to tubular joints ungrouted and unstiffened, the SCF are calculated using Efthymiou equations as described in equations from 4.123–4.155.

In lieu of a more accurate procedure for analysis, ring stiffened joints may be checked as for simple joints but using modified chord thickness.

4.8 Topside design

In general, for offshore structure design, major rolled shapes should be compact sections as defined by AISC. The minimum thickness of structural plate or section should be 6 mm. The minimum diameter to thickness ratio of tubular members should be no less than 20, where the diameter is based on the average of the tubular outside and inside diameters.

For connections, the design is to comply with the codes listed with the following minimum requirements:

1. The minimum fillet weld should be 6 mm.
2. Wherever possible, joints should be designed as simple joints with no overlap.
3. Tubular joints should be designed in accordance with API RP2A.

The deflection, as discussed in Chapter 2, should match the codes and definition in the owner specification:

1. Deflections should be checked for the actual equipment, live loads, and casual area live loads. Pattern loading should be considered.
2. Deflection of members supporting deflection-sensitive equipment should be no greater than $L/500$ for beams and $L/250$ for cantilevers under live loads.
3. Deflection of beams in the workrooms and living quarters should be no greater than $L/360$ for beams and $L/180$ for cantilevers under live loads
4. Deflection limits for other structures should be $L/250$ for beams and $L/125$ for cantilevers under live loads.

4.8.1 Topside structure analysis

In performing the structural analysis by any software, SACS for example, it is better to define the direction of the model and in most cases the following directions are best followed:

- +X axis aligned with platform east.
- +Y axis aligned with platform north.
- +Z axis aligned vertically upward.
- The datum for the axes should be at chart datum.

In general, only the primary structural steel should be modeled. However, secondary members should be modeled where they are necessary for the structural integrity or to facilitate load input. The deck plate should be modeled as shear panel elements. Joint eccentricities should be modeled using discrete elements rather than using the "offset" facility of SACS. By using individual elements, joint forces can be more easily extracted from the output.

All the differing analyses (in-place, lift, loadout, etc.) should utilize the same base model. That is, the in-place model should form the basis for all the other analyses to be performed.

The in-service analyses should include a basic model of the jacket structure to ensure the correct stiffness interaction between the jacket and the topside structures. Pile foundations to the jackets need not be considered for the analysis and design of the topsides, simple supports are sufficient.

4.8.2 Deck design to support vibrating machines

In most platforms decks carry machines such as pumps, compressors, or generators. So the dynamic effect of this machine should be considered in the design, and a local check must be made of the steel structure of the deck.

To design the steel structure's main and secondary beams to support machines with vibrations, the following data and information should be delivered from the manufacturer:

- Skid dimensions
- Machine dry weight.
- Machine wet weight.

Then, the natural frequency of the steel structure xan be calculated in the following steps:

- $m = W/g$
- $T_n = 2\pi(m/K)^{0.5}$ (s)
- $F_n = 1/T_n$ (Hz)
- $\omega_n = 2\pi f_n$ (rpm)
- The machine frequency $= \omega$
- $1.2 < \omega/\omega_n < 0.8$

Example 4.11

Check steel frame support gas generator package with the following data:

- Dry weight = 12.38 ton
- Wet weight = 13.02 ton
- Skid dimension 5842 mm × 1067 mm
- Machine has 900 rpm
- Beam section W10 × 22 with steel ASTM A36 with Fy equal to 248 N/mm^2

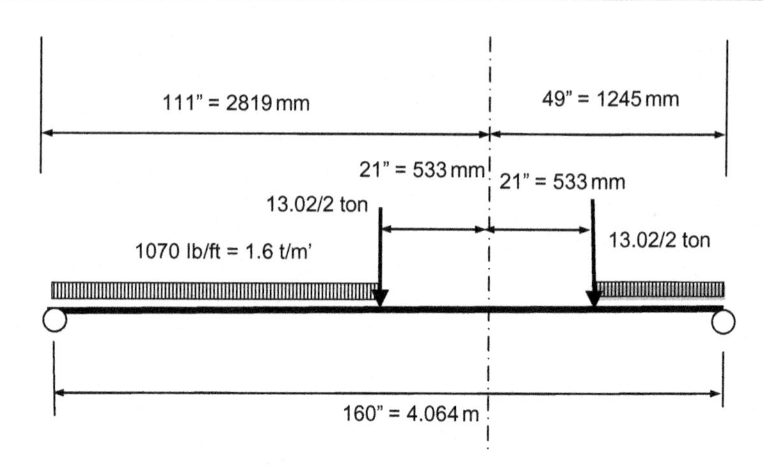

Figure 4.49 Example 4.11 solution.

Solution See Figure 4.49. In addition,

Maximum beam design moment = 4.07 *mt*
Maximum beam stress = $4.07/3.8 \times 10^{-4} = 10705.5\ t/m^2 = 105\ \text{N/mm}^2$
Allowable beam stress = $0.66 \times 240 = 158\ \text{N/mm}^2$
Unity check ratio (UC) = $105/158 = 0.66 < 1.0$ *OK*

A local check has been performed to ensure no resonance will occur due to setting down the new gas generator package over the deck:

$$\Delta = \frac{Pa^2b^2}{3EIL}$$

Beam stiffness $(K) = P/\Delta = 3EIL/a^2b^2$
$= 3 \times 29 \times 10^6 \times 118 \times 160/(111)^2\ (49)^2$
$= 55525$ lb/in. $= 9.7$ KN/mm
Mass $(M) = W/g = 13.02\ \text{KN} \times \text{s}^2/\text{m}$
$T_n = 2\pi\ (m/K)^{0.5}$
$f_n = 1/T_n = 4.35$ Hz
$\omega_n = 2\pi f_n = 27.32$ rad/s $= 1640$ rpm
The new generator frequency $(\omega) = 900$ rpm

The frequencies ratio is

$\omega/\omega_n = 900/1640 = 0.55 < 0.8$ O.k

4.8.3 Grating design

Grating is the traditional member used in covering the platform floor or in the onshore facilities. Grating should have 1-inch (25-mm) minimum bearing on supporting steel. Where grating areas are shown as removable on the drawings, the weight of fabricated grating sections for such areas should not exceed 160 kg (350 lbs).

Most grating and expanded metal needs to be supported in a specific way. The direction in which the load bars run is the important direction for grating and is usually referred to as the span.

For expanded metal the span is in the direction of the strands. The span is always the least dimension given when referencing a panel size.

In most cases. the grating has to be supported only in the span direction, and it does not require support on all four sides, as is the case with floor plate. The grating space and its types are presented in Figures 4.50 and 4.51.

There is a relation between the span and the deflection. When selecting the grating, you should refer to the manufacturer's type and calculation. Table 4.18 is an example of a grating sheet based on its type and loads.

Table 4.18 has been arranged in increasing strength order. The load, L, is a safe, superimposed, uniformly distributed load in KPa (100 kg/m^2 = 0.98 KPa), where Δ is the deflection in millimeters for the load L. Loads are calculated in accordance with an allowable bending stress of 171.6 MPa. Note that the load bars are assumed to be simply supported.

Example 4.12

Assume the distance between the secondary beams is 1.2 m and the total load of dead and live load is 20 KN/m^2, can you choose the best grating type.

Solution Based on Table 4.18, choose type 3.

4.8.4 Handrails, walkways, stairways, and ladders

Handrails, walkways, stairways, and ladders should be designed in accordance with OSHA 3124. Handrails should be provided around the perimeter of all open decks and on both sides of stairways. All handrails shall be 1.10 m high and made removable in panels no more than 4.5 m long. Handrail posts should be spaced 1.5 m apart. The gap between panels should not exceed 51 mm. Handrails in the wave zone should also be designed to withstand extreme storm maximum wave loadings.

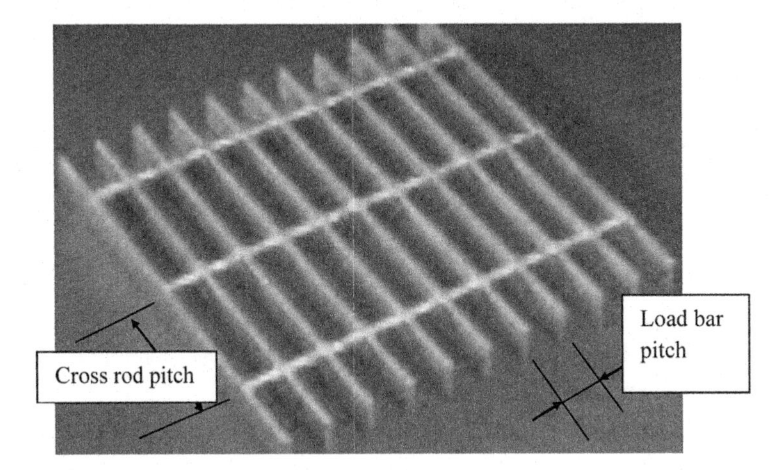

Figure 4.50 Grating dimensions.

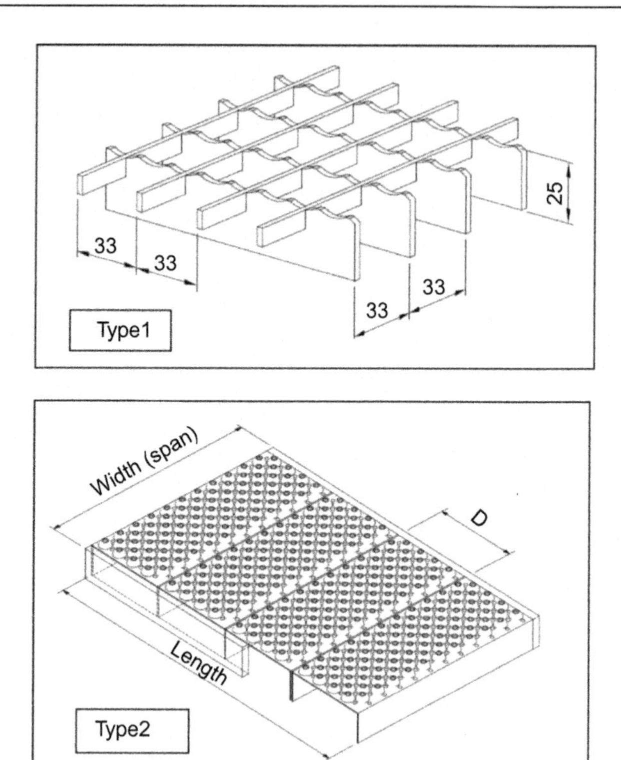

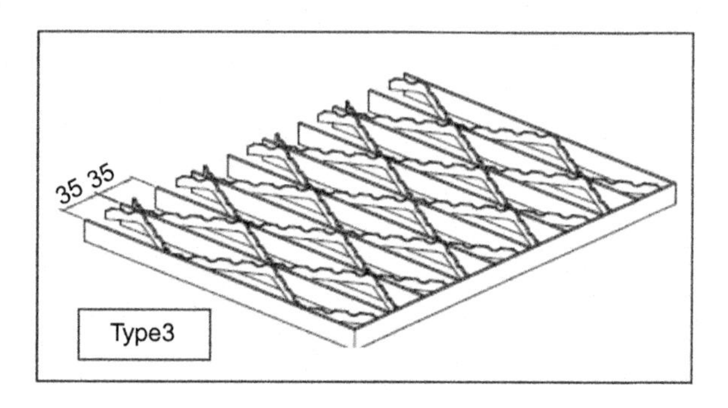

Figure 4.51 Different types of grating.

Walkways, stairways, and landings should be designed for the following load combinations:

1. Dead load plus live loads.
2. Dead loads plus extreme storm 3-second wind gusts or extreme storm maximum wave, whichever is applicable.

Stairways should be of structural steel, double runner with serrated bar grating treads and handrails.

Table 4.18 **Relationship between grating dimensions, maximum span, and maximum load**

Type	Load bar spacing, mm	Cross bar pitch	Weight kg/m^2	Bar size	Load, KPa	Spacing between supports, in mm								
						450	600	750	900	1050	1200	1500	1800	2100
1	40	100	17.5	25 × 3	L	53	30	19	13	10	7	5	3	2
					Δ(mm)	1.4	2.6	4	5.8	7.8	10.2	16	23.1	31.5
2	60	50	22.3	25 × 5	L	56	32	20	14	10	8	5	3	2
					Δ(mm)	1.4	2.6	4	5.8	7.8	10.2	16	23.1	31
3	40	100	26.9	25 × 5	L	70	39	24	17	13	9	6	4	3
					Δ(mm)	1.6	2.9	4.5	6.5	8.8	11.5	18.0	25.9	35.3
4	30	100	22.8	25 × 3	L	70	39	25	17	13	10	6	4	3
					Δ(mm)	1.4	2.6	4	5.8	7.8	10.3	16.0	23.1	31.5
5	60	50	26.4	32 × 5	L	76	43	27	19	14	11	7	5	3
					Δ(mm)	1.2	2.2	3.4	4.9	6.7	8.7	13.6	19.7	26.8
6	30	100	34.7	25 × 5	L	91	51	33	23	17	13	8	6	4
					Δ (mm)	1.6	2.9	4.5	6.5	8.8	11.5	18.0	25.9	35.3
7	40	100	34	32 × 5	L	120	67	43	30	22	17	11	7	5
					Δ(mm)	1.2	2.2	3.4	4.9	6.7	8.7	13.6	19.7	26.8
8	30	100	28.4	32 × 3	L	114	64	41	28	21	16	10	7	5
					Δ(mm)	1.1	2	3.1	4.5	6.1	8.0	12.5	18.1	24.6
9	40	100	42.1	40 × 5	L	226	127	81	56	41	31	20	14	10
					Δ(mm)	0.9	1.6	2.5	3.6	4.9	6.4	10.0	14.4	19.7
10	30	100	62.9	45 × 5	L	377	212	135	94	67	52	33	23	17
					Δ(mm)	0.8	1.4	2.2	3.2	4.3	5.7	8.9	12.8	17.5

4.9 Bridges

In some cases where two or more platforms are forming a complex or in case where separate installations are built to support a helideck or a flair, bridges may be required to connect between the different structures.

The bridge should be designed to resist the following loads:

1. Its own weight.
2. Uniformly distributed live load equal to 250 Kg/m^2 of the walk way area.
3. All piping loads carried by the bridge, if any.
4. Wind loads acting directly on the bridge.
5. Maximum imposed relative displacement between the bridge ends due to environmental-level loads acting on the two structures connected by the bridge.
6. Thermal effect due to temperature changes.

The bearings of the bridge should be designed to allow for the expected displacements and rotations. Normally, the bearings at one end are hinged allowing only for rotation. The bearings at the other end are free for sliding and rotation. Flurogold slide bearings are adequate to specify the slide bearing.

The design of the bridge should account for a span tolerance of ± 1.0 m liable to result from possible mislocation of the supporting structures. Span length rectification, in this case, should be accounted for by a possible increase or decrease of the theoretical bridge span or by availing relocation of the bearings on the structure deck.

The bridge shall have an upward camber equal to the deflection supposed to happen under dead loads.

Figure 4.52 presents the hinge support to the bridge, as it can be seen that the axial movement is prevented, This is opposite to Figure 4.53, which permits the axial movement so this support acts as roller support.

Figure 4.52 Hinge support for the bridge.

Figure 4.53 Roller support for the bridge.

4.10 Vortex-induced vibration

In fluid dynamics, vortex-induced vibrations (VIVs) are motions induced on bodies interacting with an external fluid flow, produced by periodic irregularities on this flow.

A classic example is the VIV of an underwater cylinder. You can see how this happens by putting a cylinder into the water and moving it through the water in the direction perpendicular to its axis. Since real fluids always present some viscosity, the flow around the cylinder is slowed down while in contact with its surface, forming the so-called boundary layer. At some point, however, this boundary layer can separate from the body because of its excessive curvature. Vortices are then formed, changing the pressure distribution along the surface. When the vortices are not formed symmetrically around the body (with respect to its midplane), different lift forces develop on each side of the body, leading to motion transverse to the flow. This motion changes the nature of the vortex formation in such a way that leads to a limited motion amplitude.

The tubular members of the flare/vent booms should be checked for vortex-induced vibration. If the members and booms are found to be dynamically sensitive, they should be checked during detailed design for fatigue.

Vortex-induced vibration is an important source of fatigue damage to offshore platforms especially for oil exploration and production risers. These slender structures experience both current flow and top-end vessel motions, which give rise to the flow-structure relative motion and cause VIV. The top-end vessel motion causes the riser to oscillate and the corresponding flow profile appears unsteady.

The possibility of VIV due to the design current velocity profiles should be considered for all appurtenances including risers, sump pipes, caissons, and any individual members considered potentially susceptible.

One of the classical open-flow problems in fluid mechanics concerns the flow around a circular cylinder or, more generally, a bluff body. At very low Reynolds numbers, according to the diameter of the circular member, the streamlines of the resulting flow is perfectly symmetric as expected from potential theory.

The Strouhal number relates the frequency of shedding to the velocity of the flow and a characteristic dimension of the body (diameter in the case of a cylinder). It is defined as $St = f_{st}D/U$ and is named after Čeněk (Vincent) Strouhal (a Czech scientist). In the equation f_{st} is the vortex shedding frequency (or the Strouhal frequency) of a body at rest, D is the diameter of the circular cylinder, and U is the velocity of the ambient flow. The Strouhal number for a cylinder is 0.2 over a wide range of flow velocities. The phenomenon occurs when the vortex shedding frequency becomes close to a natural frequency of vibration of the structure. When this happens large and damaging vibrations can occur.

In general, wind, current, or any fluid flow past a structural component may cause unsteady flow patterns due to vortex shedding. This may lead to oscillations of slender elements normal to their longitudinal axis. Such vortex-induced oscillations (VIO) should be investigated.

Important parameters governing vortex-induced oscillations follow:

- Geometry (L/D).
- Mass ratio ($m^* = m/(\frac{1}{4}\pi\rho D^2)$).
- Damping ratio (ζ).
- Reynolds number ($Re = uD/\nu$).
- Reduced velocity ($VR = u/fnD$).
- Flow characteristics (flow profile, steady/oscillatory flow, turbulence intensity ($\sigma u/u$), etc.).

where

L = member length (m)
D = member diameter (m)
m = mass per unit length (kg/m)
ζ = ratio between damping and critical damping
ρ = fluid density (kg/m^3)
ν = fluid kinematic viscosity (m^2/s)
u = (mean) flow velocity (m/s)
f_n = natural frequency of the member (Hz)
σ_u = standard deviation of the flow velocity (m/s)

For rounded hydrodynamically smooth stationary members, the vortex shedding phenomenon is strongly dependent on the Reynolds number for the flow, as given in Table 4.19.

For rough members and smooth vibrating members, the vortex shedding shouldl be considered strongly periodic in the entire Reynolds number range.

The vortex shedding frequency in steady flow or flow with KC numbers greater than 40 may be calculated as follows:

$$f_s = \mathrm{St}\frac{u}{D}$$

where

f_s = vortex shedding frequency (Hz)
St = Strouhal number
u = fluid velocity normal to the member axis (m/s)
D = member diameter (m)

Vortex shedding is related to the drag coefficient of the member considered. High drag coefficients usually accompany strong regular vortex shedding and vice versa.

For a smooth stationary cylinder, the Strouhal number is a function of Reynolds number (Re). The relationship between St and Re for a circular cylinder is given in Figure 4.54.

Table 4.19 Relation between type of shedding and Reynolds number

Periodic shedding	$10^2 < \mathrm{Re} < 0.6 \times 10^6$
Wide-band random shedding	$0.6 \times 10^6 < \mathrm{Re} < 3.0 \times 10^6$
Narrow-band random shedding	$3.0 \times 10^6 < \mathrm{Re} < 6.0 \times 10^6$
Quasi-periodic shedding	$\mathrm{Re} > 6.0 \times 10^6$

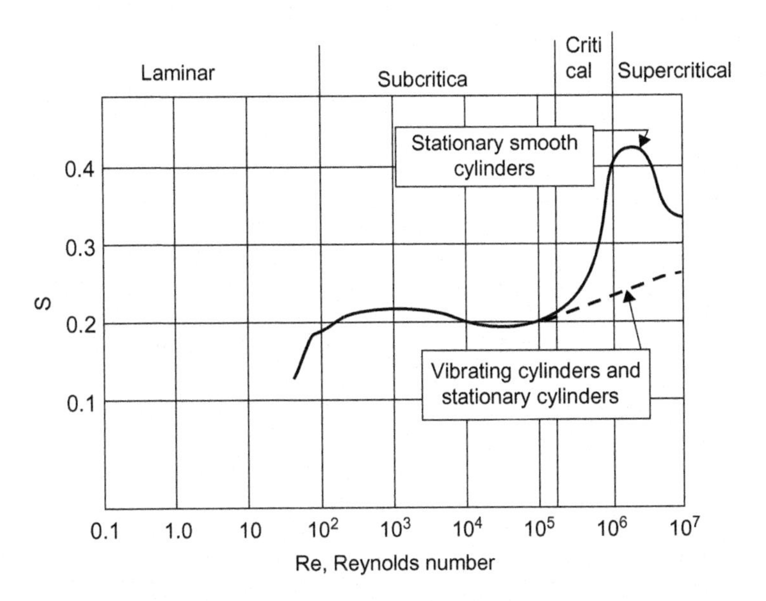

Figure 4.54 Relation between Reynolds number and Strouhal number for a circular cylinder.

Rough surfaced cylinders or vibrating cylinders (both smooth and rough surfaced) have Strouhal numbers that are relatively insensitive to the Reynolds number.

For cross sections with sharp corners, the vortex shedding is well defined for all velocities, giving Strouhal numbers that are independent of flow velocity (Re).

Structural damping is due to internal friction forces of the member material and depends on the strain level and associated deflection. For wind-exposed steel members, the structural damping ratio may be taken as 0.0015, if no other information is available. For slender elements in water, the structural damping ratio at moderate deflection typically ranges from 0.005 for pure steel pipes to 0.03–0.04 for flexible pipes.

Damping ratios for several structures and materials can be found in Blevins (1990).

Example 4.13

Calculate the vortex-induced vibration for a steel pipe in wind as follows and the Excel sheet is available on the site www.elreedyma.comli.com. The input data are

Outer diameter, $D_o = 273$ mm
Inner diameter, $D_i = 254.3$ mm
Outer coating thickness, $t_c = 0.6$ mm
Span length, $L = 9$ m

The two ends are fixed-fixed.

Structural damping ratio, $\xi = 0.001$
Density of steel, $\rho_s = 7850$ kg/m^3
Density of air, $\rho_a = 1.21$ kg/m^3
Kinematic viscosity of air, $\nu = 1.52 \times 10^{-5}$
Midspan velocity; u = 42 m/s

The structural member properties are

Wall thickness, $t_w = 9.35$ mm
Moment of inertia, $I = 6.738 \times 10^{-5}$ m^4
Section modulus, $z = 4.936 \times 10^{-4}$ m^3
Cross-sectional area, $A_s = 0.008$ m^2
Mass/unit length = $\rho_s x A_s = 60.794$ kg/m
Actual out diameter, = $D_o + 2t_c = 0.274$ m
Effective mass calculation, $m_s = 60.794$ kg/m, neglecting the coating density

Table 4.20 shows the fixity status of the steel pipe.

Stiffness, $k = (a_1)^2\ E\ I/(L)^4 = 1.07 \times 10^6$ (N/m^2)
Natural frequency, $f_n = 1/2\pi(k/m)^{0.5} = 21.08$ Hz
Natural period, $T_n = 0.05$ s
Reynolds number, $\mathrm{Re} = uD/\nu = 7.52 \times 10^{-5}$
Reduced velocity parameter, $V_r = u/D\ f_n = 7.30$
Mass –damping parameter, $K_s = 4\ \pi\xi\ m/\rho_a\ (D)^2 = 8.47$
In-line VIV response in case of $1.7 < V_r < 3.2$
Cross-flow VIV response $4.25 < V_r < 8.0$ (It is in our case)

Strouhal number, St = 0.20
Approximate reduced velocity at resonance, $V_{Re} = 1/S = 5.00$
$V_r > V_{Re}$, resonance will occur
Flow velocity at resonance, $ur = V_{Re} f_n D = 28.90$ m/s
Reynolds number at resonance, Re(res) = 5.2×10^5
Base lift coefficient, $L_c = 0.29$ if Re(res) > 400,000 = 0.42 if Re(res) ≤ 400,000
Turbulence intensity, $T_i = 1.02$

The turbulence intensity is obtained from the relation between turbulence ratio to the reduced amplitude, as shown in Table 4.21.

Table 4.20 Fixity status of the steel pipe in Example 4.13

Fixity status	Restraint factor, a_1	Correlation length factor, L_f	Restraint constant, K_n
Fixed-free	3.52	0.66	2
Pinned-pinned	9.87	0.63	4.8
Fixed-pinned	15.4	0.58	11.57
Fixed-fixed	22.4	0.52	16

Table 4.21 Turbulence intensity

Reduced amplitude	Turbulence intensity
0.000	1.00
0.025	1.01
0.050	1.04
0.075	1.14
0.100	1.40
0.125	1.58
0.150	1.68
0.175	1.76
0.200	1.81
0.225	1.85
0.250	1.88
0.275	1.90
0.300	1.91
0.325	1.92
0.350	1.91
0.375	1.89
0.400	1.86
0.425	1.83
0.450	1.80
0.475	1.76
0.500	1.72

The value of the reduce amplitude $= L_c L_f (V_r)^2 / 4\pi K_s$
Fluctuating cross-flow force coefficient, $F_c = L_c L_f T_i = 0.15$
Individual shedding frequency $= 2S.u/D = 61.269$ Hz
Reduced amplitude response, $A_{mr}/D = \overline{\text{DAF}}(F_c)\ (V_r)^2/(4\pi\ ks) = 3.62 \times 10^{-2}$
Amplitude of response, $A_m = A_{mr} D = 10$ mm
Maximum bending moment, $M = K_n\ Z\ E\ D_o\ A_m / L^2 = 54.39$ kN m
Maximum bending stress, $\sigma = M/Z = 1\ 10.20$ MPa

Example 4.14

Calculate the same steel pipe as in the previous example but with span length 5 m. The input data are

Outer diameter, $D_o = 273$ mm
Inner diameter, $D_i = 254.3$ mm
Outer coating thickness; $t_c = 0.6$ mm
Free span length, $L = 5$ m

The two ends are fixed-fixed.

Structural damping ratio, $\xi = 0.001$
Density of steel, $\rho_s = 7850$ kg/m^3
Density of air, $\rho_a = 1.21$ kg/m^3
Kinematic viscosity of air, $\nu = 1.52 \times 10^{-5}$
Midspan velocity, $u = 42$ m/s

The structural member properties are

Wall thickness, $t_w = 9.35$ mm
Moment of inertia, $I = 6.738 \times 10^{-5}$ m^4
Section modulus, $z = 4.936 \times 10^{-4}$ m^3
Cross-sectional area, $A_s = 0.008$ m^2
Mass/unit length $= \rho_s x A_s = 60.794$ kg/m
Actual out diameter, $= D_o + 2t_c = 0.274$ m
Effective mass calculation, $m_s = 60.794$ kg/m, neglecting the coating density

Table 4.22 shows the fixity status of the steel pipe.

Stiffness, $k = (a_1)^2\ E\ I/(L)^4 = 1.12 \times 10^7$ (N/m^2)
Natural frequency, $f_n = 1/2\pi(k/m)^{0.5} = 68.302$ Hz

Table 4.22 Fixity status of the steel pipe in Example 4.14

Fixity status	Restraint factor, a_1	Correlation length factor, L_f	Restraint constant, K_n
Fixed-free	3.52	0.66	2
Pinned-pinned (guided-guided)	9.87	0.63	4.8
Fixed-pinned (guide-fixed)	15.4	0.58	11.57
Fixed-fixed	22.4	0.52	16

Natural period, $T_n = 0.01$ s
Reynolds number, $\mathrm{Re} = uD/\nu = 7.52 \times 10^{-5}$
Reduced velocity parameter, $V_r = u/D\, f_n = 2.25$
Mass-damping parameter, $K_s = 4\pi\, \xi m/\rho_a\, (D)^2 = 8.47$
In-line VIV response in case of $1.7 < V_r < 3.2$ (it is in our case)
Cross-flow VIV response $4.25 < V_r < 8.0$
Strouhal number, $\mathrm{St} = 0.20$
Approximate reduced velocity at resonance, $V_{\mathrm{Re}} = 1/S = 5.00$
Flow velocity at resonance, $V_r = 2.25$ m/s
Reynolds numbet at resonance, $\mathrm{Re(res)} = 7.52 \times 10^5$
Base lift coefficient, $L_c = 0.29$, if $\mathrm{Re(res)} > 400{,}000$
$= 0.42$ if $\mathrm{Re(res)} \leq 400{,}000$
Turbulence intensity, $T_i = 1.00$, from Table 4.21
The value of the reduce amplitude $= L_c\, L_f (V_r)^2/4\pi \mathrm{K_s}$
Fluctuating cross-flow force coefficient, $F_c = L_c L_f T_i = 0.15$
Individual shedding frequency, $f_v = 2S_u/D = 61.269$ Hz
Frequency ratio, $f_v/f_n = 0.9$
Dynamic amplification factor, $\mathrm{DAF} = 1/\{([1-(f_v/f_n]^2)^2 + [2\xi(f_v/f_n)]^2\}^{0.5} = 5.12$
Reduced amplitude response, $A_{mr}/D = (C_1)\,(V_r)2/4\pi[2\pi m/(\rho_a\, D_o{}^2)] = 7.38 \times 10^{-5}$
Amplitude of response, $A_m = A_{mr} \cdot D = 0.02$ mm
Maximum bending moment, $M = K_n\, Z\, E\, D_o\, A_m/L^2 = 0.36$ kN m
Maximum bending stress, $\sigma = M/Z = 0.73$ MPa

Example 4.15

A riser ose under VIV. As in most cases, it was designed by the piping engineer and the responsibility of the structure engineer is to check the design of clamp fixation. The data of the pipe are as follows:

Outer diameter, $D_o = 254$ mm
Inner diameter, $D_i = 228.6$ mm
Marine growth thickness, $t_m = 50$ mm
Marine growth density = 1400 kg/m^3
Effected diameter, $D_e = D_o + 2t_m = 0.345$ m
Free span length, $L = 8.8$ m

The two ends are fixed-fixed.
Calculated as follows:

Structural damping ratio, $\xi = 0.02$
Density of steel, $\rho_s = 7850$ kg/m^3
Density of water, $\rho_w = 1030$ kg/m^3
Maximum particle horizontal velocity, $u = 1.5$ m/s (from SACS output or from the wave meteorological data)
Wave period, $T = 8$ s

The member is not flooded.
The structural member properties follow:

Wall thickness, $t_w = 12.7$ mm
Moment of inertia, $I = 7.0465 \times 10^{-5}$ m^4

Mass/unit length of pipe, $m_s = \rho_s x A_s = 75.58$ kg/m'
Marine growth unit mass, $m_m = 66.85$ kg/m'
Displaced unit mass, $m_d = \pi\rho_w (D_o + 2t_m)^2/4 = 101.38$

Note: if the riser is flooded, the mass of the liquid should be include in the mass value.

Effective mass calculation, $m = m_s + m_m + m_d = 243.810$ kg/m
Stiffness, $k = (a_1)^2 E I/(L)^4 = 1.0 \times 10^6$ (N/m^2)
Natural frequency, $f_n = 1/2\pi(k/m)^{0.5} = 11.222$ Hz
Natural period, $T_n = 0.09$ s
Reduced velocity parameter, $V_r = u/D_e f_n = 0.38$
Mass-damping parameter, $K_s = 4\ \pi\xi\ m/\rho_w\ (D_e)^2 = 0.47$
Keulegan-Carpenter number; $K_c = uT/D_e = 33.9$; $K_c < 40$
Cross-flow VIV response $3 < V_r < 9$

In-line VIV response in case of $V_r > 1.0$ (it is in our case). In this case, $V_r < 1.0$, resonance will not occur.

Example 4.16

The problem and input data are the same as in Example 4.15 with maximum particle horizontal velocity equal to 5 m/s.

Solution All calculation is the same as in Example 4.15, except

Reduced velocity parameter; $V_r = u/D_e f_n = 1.26$
Calculate Keulegan-Carpenter number; $K_c = uT/D_e = 112.99$; in this case $K_c > 40$
In n line excitation $1.0 < V_r < 4.5$, in this case
In cross-flow excitation $3.0 < V_r < 9$

So, check the stability number, $K_s = 0.47 < 1.8$ and $V_r = 1.26$. In $1.0 < V_r < 3.5$, resonance will occur.

The action is to reduce the distance between the clamps or by another means reduce the free span length.

Example 4.17

The problem and input data are the same as in Example 4.15 but the span is reduced to 5.0 m.

Solution Members properties as in Example 4.16.

Stiffness, $k = (a_1)^2 E I/(L)^4 = 1.0 \times 10^7$ (N/m^2)
Natural frequency, $f_n = 1/2\pi(k/m)^{0.5} = 34.75$ Hz
Natural period, $T_n = 0.029$ sec
Reduced velocity parameter, $V_r = u/D_e f_n = 0.41$
Mass–damping parameter, $K_s = 4\ \pi\xi\ m/\rho_w\ (D_e)^2 = 0.47$
Calculate Keulegan-Carpenter number, $K_c = uT/D_e = 112.99$; $K_c > 40$
Cross-flow VIV response, $3 < V_r < 9$

In this case, $V_r < 1.0$, resonance will not occur.

Further reading

AISC. Specification for the Design, Fabrication, and Erection of Structural Steel for Buildings.

American Welding Society, 1972. AWS D1.1, Structural Welding Code, first ed. American Welding Society, USA.

API RP2A, 2007. Recommended Practice for Planning, Designing, and Constructing Fixed Offshore Platforms, twentieth ed. American Petroleum Institute, Washington, DC.

Dier, A.F., Lalani, M., 1995. Strength and stiffness of tubular joints for assessment/design purposes. Paper OTC 7799. Offshore Technology Conference, Houston, TX, May.

Det Norske Veritas, 2008. Structure Analysis for Piping System. Recommended practice DNV-RP-D101. DNV, Norwige, October.

Healy, R.E., Buitrago, J., 1994. Extrapolation procedures for determining SCFs in mid-surface tubular joint models. Sixth International Symposium on Tubular Structures. Monash University, Melbourne, Australia.

International Organization for Standardization, 2004. Petroleum and Natural Gas Industries – Offshore Structures. Part 2. Fixed Steel Structures. ISO, Switzerland, ISO/DIS 19902:2004.

Kuang, J.G., Potvin, A.B., Leick, R.D., 2003. "Stress concentration in tubular joints," 1975 OTC proceedings, paper 2205 References EDI, SACS IV, User's Manual, release 5.1.

Lloyd's Register of Shipping, 1988. Stress Concentration Factors for Tubular Complex Joints. Complex Joints JIP. Final Report No.3 of 5 of Simple Unstiffened Joint SCFs, March 8.

Marshall, P.W., 1989. Recent Developments in Fatigue Design Rules in the U.S.A. In: Fatigue Aspects in Structural Design, Delft University Press, Delft, the Netherlands.

Marshall P.W., Bucknell, J., Mohr W.C., 2005. Background to new RP 2A-WSD fatigue provision. OTC 17295, Proceedings of the Offshore Technical Conference, OTC, Houston, Tx, USA. May.

Marshall, P.W., Toprac, A.A., 1974. Basis for tubular joint design. Welding J. 53 (5), (May).

MSL Engineering Limited, 1996. Assessment Criteria, Reliability and Reserve Strength of Tubular Joints. MSL, Doc. Ref. C14200R018, Ascot, England, March.

Niemi, E., et al., 1995. Stress Determination for Fatigue Analysis of Welded Components. IIW-1221-93. Abington Publishing, Cambrudge, UK.

Proposed Revisions for Fatigue Design of Welded Connections for adoption by IIW (International Institute of Welding), Eurocode 3

Yettram, A., Husain, H., 1966. Plane framework methods for plates in extension. J. Eng. Mech. Div. ASCE. (February).

Helidecks and boat landing design

5.1 Introduction

The design of the auxiliary structure on the platform is mandatory, as the helideck and boat landing are essential for any offshore structure and their design depends mainly on the information that obtained from operations about the type of helicopter that is used now and will be future. For the design of the boat landing, it is important to define which boats that will be used on a regular basis and their weight.

5.2 Helideck design

The helideck design is critical, as the helideck layout should match the ICAO or CAP requirements, the last edition of these regulations, and the owner's safety policy. After that, the structure analysis is based on the loads that affect the helideck. The transportation, sea fastening, and lifting of the helideck follow the same steps and assumptions as the topside.

5.2.1 Helicopter landing loads

The maximum dynamic local actions from an emergency landing may be determined from the collapse load of the landing gear. This should be obtained from the helicopter's manufacturer.

Alternatively, default values may be used for design by considering an appropriate distribution of the total impact load of 2.5 times the maximum takeoff weight (MTOW).

The local loads used in the design should correspond to the configuration of the landing gear. A single main rotor helicopter may be assumed to land simultaneously on its two main undercarriages or skids. A tandem main rotor helicopter may be assumed to land on the wheels of all main undercarriages simultaneously.

For a single main rotor helicopter, the total loads imposed on the structure should be taken as concentrated loads on the undercarriage centers of the specific helicopter, divided equally between the two main undercarriages. For tandem main rotor helicopters, the total loads imposed on the structure should be taken as concentrated loads on the undercarriage centers of the specific helicopter and distributed among the main undercarriages in the proportion in which they carry the maximum static loads.

Marine Structural Design Calculations. DOI: http://dx.doi.org/10.1016/B978-0-08-099987-6.00005-2

The concentrated undercarriage loads should normally be treated as point loads, but where it is advantageous, a tire contact area may be assumed in accordance with the manufacturer's specification. The MTOW and undercarriage centers for which the platform has been designed and the maximum size and weight of helicopters for which the deck is suitable should be recorded. Information on the dimensions and MTOW of specific helicopters is given in Table 5.1.

Based on CAP 437, the takeoff and landing area should be designed for the heaviest and largest helicopter anticipated to be used at the facility, as shown in Table 5.1. Helideck structures should be designed in accordance with the International Civil Aviation Organization (ICAO) requirements in the *Heliport Manual*, International Standards Organization (ISO) codes for offshore structures, and for a floating installation, the relevant International Maritime Organization (IMO) code. The maximum size and mass of helicopter for which the helideck has been designed should be stated in the installation or vessel operations manual and verification and classification documents.

5.2.1.1 Loads for helicopter landings

The helideck should be designed to withstand all the forces likely to act when a helicopter lands, including

1. **Dynamic loads due to impact landing.** This should cover both a heavy normal landing and emergency landing. For the former, an impact load of 1.5 × maximum takeoff mass (MTOM) of the helicopter should be used. This should be treated as an imposed load, applied together with the combined effects of (2) to (6) in any position on the safe landing area, so as to produce the most severe landing condition for each element concerned. For an emergency landing, an impact load of 2.5 × MTOM should be applied in any position on the landing area together with the combined effects of (2) to (6). Normally, the emergency landing case governs the design of the structure.
2. **Sympathetic response of landing platform.** The preceding dynamic load should be increased by a structural response factor, depending on the natural frequency of the helideck structure, after considering the design of its supporting beams and columns and the characteristics of the designated helicopter. It is recommended that a structure response factor of 1.3 be used, unless further information is available to allow a lower factor to be calculated. Information required to do this include the natural periods of vibration of the helideck and dynamic characteristics of the designated helicopter and its landing gear.
3. **Overall superimposed load on the landing platform.** To allow for snow, personnel, and other live loads in addition to wheel loads, an allowance of 0.5 kN/m^2 should be added over the whole area of the helideck.
4. **Lateral load on landing platform supports.** The landing platform and its supports should be designed to resist concentrated horizontal imposed loads equivalent to 0.5 × MTOM of the helicopter, distributed among the undercarriages in proportion to the applied vertical loading in the direction that will produce the most severe loading on the element being considered.
5. **Dead load of structural members.**
6. **Wind loading.** Wind loading should be allowed for in the design of the platform. This should be applied in the direction that, together with the imposed lateral loading, will produce the most severe loading condition on each element.

Table 5.1 Weights, dimensions and D value for different types of helicopters

Type	*D* value[a] (m)	Perimeter *D* marking	Rotor height (m)	Rotor diameter (m)	Max. weight (kg)	*t* value[c]	Landing net size
Bolkow bo 105D	11.81	12	3.80	9.90	2,300	2.4 t	Not required
Bolkow 117	13.00	13	3.84	11	2,300	3.2 t	Not required
Augusta A109	13.05	13	3.30	11	2,600	2.6 t	Small
Dauphin SA365N2	13.68	14	4.01	11.93	4,250	4.3 t	Small
Sikorsky S76B&C	16.00	16	4.41	13.40	5,307	5.3 t	Medium
Bell 212	17.46	17	4.80	14.63	5,080	5.1 t	Not required
Super Puma AS 332L2	19.50	20	4.92	16.20	9300	9.3 t	Medium
Super Puma AS 332L	18.70	19	4.92	15	8,599	8.6 t	Medium
Bell 214ST	18.95	19	4.68	15.85	7,936	8.0 t	Medium
Sikorsky S61N	22.20	22	5.64	18.90	9,298	9.3 t	Large
EH101	22.80	23	6.65	18.60	14,600	14.6 t	Large
Boeing BV234LR Chinook[b]	30.18	30	5.69	18.29	21,315	21.31 t	Large

Note: For skid-fitted helicopters, the maximum height may be increased if ground-handling wheels have been fitted.
[a]The *D* value is the largest overall dimension of the helicopter when rotors are turning. This dimension normally is measured from the most forward position of the main rotor tip path plane to the most rearward position of the tail rotor tip path plane or the most rearward extension of the fuselage in the case of Fenestron or Notar tails. The *D* circle is a circle, usually imaginary unless the helideck itself is circular, the diameter of which is the *D* value of the largest helicopter the helideck is intended to serve.
[b]The BV234 is a tandem rotor helicopter; and in accordance with ICAO, the helicopter size is 0.9% of the helicopter *D* value, that is, 27.16 m.
[c]The notation t indicates the allowable helicopter weight in tons.

7. **Punching shear.** A check should be made for the punching shear from an undercarriage wheel with a contact area of $65 \times 103\ mm^2$ acting in any probable location. Particular attention to detail should be taken at the junction of the supports and the platform deck.

5.2.1.2 Loads for helicopters at rest

The helideck should be designed to withstand all the applied forces that could result from a helicopter at rest. The helideck components should be designed to resist the following simultaneous actions in normal landing and at-rest situations:

- Helicopter static loads (local landing gear, local patch loads).
- Area load.
- Helicopter tie-down loads, including wind loads from a secured helicopter.
- Dead loads.
- Helideck structure and fixed appurtenances self-weight.
- Wind loading.
- Installation motion.

5.2.1.3 Helicopter static loads

All parts of the helideck accessible to a helicopter should be designed to carry an imposed load equal to the MTOW of the helicopter. This should be distributed at the landing gear locations in relation to the position of the center of gravity of the helicopter, taking into account different orientations of the helicopter with respect to the installation.

5.2.1.4 Area load

To allow for personnel, freight, refueling equipment and other traffic, snow and ice, rotor downwash, and other loads, a general area load of $2.0\ kN/m^2$ should be included.

5.2.1.5 Helicopter tie-down loads

Each tie-down should be designed to resist the total wind load imposed by "100-year storm" winds on the helicopter.

Sufficient flush-fitting tie-down points should be provided for securing all the helicopter types for which the landing area is designed. They should be located and be of such construction as to secure the helicopter when subjected to weather conditions of a severity unlikely to be exceeded in any one year. They should also take into account any recommendations made by the aircraft manufacturer, and where significant, the inertial forces resulting from movement of floating platforms. Figure 5.1 presents a suitable tie-down configuration.

5.2.1.6 Wind loading

Wind loading on the helideck structure should be applied in the direction that, together with the horizontal imposed loading, will produce the most severe loading condition for the element considered. Consideration should also be given to the additional wind loading from a secured helicopter, as noted previously.

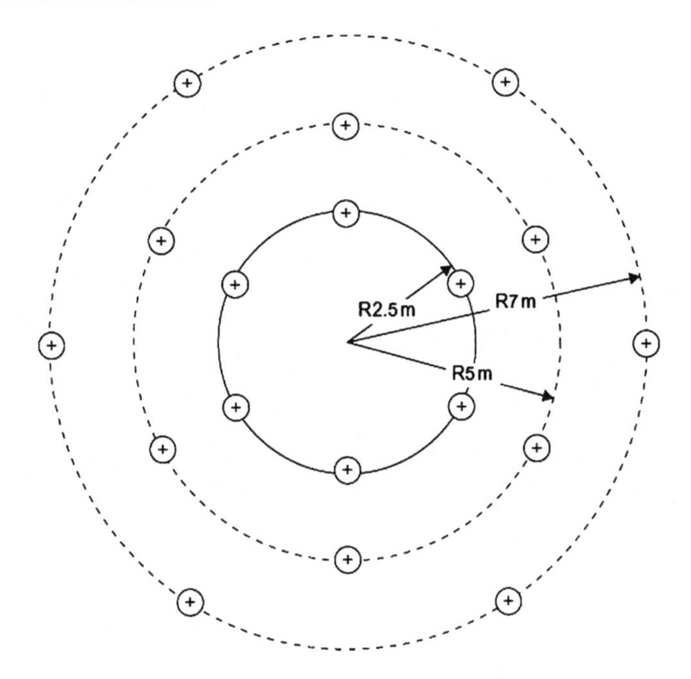

Figure 5.1 Tiedown location.

5.2.1.7 Installation motion

The effect of acceleration forces and other dynamic amplification forces arising from the predicted motions of the installation under the 100-year storm condition should be considered.

5.2.2 Safety net arms and framing

Safety nets (Figure 5.2) for the protection of personnel should be installed around the landing area except where structural protection exists. The netting used should be flexible and manufactured from nonflammable material, with the inboard edge fastened level with, or just below, the edge of the helicopter landing deck. The net itself should extend at least 1.5 m in the horizontal plane and be arranged so that the outboard edge is slightly above the level of the landing area but not by more than 0.25 m, so that it has an upward and outward slope of at least 10°. The supporting structure associated with the safety net should be capable of withstanding, without damage, a 75 kg weight being dropped from a height of 1 m onto an area of 0.25 m^2. The net should be strong enough to withstand and contain, without damage, a 100 kg weight being dropped from a height of 1 m.

Based on the API RP2L Recommendation, the safety net designed to meet these criteria should not act as a trampoline, giving a "bounce" effect. Where lateral or

Figure 5.2 Main configuration of a safety net.

longitudinal center bars are provided to strengthen the net structure, they should be arranged and constructed to avoid causing serious injury to persons falling on to them. The ideal design should produce a “hammock” effect, which should securely contain a body falling, rolling, or jumping into it without serious injury. When considering securing of the net to the structure and the materials used, care should be taken that each segment meets adequacy of purpose considerations. Polypropylene deteriorates over time; various wire meshes have been shown to be suitable if properly installed.

The entire helideck, including any separate parking or runoff area, should be designed to resist an imposed load equal to the MTOM of the helicopter. This load should be distributed among all the undercarriages of the helicopter. It should be applied in any position on the helicopter platform so as to produce the most severe loading condition for each element considered.

The values for these overall superimposed loads, dead loads, and wind loads should be considered to act in combination with the dynamic load impact, as discussed previously. Consideration should also be given to the additional wind loading from any parked or secured helicopter.

Based on the American Petroleum Institute (API) RP2L design for heliports, the flight deck, stiffeners, and supporting structure should be designed to withstand the helicopter landing load encountered during an exceptionally hard landing, after power failure while hovering. Examples of the helicopter parameters that should be obtained are shown in Table 5.2, and it is recommended that parameters similar to those given in the table be obtained from the manufacturer of any helicopter considered in the helideck design.

The maximum contact area per landing gear used to design deck plate bending and shear should conform to the manufacturer’s values given in Table 5.2. For

Table 5.2 Technical parameters for helicopters, as per API

Manufacturer model	Common name	Gross weight, kg	Rotor diameter, m	Overall length, m	Type	Number		Contact area, cm^2		Percentage of gross weight		Distance between fore and aft gears, m	Width between gears, m
						Fore	Aft	Fore	Aft	Fore	Aft		
Aerospatiale													
315B	Lama	2305	11.0	12.9	Skid					38	62		2.4
316B		2205	11.0	12.9	Wheel	1	2	297	594	28	72	3.1	2.6
318C	Alouette II	1656	10.2	12.1	Skid								2.3
319B	Alouette III	2250	11.0	12.9	Wheel	1	2	297	594	34	66	4.1	2.6
330J	Puma	7400	15.1	18.2	Wheel	2	4	1200	2142	36	64	5.3	3.0
332L	Super Puma	8351	15.6	18.7	Wheel	2	2	465	735	40	60	4.5	3.0
332C	Super Puma	8351	15.6	18.7	Wheel	2	2	465	735	33	67		2.0
341G	Gazelle	1800	10.5	12.0	Skid					51	49		2.1
350B/d	ASTAR	1950.5	10.7	13.0	Skid					51	49		2.1
355F	Twin Star	2305	10.7	13.0	Skid								
360	Dauphin	2799	11.5	13.4	Wheel	2	1					7.2	2.0
360C		2994	11.5	13.4	Wheel	2	1	213	123	84	16	3.32	2.4
360C		2994	11.5	13.4	Skid					84	16	3.32	2.3
365C		3401	11.5	13.4	Wheel	2	1	213	123	84	16	3.32	2.4
365C		3401	11.5	13.4	Skid								2.3
365N	Dauphin 2	3850	11.9	13.5	Wheel	2	2	245	426	22	78	3.6	2.0
Augusta/ Atlantic													
A109	Hirando	2450	11.0	13.1	Wheel	1	2	129	129			3.5	2.3
A-19A	Mark II	2600	11.0	13.1	Wheel	1	2	46	284	23	77	3.5	2.5
Bell Helicopter													
47G		1338	11.6	13.3	Skid			174	174			1.6	2.3
205A-1		4309	14.7	17.4	Skid			310	310			2.3	2.7

(Continued)

Table 5.2 **(Continued)**

Manufacturer model	Common name	Gross weight, kg	Rotor diameter, m	Overall length, m	Type	Number		Contact area, cm^2		Percentage of gross weight		Distance between fore and aft gears, m	Width between gears, m
						Fore	Aft	Fore	Aft	Fore	Aft		
206B	Jet Ranger	1451	10.2	12.0	Skid			174	174	19	81	1.4	1.8
206L	Lone Ranger	1882	11.3	13.0	Skid			174	174	29	71	2.1	2.2
212	Twin	5080	14.6	17.5	Skid			310	310	22	78	2.3	2.5
214B	Big Lifter	7257	15.2	19.0	Skid								2.6
214ST	Super Transport	7938	15.9	19.0	Skid			319	319	22	78	2.3	2.5
214ST	Super Transport	7938	15.9	19.0	Wheel	2	2	247	581	22	78	4.8	2.8
222		3561	12.1	14.5	Wheel	1	2	122	410	19	81	3.7	2.8
222B		3742	12.8	15.3	Wheel	1	2	123	413	19	81	3.7	2.8
222UT		3742	12.8	15.3	Skid			310	310	32	68	2.4	2.4
412		5262	14.0	17.1	Skid			310	310	20	80	2.4	2.5
Boeing vertol													
BO105C		2300	9.8	11.8	Skid								2.6
BO105CBS		2400	9.8	11.9	Skid			181	181	36	64		2.5
BK117		2850	11.0	13.0	Skid			206	206	34	66		2.5
234		21,900	18.3	30.2	Wheel	4	2	2529	1600	58	42	7.9	3.4
CH47234		22,680	18.3	30.2	Wheel	4	2	1007	503			6.9	3.4
107-11		10,030	5.2	25.3	Wheel	2	4	323	323			2.6	3.9
179		8482	14.9	18.1	Wheel	2	2	1058	529			4.7	2.7

Fairchild													
FH-1100		1247	10.8	12.7	Skid								2.2
Hiller													
UH-12L4		1406	10.8	12.4	Skid								2.3
UH12E/E4		1270	10.8	12.4	Skid								2.3
Hughes													
269A/B		758	7.7	8.8	Skid								2.0
269C		930	8.2	9.4	Skid			71	71	41	59		2.0
369HS		1158	8.0	9.2	Skid								2.1
369D		1361	8.0	9.3	Skid			194	242	33	67		2.1
Sikorsky													
S55T		3266	16.2	19.0	Wheel	2	2	258	258			3.2	3.4
S58T		5897	17.1	20.1	Wheel	2	1	1032	290	88	12	8.6	3.7
S61NL		9299	18.9	22.3	Wheel	2	1	1497	277	85	15	7.2	4.3
S62		3583	16.2	19.0	Wheel	2	1	697	348	87	13	5.4	3.7
S64		19,050	22.0	27.0	Wheel	1	2					7.4	6.0
S65C		19,050	22.0	26.9	Wheel	2	4	994	994	25	75	8.2	4.0
S76		4672	13.4	16.0	Wheel	1	2	123	310			5.0	2.4
S78C		9072	16.4	19.8	Wheel	2	1	471	471			8.8	2.7

multiwheeled landing gear, the value of the contact area is the sum of the areas for each wheel. The contact area for float or skid landing gear is the area of the float or skid around each support strut. The load distribution per landing gear, in terms of percentage of gross weight, is given in Table 5.2.

The design landing load is the landing gear load based on a percent of a helicopter's gross weight times an impact factor of 1.5. For percentages and helicopter gross weight, see Table 5.2.

5.3 Design load conditions

The heliport should be designed for at least the following combinations of design loads:

1. Dead load plus live load.
2. Dead load plus design landing load. If icing conditions are prevalent during normal helicopter operations, superposition of an appropriate live load should be considered.
3. Dead load plus live load plus wind load.

Most classification societies with jurisdiction in the U.K. sector of the North Sea have classification notes available that include helideck specifications. Because the main duties of some of these authorities is primarily classification of floating structures, the helideck specifications generally only appear within rules for mobile offshore units and the like. Two, however, Lloyd's Register of Shipping and Det Norske Veritas, have adopted identical specifications for fixed platforms.

The authorities relevant to this current investigation are

- American Bureau of Shipping (ABS).
- Bureau Veritas (B.V.).
- Det Norske Veritas (DNV).
- Germanischer Lloyd (GL).
- Lloyd's Register of Shipping (LRS).

Tables 5.3 and 5.4 summarize the loading requirements for the most regular specifications. Table 5.3 is concerned with the load specification during helicopter landings while Table 5.4 relates to helicopters in the stowed position. From both tables, it is seen that a considerable variation of requirements exists between all the specifications with variations particularly on

- The factor on MTOW (M) for emergency landing conditions.
- Whether a deck response factor is considered.
- The level of superimposed load considered simultaneously or separately.
- Whether a lateral load is considered simultaneously with the emergency-landing load.

Example 5.1

The helicopter loads presented in Table 5.3 are used here for the skid loading model. In the SACS input loading menu for a Bell-212 helicopter, the MTOW = 50 kN (5 tons).

Table 5.3 Comparison among authorities for helicopter landing loading specifications

	Authority							
	ISO	**CAP**	**HSE**	**ABS**	**B.V.**	**DNV**	**GL**	**LRS**
Heavy landing	—	1.5*M*						
Emergency landing	2.5*M*	2.5*M*	2.5*M*	1.5*M*[a]	3.0*M*	2.0*M*	1.5*M*	1.5*M*[b] 2.5*M*[c]
Deck response factor	1.3	1.3[d]	1.3[d]	—	—	—	—	—
Superimposed load, kN/m^2	0.5	0.5	0.5	2.0[e]	2.0[e]	As normal class	0.5	—[b] 0.2[c]
Lateral load	0.5*M*	0.5*M*	0.5*M*	—	—	0.4*M*	—	0.5*M*
Wind load	Max. operation			Normal design		$V_w = 30$ m/s	$V_w = 25$ m/s	—

Note: *M* is the maximum takeoff weight; V_w is the wind velocity.
[a]Or manufacture's recommend wheel impact loads.
[b]For design of plating.
[c]For design of stiffening and supporting structure.
[d]Additional frequency-dependent values given for Chinook helicopter.
[e]Considered independently.

Table 5.4 Comparison among authorities for helicopter at-rest loading specifications

	Authority							
	ISO	**CAP**	**HSE**	**ABS**	**B.V.**	**DNV**	**GL[a]**	**LRS**
Self-weight	M	M	M	M	M	M	1.5 M	M
Superimposed load, kN/m^2	2.0	0.5	0.5	0.49	0.5	As normal class	2.0	2.0
Wind load	100-year storm	As HSE	As HSE	Normal design	Normal design	$V_w = 55$ m/s	$V_w = 50$ m/s	—

Note: M is the maximum takeoff weight; V_w is the wind velocity.
[a]Fixed platforms; for floating platforms, also include lateral load of $0.6(M + W)$, where W is the deck weight in place of platform motion.

The helicopter has two skids. Each skid's geometrical dimensions are assumed to be 3.68 (L) × 2.856 (W) meters, with the center of gravity (CoG.) point of application at the middle of the skid. The height of the helicopter fuselage is taken as 2 m.

1. CAP-437 imposed live load = 0.5 kN/m^2.
2. At-rest condition = 1 × MTOW = 50.0 kN.
3. Normal operating condition = 1.5 × 1.3 × MTOW = 97.50 kN.
4. Emergency landing condition = 2.5 × 1.3 × MTOW = 162.50 kN.

The landing conditions, either normal operating or emergency, are combined with lateral horizontal force = 0.5 × MTOW = 25.0 kN applied on both skids.

5.3.1 Helideck layout design steps

The loads affecting the helideck are discussed in previous sections.

- Helidecks are designed in accordance with API RP2A and API RP2L.
- Layout of the helideck should be sufficient for one helicopter.
- One stair for primary access and a ladder for secondary access are required.
- Safety netting should surround the helideck completely, at a minimum width of 1500 mm.
- Paint markings, sizes, and colors are in accordance with API RP2L.

The helideck may have a separate platform and may be connected to the main platform by a bridge.

The helicopter landing area should be clear, with no obstacle, and for obstacles below the landing area, there are some limits from CAP437, as shown in Figure 5.3.

According to CAP437, the following dimensions should be considered. For the helideck markings, the color of the helideck should be dark green or gray, as shown in Figure 5.4. In addition, the perimeter should be clearly marked with a white painted line 300 mm wide and also presents the dimensions of the yellow circle and the H marking, painted white, on the helideck.

Based on CAP 437, the helideck netting dimension depends on the landing net size, which, in turn, depends on the helicopter type, the D value and the maximum weight (see Table 5.5).

The helicopter should be tied to the helideck. This tie-down point should be located and be of such strength and construction to secure the helicopter when subjected to weather conditions pertinent to the installation design considerations.

The tie-down rings should match the tie-down strap attachments. Note that the maximum bar diameter of the tie-down ring should be 22 mm in order to match the strap hook dimension of the tie-down strap in most helicopters.

Advice on the recommended safe working load should be obtained from the helicopter operator.

Note that the outer circle is not required if the D value is less than 22.2 m. The centers of all the circles coincide with the center of the marking circle.

Based on CAP437, the dimensions in Table 5.6 should be considered.

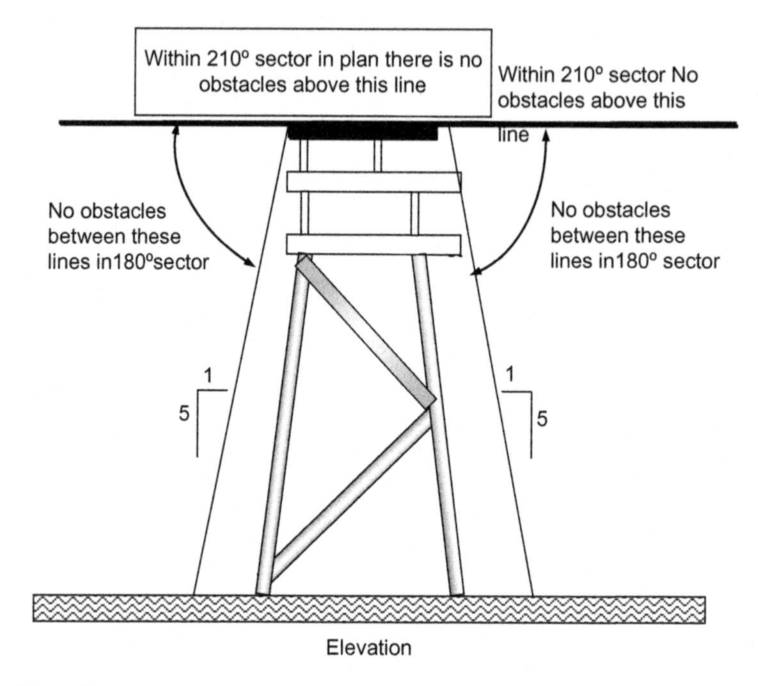

Figure 5.3 Helideck marking specifications and dimensions.

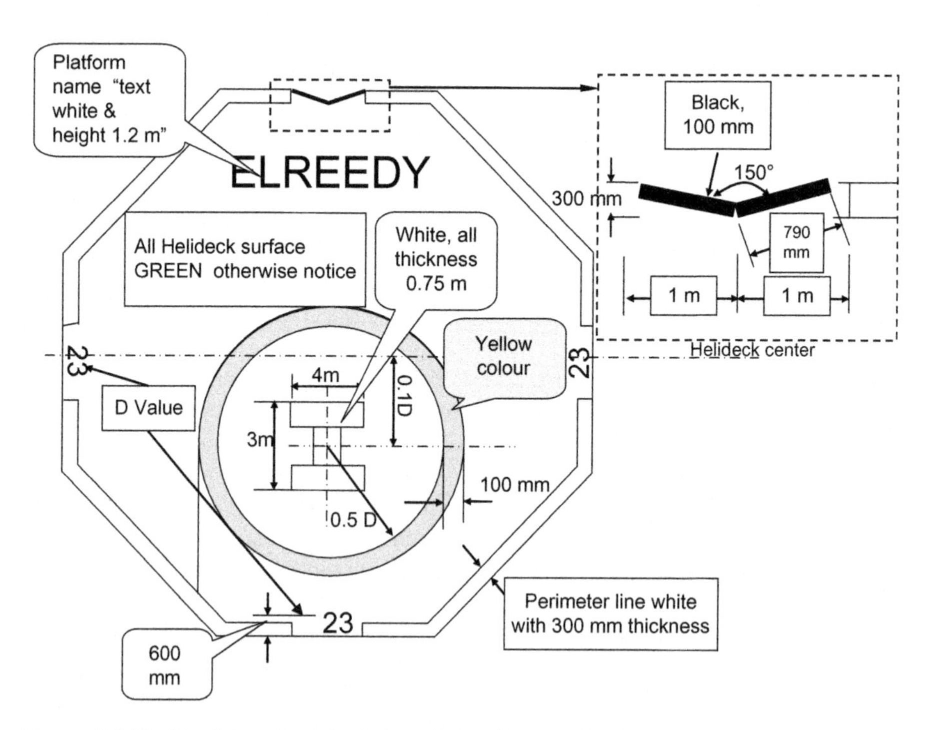

Figure 5.4 Limits of free obstacles below the landing area level.

Table 5.5 **Weights, dimensions, and D values for different types of helicopters, as per CAP437**

Type	*D* value, m	Perimeter *D* marking	Max. weight, tons	*t* value	Landing net size	Landing minimum size, m
Bolkow bo 105D	11.81	12	2.300	2.4 t	Not required	—
Bolkow 117	13.00	13	2.300	3.2 t	Not required	—
Augusta A109	13.05	13	2.600	2.6 t	Small	9 m × 9 m
Dauphin SA365N2	13,68	14	4.250	4.3 t	Small	9 m × 9 m
Sikorsky S76B&C	16	16	5.307	5.3 t	Medium	—
Bell 212	17.46	17	5.080	5.1 t	Not required	—
Super puma AS 332L2	19.50	20	9.300	9.3 t	Medium	12 m × 12 m
Super Puma AS 332 L	18.70	19	8.599	8.6 t	Medium	12 m × 12 m
Bell 214ST	18.95	19	7.936	8.0 t	Medium	12 m × 12 m
Sikorsky S61N	22.20	22	9.298	9.3 t	Large	15 m × 15 m
EH101	22.80	23	14.600	14.6 t	Large	15 m × 15 m
Boeing BV234LR Chinook[a]	30.18	30	21.315	21.31 t	Large	15 m × 15 m

Note: With a skid-fitted helicopter, the maximum height may be increased with ground handling wheels fitted.
[a]The BV234 is a tandem rotor helicopter and in accordance with ICAO the helicopter size is 0.9 of the helicopter D value, that is, 27.16 m.

Table 5.6 **Helicopter deck netting standard dimensions**

Small	9 × 9 m
Medium	12 × 12 m
Large	15 × 15 m

5.3.2 Plate thickness calculation

If the helicopter has tires, calculation is needed to check the plate thickness based on Lloyd's Register of Shipping, *Rules and Regulations for the Classification of Fixed Offshore Installations*, Part 4, "Steel Structures December 1989," Chapter 2, "Topside Items and Fittings," Section 2", Helicopter Landing Area." The following example presents these equations.

Example 5.2

A Sikorisky model S-76 helicopter with a single main rotor has the following input data:

Maximum helicopter weight, $P = 4.670$ tons
Contact area per wheel (main gear), $A = 15484\ \text{mm}^2$
Panel short dimension, $s = 914$ mm
Panel long dimension, $1 = 4572$ mm
Steel yield strength, $F_y = 248$ MPa
Tire print dimension, $u = 176$ mm
Tire print dimension, $v = 88$ mm
Tire print dimension ratio, $u/v = 2.0$
Manned spaced factor, $f = 1$
Location factor, $g = 0.6$

The calculation is as follows:

Higher steel tensile factor, $k = 245/F_y = 0.987$
Lesser of v and s, $v_1 = 88$ mm
Patch aspect ratio correction factor, $\phi_1 = (2v_1 = 1.1s)/(u + 1.1s)$
$\phi_1 = 1$ In this example
Patch aspect ratio correction factor; $\phi_2 = 1$ if $u \leq (a - s)$
$= 1/[1.3 - (0.3/s(a - u)]$ if $u \leq a$
$= 0.77a/u$ if $u > a$
$\phi_2 = 1$ In this example
Patch aspect ratio correction factor, $\phi_3 = 1$ if $v/s < 1$
$\phi_3 = 0.6\ s/(v + 0.4)$ if $v/s < 1.5$
$\phi_3 = 1.2\ s/v$ if $v/s \geq 1.5$
$\phi_3 = 1$ in this example
Landing load on tire print, $P_w = P/2 = 2.335$ tons
Corrected patch load, $P_1 = 2.5\phi_1\phi_2\phi_3 f\, g\, P_w = 3.503$ tons
Tire print coefficient; $\beta = \log P_1\ (k)^2 \times 107/(s)^2$

By knowing the v/s ratio from Table 5.7 the thickness coefficient α, is obtained.

Table 5.7 **Lookup table for α (thickness coefficient)**

β	v/s										
	0.0	**0.1**	**0.2**	**0.3**	**0.4**	**0.5**	**0.6**	**0.7**	**0.8**	**0.9**	**1.0**
0.8	5	5	4	4	x	x	x	x	x	x	x
1.0	6	6	5	5	5	4	4	x	x	x	x
1.2	8	7	7	6	6	5	5	5	4	4	X
1.4	10	9	8	8	7	7	6	6	5	5	5
1.6	12	11	10	9	8	8	7	7	6	6	5
1.8	15	13	12	11	10	9	9	8	7	7	6
2.0	18	16	15	13	12	11	10	9	9	8	7
2.2	23	20	18	16	14	13	12	11	10	9	9
2.4	30	26	23	20	18	16	14	13	12	11	10
2.6	37	33	29	26	22	20	18	16	15	14	13
2.8	x	40	36	32	28	25	23	20	18	17	15
3.0	x	x	x	40	36	32	29	26	24	21	19
3.2	x	x	x	x	x	40	36	33	30	28	25
3.4	x	x	x	x	x	x	x	40	38	35	32
3.6	x	x	x	x	x	x	x	x	x	x	40

The thickness coefficient, $\alpha = 11$
The deck plate thickness $= [\alpha s\,(k)^{0.5}/100] + 1.5$ (corrosion allowance) $= 11.5$ mm

5.3.3 Aluminum helideck

A new trend is to use an aluminum helidecks, and they are used successfully in the Middle East and on the some barges. The main benefit of this type of helideck is its weight, which is much than conventional steel, about 40–60% less than the steel weight. This load saving is essential for mature marine structures.

The aluminum helideck remains free from corrosion, so it is cheap to maintain, with a long life. In addition, as the structures mature, they carry a reduced load and gain a lot in structural integrity.

5.4 Boat landing design

The boat landing is designed mainly to handle the impact load from the vessel or the boat to the offshore structure. To absorb the impact load, a fix a fender is usually attached to the boat landing, which is a car tire or a special tire. The connection between the boat landing frame structure and the platform jacket is through a shock absorber, like ae piston. Figure 5.5 presents a boat landing and its connection to the offshore structure leg. The connection between the boat landing and the leg is shown in Figure 5.6, and Figure 5.7 presents the chock cell.

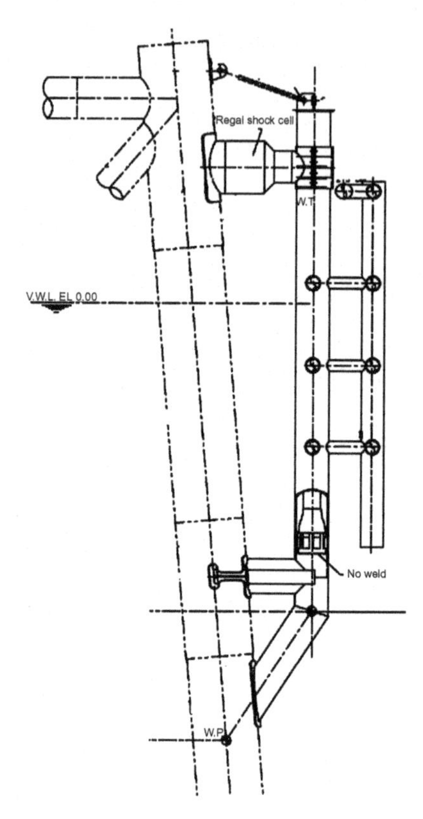

Figure 5.5 Boat landing support.

The barge bumpers and their associated connections to the jacket should be designed for a loading in which the vessel impact is directly in the middle one-third height of the post. The energy to be absorbed in the system is 560 kJ. Barge bumpers aree designed to be easily replaced in case of damage. Details should consider jacket elevation tolerances during the detailed design phase. In lieu of specific data, a minimum installation tolerance on the jacket elevation of $\pm$ 0.3 m is adopted.

The following should be considered regarding stab-in guides and bumpers:

1. The aids should be designed such that they fail prior to permanent deformation of any part of the permanent structure. The permanent structural members should be designed such that they can withstand significantly more load than the aids.
2. Any deflections must be within the elastic limit of the material.
3. A 33% overstress increase in allowable member stresses is permitted.

5.4.1 Boat landing calculation

The transfer of loads from the vessel to the boat landing is presented by the following example.

Figure 5.6 Connection between the leg and the boat landing.

Figure 5.7 The shock cell.

Example 5.3

We want to calculate the collision force based on a 1000-ton boat impacting at 0.5 meter per second velocity on a landing using a Regal shock cell model SC1830. Note that we must define the force that applies to the boat landing frame then start to calculate the stresses on the frame using software to start the design.

$m = (1000 \times 9810) \times (9.81 \times 1000) = 1100 \text{ N sec}^2/\text{mm}$
$\nu = \text{approaching velocity} = 500 \text{ mm/s}$
$E = ½ \times 1100 \times (500)^2 = 13.75 \times 10^7 \text{ N mm}$

From the shock cell type, choose the suitable curve for the relation between the energy versus deflection curve, as shown in Figure 5.8:

$\delta = 360 \text{ mm for } E = 13.75 \times 10^7 \text{ N mm}$

The force versus deflection is based on the model type curve shown in Figure 5.9:

$F = 70 \text{ tons} = 686.6 \text{ kN}$

Some of the energy is absorbed by the vessel. (Assume 30% is absorbed by the vessel and 70% is absorbed by the structure.)

$E = 13.75 \times 10^7 \times 0.7 = 9.63 \times 10^7 \text{ N mm}$
$\delta = 300 \text{ mm}$
$F = 510 \text{ kN}$

This force is concentrated at one point because collision occurs on a considerable area, depending on the dimensions of vessel and its position during impact, and because the fender system distributes the load.

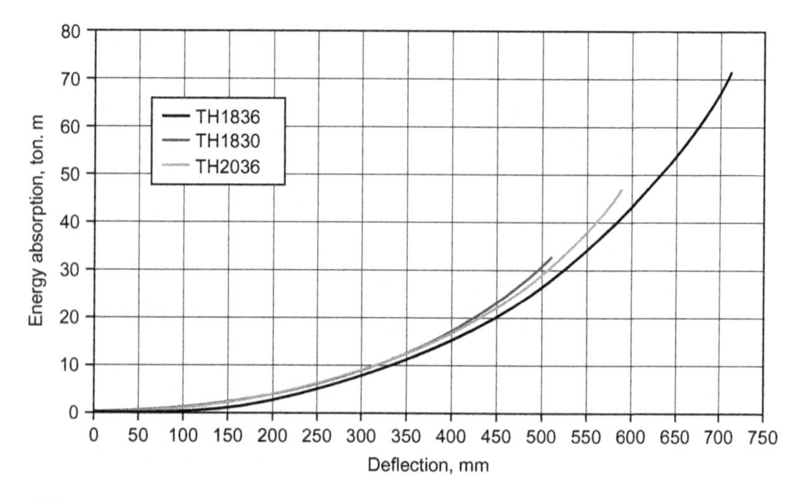

Figure 5.8 Energy absorption versus deflection.

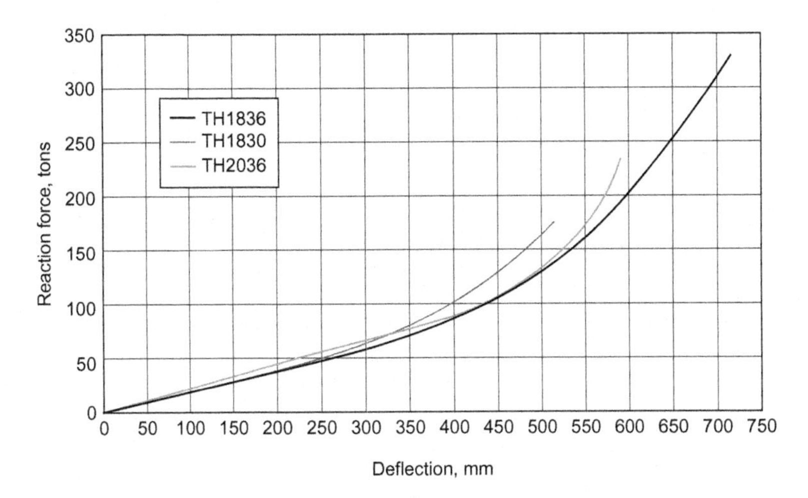

Figure 5.9 Reaction force versus deflection.

5.4.1.1 *Cases of impact load*

$F = 9.81 \times 10^5$ N
$L = 5560$ mm (assumed)

In Case 1, a uniform load is at midspan at an elevation of +300, as in Figure 5.10(a):

Uniform load = $F/L = 510/6000 = 85$ kN/m′

In Case 2, The uniform load at mid-span is at an elevation of −900. The same values apply as in Case 1 but at level −900.

In Case 3, the uniform load at mid-span is at elevations of +300 and −900, as shown in Figure 5.10(b):

Uniform load = $0.5F/L = 42.5$ kN/mm′

In Case 4, assume the impact load is distributed as a concentrated load at mid-span at elevations of +300 and −900 (six concentrated loads):

Force at each joint = $F/6 = 510/6 = 85$ kN

5.4.2 Boat landing design using a nonlinear analysis method

Boat landings and riser guards have been studied using a nonlinear analysis method that depends on the strain and denting that would occur on the boat landing member. This analysis is usually performed by special nonlinear analysis software to reduce the member size, which reduces the wave load effect on the platform.

A nonlinear analysis can be performed to study the behavior of the platform due to the impact of a 3500 ton vessel with 1 knot velocity. This was done by calculating the impact energy and analyzing the model with software. By resizing the boat

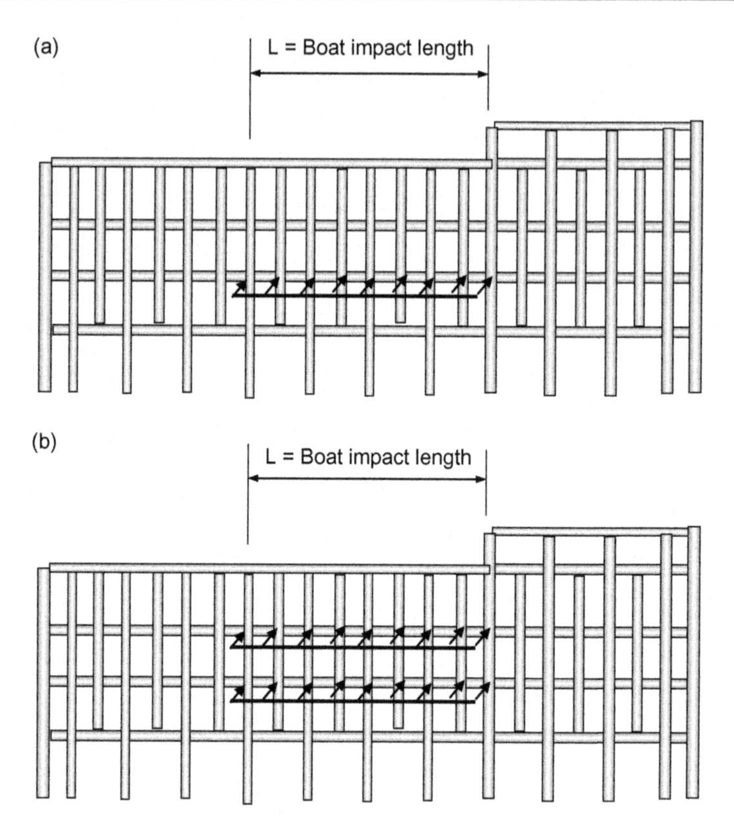

Figure 5.10 (a) Boat impact length; (b) boat impact length.

landing members, the impact energy was absorbed and the results show the impact force and members with plastic hinges.

In addition to that, a dynamic boat impact analysis can be carried also by software. The comparison between these two methods shows that dynamic boat impact and static boat impact analysis give us a similar impact force for a fixed offshore jacket and a static boat impact analysis is sufficient for the jacket boat impact analysis.

5.4.3 Boat impact methods

The static method utilizes the impact energy to calculate an impact load that is incremented until the impact energy has been dissipated. The fracture control is used in the software is to monitor the strain in the members as energy is progressively applied. When the specified strain of 15% is reached in a member, that member is removed and the loading is redirected to other members. If no other load path can be found, the analysis is terminated.

In the dynamic method, the ship is modeled as a mass point connected to the platform through a nonlinear spring. The mass representing the ship is given an initial velocity corresponding to the impact speed, and the analysis is carried out as a free vibration problem. The ship force unloads once the spring starts to elongate; that is, the ship and vessel go away from each other. When the contact force has vanished, the ship mass is disconnected from the model. The dynamic impact analysis is a nonlinear, step-by-step analysis.

For dynamic analysis, the vessel mass and velocity are modeled and analyzed. At first, the results of the analysis show a higher (20–30%) impact load in dynamic impact analysis than the static impact condition. By referring to the static impact results, it can be found that a major portion of the impact energy is absorbed by the denting of the tubular member, and the software usually automatically models the denting progress of a pipe during a static impact.

The effect of denting in a dynamic boat impact analysis is very important. As in an impact between two bodies, the duration of the impact directly affects the impact load. Which mean that if the impact time increase the impact force will reduce. Denting behavior is like a soft spring and makes the impact condition a "soft impact" case so it reduces the impact load.

After the modeling the denting springs, the impact loads in the dynamic analysis are reduced and are slightly around 5% less than static impact load. In some cases, the impact analysis was done for 12-m and 55-m water depth jackets and, after modelling the denting, the dynamic impact analysis gives us impact loads similar to the static impact analysis.

5.4.4 Tubular member denting analysis

API presents a formula for modeling a denting spring. According to the API RP2A WSD 21st edition, we have

$$P_d = 40F_Y t^2\ (X/D)^{0.5} \tag{5.1}$$

where P_d is the denting force; X is the denting depth; F_y, t, and D are the yield stress, wall thickness, and diameter of the denting member, respectively.

The simplified dynamic impact model follows. Denting a tubular member is a complicated behavior, and few formulas describe the relation between denting force, denting depth, and denting energy, some of which are presented in the API RP2A standard. The critical zone with high stress and strain is shown in Figure 5.11.

For a simplified calculation of the strain at the critical zone of the oval, we need to find the geometry and the minimum radius of the oval. The main assumption is "the circumference of the tubular circle and deformed ellipsis are similar." After finding the oval minimum radius, the strain can be checked against a 15% strain to calculate the denting limit. Calculation of the simplified denting method is presented in the following subsection.

This method is valid only for boat landing members with bending behavior. For checking the accidental impact on jacket members with axial loads, the simplified method is not valid and a more detailed analysis need to be done.

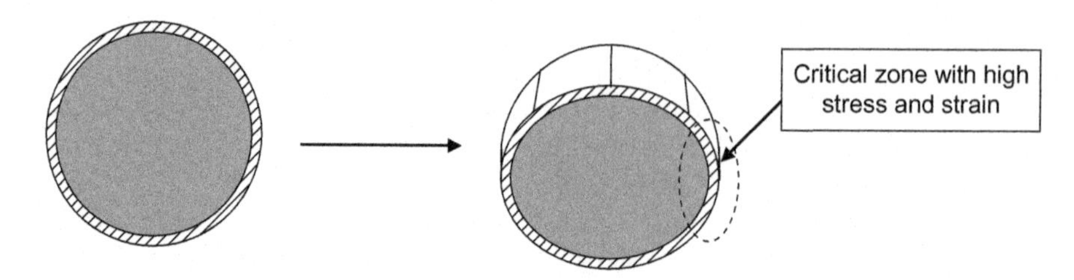

Figure 5.11 Critical zone for member being dented.

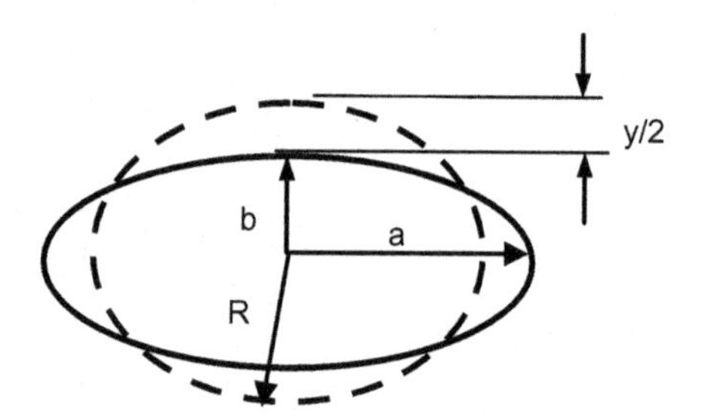

Figure 5.12 Calculation parameters.

5.4.4.1 *Simplified method for denting limit calculation*

The circle perimeter is as in Figure 5.12:

$$P_c = 2\pi R \tag{5.2}$$

The ellipse parameter has a different equation, and the following equation has more accuracy:

$$P_e \cong 2\pi\left(\frac{a^{1.5}+b^{1.5}}{2}\right)^{2/3} \tag{5.3}$$

Noting that, in case of denting, the cylinder section the perimeter of the circle is the same after denting but shaped like the ellipse in Figure 5.13.

$$\begin{aligned} 2\pi R &= 2\pi\left(\frac{a^{1.5}+b^{1.5}}{2}\right)^{2/3} \\ a &= (2R^{1.5}-b^{1.5})^{2/3} \end{aligned} \tag{5.4}$$

where y is the dent distance,

$$b = R - y/2 \tag{5.5}$$

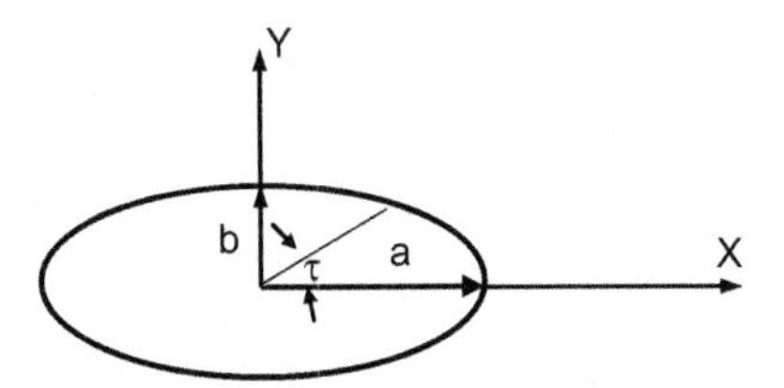

Figure 5.13 Obtaining a and b.

By knowing the dent distance and the radius of the section, R, we can obtain b and a from equations (5.4) and (5.5), respectively.

By assuming that coordinate system has the origin at ellipse's center,

$$\frac{x^2}{a^2}+\frac{y^2}{b^2}=1$$

We need the radius of curvature at $(x, y) = (a, 0)$. This is actually a question that is found using calculus. The radius of curvature is

$$R=\frac{[(x')^2+(y')^2]^{1.5}}{x'y''-y'x''} \tag{5.6}$$

where the x and y coordinates can be parameterized as

$$x(\tau)=a\cos(\tau),\ y(\tau)=b\sin(\tau)$$
$$x'(\tau)=-a\sin(\tau),\ y'(t)=b\cos(\tau)$$
$$x''(\tau)=-a\cos(\tau),\ y''(\tau)=-b\sin(\tau)$$

and plugging these into the expression for R gives us

$$R=\frac{[a^2\sin^2\tau+b^2\cos^2\tau]^{1.5}}{ab[\sin^2\tau+\cos^2\tau]}$$

$$\text{at } \tau=0\ R_{\min}=b^2/a$$

Figure 5.14 presents the strain ε that happens in the tubular section due to denting:

$$\varepsilon=\frac{\phi t/2}{R_1\theta/2}=\frac{t}{R_1}\cdot\frac{\phi}{\theta}$$

$$R_1\theta=2(\theta/2+\phi)R_2$$

$$\phi=\theta/2\left(\frac{R_1}{R_2}-1\right) \tag{5.7}$$

$$\varepsilon=\frac{t}{2R_1}\left(\frac{R_1}{R_2}-1\right)$$

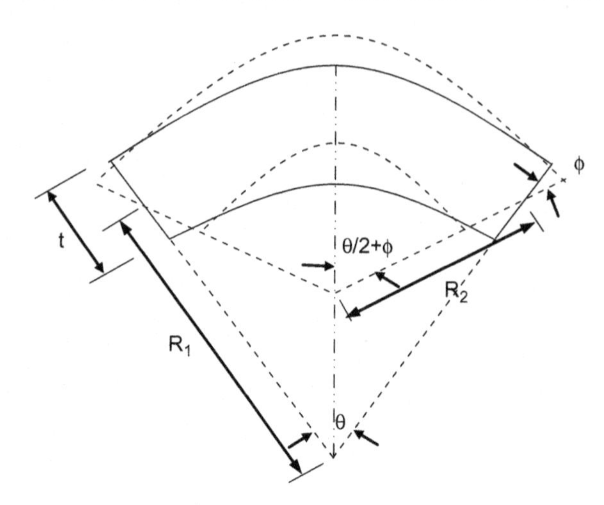

Figure 5.14 Strain to the denting member.

5.5 Riser guard

Riser guards consist of a tubular steel space frame provided to the jacket face between elevations +2.5 m (LAT) and −2.5 m (LAT) to protect the riser from boat collisions or any accidents that may occur. The riser guard is designed to resist a collision of an equivalent static force acting anywhere on the frame.

The said equivalent static force has to be defined by the client or by the engineer and approved by the client. This force has to be mentioned in the project premises specification.

A static in-place analysis of the riser guard is performed using a structural computer program. The model geometry should include all the structural members of the riser guard.

The structural dead loads are generated by the computer program based on geometry and member properties.

Several load cases are investigated in the analysis for the equivalent static force. For each case, the program calculates reactions, deformations, member forces, and checks all members for compliance with AISC Code. A 33% stress increase is allowed for all load cases.

5.5.1 Riser guard design calculation

The riser guard design is the same as the boat landing design but it need not have a shock cell, as the boats rarely collide into the riser guard. So, the design is based on the member of the riser guard reaching plasticity in case of an accident and any effect the deformation of the riser guard has on the risers. The tubular member reaches plasticity as shown in Figure 5.15.

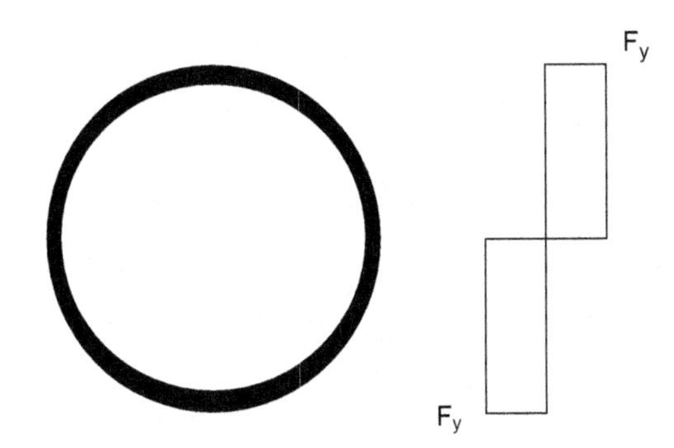

Figure 5.15 Member at a plastic moment.

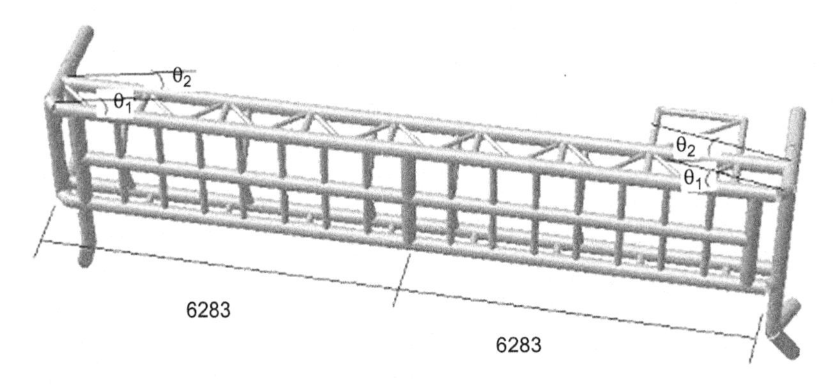

Figure 5.16 Proposed deformation of a riser guard.

Example 5.4

The riser guard configuration is as shown in Figure 5.16. Using tubular member 12.5″ × 0.5″ (323.9 × 12.7 mm),

$F_y = 240$ N/mm^2 (mild steel)
The plastic modulus = 70.2 in.^{3} = 1,150,371.9 mm^3
$M_p = 240 \times 1{,}150{,}371.9/1{,}000 \times 1{,}000 = 276.09 \times 10^6$ N mm
$P \times \delta = 5 \times M_p \times (\theta_1 + \theta_2)$
$\theta_1 = \theta_2 = \delta/6283$
$P \times \delta = 5 \times 276.09 \times 10^6 \times 2 \times \delta/6000$
$P = 4.6 \times 10^5$
$E = 0.9 \times 4.6 \times 10^5 \times \delta = 0.5\ mv^2 \times 0.7$
$V = 500$ mm/s
$m = 1100$ N s^2/mm
$E = 0.7\ (1/2)\ 1100\ (500)^2 = 0.9 \times 4.7 \times 10^5 \times \delta$

The 232-mm deflection is less than the distance between the outside diameter of the risers and the rear face of the riser guard.

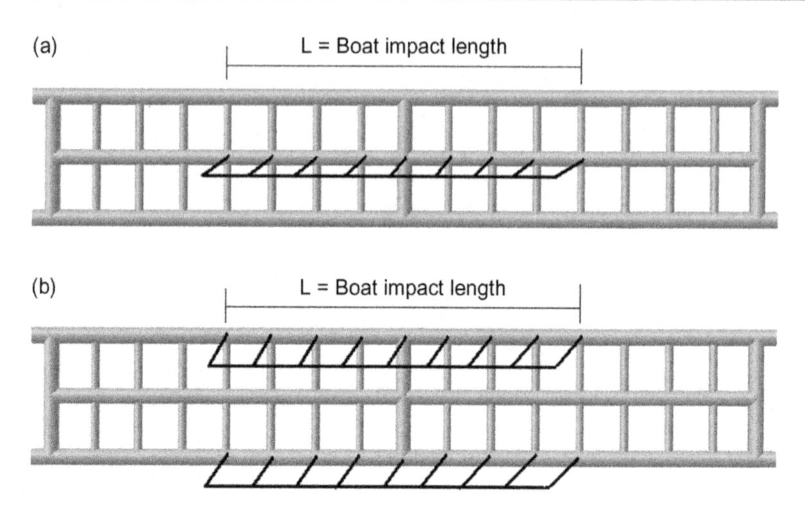

Figure 5.17 Load distributed in (a) Case 1 and (b) Case 2.

5.5.1.1 *Cases of impact load*

In Case 1, a uniform load is at midspan at an elevation of 0.0 mean water level (MWL), as in Figure 5.17(a):

L = 7000 mm (assumed)
Uniform load = 65.7 N/mm

In Case 2, a uniform load at midspan is at elevations +1200 and −1200; the total load is distributed to two levels, so the applied uniform load is as shown in Figure 5.17(b):

Uniform load = 32.9 N/mm (for each elevation)

Further reading

American Petroleum Institute. Recommended Practice for Planning, Designing and Constructing Heliports for Fixed Offshore Platforms. API-RP2L. API. Houston, USA, 1996.

CAP, 2008. 437 Offshore Helicopter Landing Areas-Guidance on Standards 2008. CAP, Civil aviation Authority.

International Civil Aviation Authority, 1995. International Standard and Recommended Practices, Annex 14, vol. 2, Heliports. ICAO, Chicago, USA.

Lloyd's Register of Shipping, 1989. Rules and Regulations for the Classification of Fixed Offshore Installations, Part 4, Steel Structures December 1989, Chapter 2 Topside Items and Fittings, Section 2, Helicopter Landing Area. Lloyd's Register of Shippin, Aberdeen, England.

Geotechnical data and piles design 6

6.1 Introduction

To begin the design of any structure, the designer must know the type of soil on which the structure will be built. So, the first step in design of an offshore structure platform is a soil investigation, but the offshore soil investigation technique is not like onshore soil investigation.

A fixed offshore platform is defined as a jacket-type structure extending above water level and supported by relatively long piles, because this type of structure has been an industry standard for many years. Of course, nowadays other types of structures are being built, many different designs are being proposed, and the soil investigation requirements and the piles' depth depends on the type of structure, since these factors could vary considerably for different types of structures.

For a typical pile-supported platform, the soil investigation usually consists of a single bored sampled to a penetration below the expected pile-tip elevation.

6.2 Geotechnical investigation

The final design of a fixed offshore platform is based on the best soil field test data that can be obtained at the exact structure location. Therefore, an accurate survey should be implemented with the soil investigation. In general, the soil data are usually needed well in advance of any construction at a site.

The marine vessel doing the soil investigation should have a mooring or positioning system and should be capable of drilling and sampling well below the maximum expected penetration of the piles. Furthermore, offshore soil investigation is characteristically more costly than similar work onshore, due to the cost of the marine vessel itself, as its fee is calculated per day, which is expensive. This is a main difference between soil investigation onshore and offshore soil investigation. Therefore, the objectives of an investigation should always be kept in mind and the planning effort and selection of equipment and techniques should be aimed at achieving the objectives at the lowest cost.

Once the desired field investigation has been completed, further investigation is needed in the laboratory to evaluate and soil properties in orderto apply the field and laboratory results to the design problem. It is appropriate to mention at this point that a fairly wide range of results can be obtained, particularly in the field,

Marine Structural Design Calculations. DOI: http://dx.doi.org/10.1016/B978-0-08-099987-6.00006-4

depending on several factors. It is essential that the factors influencing the results be evaluated in interpreting the data and applying the results to the design.

The observations made during drilling and sampling should be recorded, because they can also provide some indication of potential problems in pile installation.

6.2.1 Performing an offshore investigation

An investigation at sea must begin by identifying the location, which is already done by a survey boat. A typical vessel for soil investigation has a length in the range of 40 to 70 m. The vessel is equipped with winches, cables, and anchors for four-point mooring. Anchor lines usually approach about eight times water depth.

Pile foundations frequently penetrate 90–120 m into the sea bottom. The combination of water depth and bottom penetration means that drilling equipment for soil borings should have a depth capability of 300 m or more.

In areas of very soft, underconsolidated soils, it is necessary to exercise very careful control of drilling fluid weight to counteract the tendency of the material to squeeze into the drilled hole and up into the drill pipe. Failure to do so can result in very severe sample disturbance. Furthermore, in such areas, the problem of handling disturbance on recovered samples is also quite severe.

Once a boring has penetrated a granular soil formation, it is essential that drilling mud having suitable viscosity and gel properties, which will be used to stabilize the drilled hole and to prevent caving; commercial saltwater gel is excellent for this purpose. Particularly in glacial deposits, coarse granular material such as cobbles or boulders may be encountered and make drilling extremely difficult. The presence of rock formations within the depth of investigation requires that special tools and procedures to be used.

In most cases, gas be present within formations penetrated by soil borings and flows of water may also be encountered. The normal procedure of using a blowout preventer on cased holes is not easily applied to the wire-line method; the only reasonable protection against blowout is through a controlled mud-weight program using commercial barites.

6.3 Soil tests

The wet rotary process commonly used in advance of onshore soil boring is used also for offshore soil testing. In this respect, there is little difference between onshore and offshore practice except in some details in the way the objective is accomplished. However, the different offshore environmental conditions have necessitated changes in soil sampling procedures. An understanding of onshore sampling techniques, tools, and results will aid in understanding the required alterations, the concessions made, and the advantages of the offshore procedures.

Due to the cost of soil investigation offshore, preliminary engineering is recommended. Many sources of useful information and data about soil characteristics can be gained from the geologic information about the platform location, drilling records, and acoustic data.

Many in-situ testing devices have been developed and are used to determine soil properties and conditions in the ground. Among them are the cone penetrometer, commonly known as the *Dutch cone*; the pressure meter; and the vane shear device. None of these provides a sample for other tests; if samples are required, sampling must be done between the in-situ tests or in a companion boring.

The Dutch cone was developed principally to define granular soil strata that would serve to support point-bearing piles. The cone is modified to measure both side friction and point-bearing resistance. Experienced operators claim to be able to identify soil types by the ratios of these resistances. Modern cone equipment has been used successfully only in shallow penetrations at sea. Remotely operated sea floor equipment presently being developed in Europe may have a substantially greater depth capability.

The pressure meter is designed to determine soil behavior by using an expanding pressure cell to measure load-deformation characteristics. The pressure cell is either driven or pushed into undisturbed soil to begin testing or is installed in the bottom of a drilled hole of carefully controlled dimensions. Readings, interpreted in terms of modulus of deformation and limit pressure, are used to determine bearing capacity, settlement, and other data. This tool has seen little success in connection with deep exploration for pile foundations at sea.

The remote vane, which is the vane shear device, has been used for years to measure in-place shear strengths of soils; however, soil shear failure was produced through torque applied to rods extending to the surface.

Remotely operated vane equipment has been developed in recent years and has been used successfully in a number of offshore investigations.

Tests can be conducted fairly quickly and economically in materials having a shear strength up to about 192 kN/m^2. The present vane probe is powered electrically through a conductor cable extending to the surface but has no connection to the drill pipe. Continuous torque readings throughout a test are monitored on a readout at the surface. Development is under way on another device that has no electrical connection to the surface; readings will be stored in the down-hole device and will be read when it is brought to the surface after a test.

Results of the remote vane tests show strengths consistently higher than can be measured on samples recovered. This is to be expected because of disturbance created by sampling and sample handling and because of the pressure relief experienced by samples brought to the surface. The vane shear device probably better measures the true in-situ shear strength of a cohesive soil than any other device.

Until the principal emphasis shifts from pile-supported platforms, which is not expected any time soon, and until offshore activity extends into much deeper water, the present methods and techniques of offshore soil investigation will continue to be employed. Boat drilling and wire-line sampling methods offer the most versatile and economical means of investigation. Use of remote devices, such as the vane probe and cone penetrometer, in drilled holes, will increase and add much to our knowledge of in-situ soil properties. Dynamic positioning of surface drilling vessels or support vessels for other operations will become more important as operations move into deeper waters beyond the continental shelf. There will be much more activity in the development of remotely operated sea floor drilling and sampling devices. Manned submersibles should also play an increasingly important role with further development.

6.4 In-situ testing

As mentioned previously, offshore soil investigation is much more expensive than onshore investigation. Therefore, the Nosrok (2004) report recommends that all in-situ test equipment systems prescribed should be checked for functionality during mobilization of equipment on board the survey vessel. The functionality checks should include but not be limited to the signal response of sensors, the data-acquisition system, and a wet test of essential subsea equipment.

In-situ equipment with electronic transmission should be designed to sustain the water pressures expected in the field.

During testing, zero readings of all sensors should be recorded before and after each test. The specifications for in-situ test equipment are made for the most commonly used tests. For other in-situ testing, equipment specifications and procedures should be established prior to mobilization.

Records of experience with the use of the equipment, routines, and procedures for interpretation of measurements for assessment of soil parameters should be documented and should be made available on request

In most cases, the in-situ test tool can be inserted into the soils from the sea bed to either a preset depth or to refusal due to limitation in pushing force, capacity of load sensor(s), or other factors, which is called *seabed mode*, or from the bottom of the borehole to either a preset depth or to refusal due to limitation in pushing force, capacity of load sensor(s), or other factors, which is called *drilling mode*.

The drilling of the borehole should be carried out in such a way that the disturbance to the soil below the drill bit is minimized. To avoid any disturbed zone below the drill bit, the in-situ test tool should penetrate at least 1 m if soil strength and density allow. The disturbed zone can be assessed by continuous CPT/CPTU (see next section) penetrated to approximately 3 m below the drill bit.

6.4.1 Cone penetration test

The cone penetration test (CPT; Dutch cone) is modified to piezocone penetration depth. The piezocone test (CPT testing that also gathers piezometer data, called *CPTU testing*) is a CPT with additional measurement of the pore water pressure at one or more locations (U_1, U_2, and U_3) on the penetrometer surface, as shown in Figure 6.1.

The CPT is an in-situ testing method for determining the geotechnical engineering properties of soils and for delineating soil stratigraphy. It was initially developed in the 1950s at the Dutch Laboratory for Soil Mechanics in Delft to investigate soft

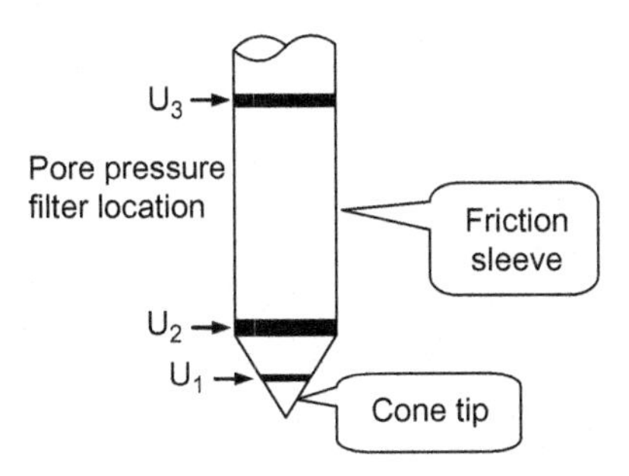

Figure 6.1 Cone penetration testing.

soils (which is why it has also been called the *Dutch cone test*). Today, the CPT is one of the most used and accepted in-situ test methods for soil investigation worldwide.

The test method consists of pushing an instrumented cone, with the tip facing down, into the ground at a controlled rate, usually 20 mm/s. The resolution of the CPT in delineating stratigraphic layers is related to the size of the cone tip, with typical cone tips having a cross-sectional area of either 1000 or 1500 mm^2, corresponding to diameters of 36 and 44 mm.

The American Society for Testing and Materials (ASTM) presents the apparatus and test procedure for the CPT and the measurement of q_c. In particular, ISO (2005) prescribes cones with a base area in the range of 500 mm^2 to 2000 mm^2 and a penetration rate of 20 $\pm$ 5 mm/s.

It is noted that the CPT-based design methods were established for cone resistance values up to 100 MPa. Caution should be used when applying the methods to sands with higher resistances.

The early applications of CPT mainly determined soil bearing capacity. The original cone penetrometers involved simple mechanical measurements of the total penetration resistance to pushing a tool with a conical tip into the soil. Different methods were employed to separate the total measured resistance into components generated by the conical tip due to the tip friction and friction generated by the rod string.

In the 1960s, a friction sleeve was added to quantify the friction component and aid in determining soil cohesive strength (Begemann, 1965). Electronic measurements began in 1948 and improved further in the early 1970s (de Reister, 1971).

Most modern electronic CPT cones now also employ a pressure transducer with a filter to gather pore water pressure data. The filter is usually located either on the cone tip (the so-called U_1 position), immediately behind the cone tip in the most common U_2 position, or behind the friction sleeve, which is the U_3 position. Pore water pressure data aid in stratigraphy and is primarily used to correct tip friction values. CPT and CPTU testing equipment generally advance the cone using hydraulic rams mounted on a heavily ballasted vehicle or using screwed-in anchors as a counterforce.

CPT for geotechnical applications was standardized in 1986 by ASTM Standard D-3441 (ASTM, 2004). ISSMGE provides international standards for CPT and

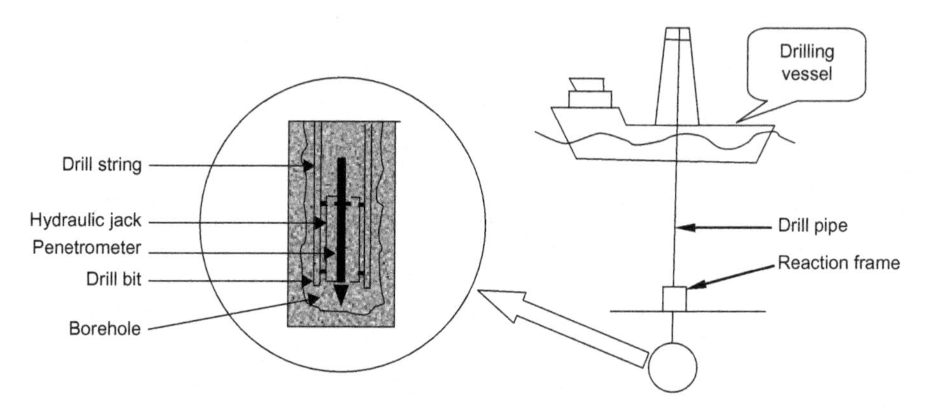

Figure 6.2 Arrangement for cone penetrometer tests.

CPTU. Later, ASTM standards addressed the use of CPT for environmental site characterization and groundwater monitoring activities.

The ability to advance additional in-situ testing tools using the CPT direct-push drilling rig, including the seismic tools, is accelerating this process.

The arrangement for the CPT on the drilling vessel is shown in Figure 6.2.

6.4.1.1 Equipment requirements

The geometry of the cone penetrometer, with tip, sleeve, and pore pressure filters, should follow IRTP (1999), from which the following is extracted: (1) The penetrometer tip and adjoining rods should have the same diameter for at least 400 mm behind the tip, and (2) the cone should have a nominal cross-section area, A_c, of 1000 mm^2, with $35.3\ \text{mm} \leq d_c \leq 36.0\ \text{mm}$ and $24.0\ \text{mm} \leq h_c \leq 31.2\ \text{mm}$, where d_c is the cone diameter and h_c is the height of the conical part.

According to the IRTP (1999), cone penetrometers with a diameter between 25 mm ($A_c = 500\ \text{mm}^2$) and 50 mm ($A_c = 2000\ \text{mm}^2$) are permitted for special purposes, without the application of correction factors. The recommended geometry and tolerances given for the 1000-mm^2 cone penetrometer should be adjusted proportionally to the diameter.

The use of accuracy classes, as required in IRTP (1999), should be adopted. Equipment and procedures to be used should be selected according to the required accuracy class given in Table 6.1. These precautions are critical, to be considered and monitored by the project quality team.

Class 1 is meant for situations where the results will be used for precise evaluation of stratification and soil type as well as parameter interpretation in profiles including soft or loose soils. Class 2 may be considered more appropriate for stiff clays and sands. For Class 3, the results should be used only for stratification and parameter evaluation in stiff or dense soils.

Table 6.1 **Accuracy classes**

Class	Measured parameter	Minimum allowable accuracy	Maximum length between measurements, mm
1	Cone resistance	50 kPa or 3%	20
	Sleeve friction	10 kPa or 10%	
	Pore pressure	5 kPa or 2%	
	Inclination	2 degrees	
	Penetration	0.1 m or 1%	
2	Cone resistance	200 kPa or 3%	20
	Sleeve friction	25 kPa or 15%	
	Pore pressure	25 kPa or 3%	
	Inclination	2 degrees	
	Penetration	0.2 m or 2%	
3	Cone resistance	400 kPa or 5%	50
	Sleeve friction	50 kPa or 15%	
	Pore pressure	50 kPa or 5%	
	Inclination	5 degrees	
	Penetration	0.2 m or 2%	

6.4.1.2 CPT results

The reporting of results from CPTs should comply with IRTP (1999).

For each cone test, the following information should be reported offshore after each test and in the field report (either in tabular form or in the CPT profiles):

- Location.
- Test number.
- Coordinates of test location.
- Date of test.
- Cone serial number.
- Cone geometry and dimensions, with position and dimensions of filter stone.
- Capacity of sensors (tip, pore pressure, and sleeve friction).
- Calibration factors used.
- Zero readings of all sensors before and after each test, either at sea floor or bottom of borehole.
- Observed wear or damage on tip or sleeve.
- Penetration rate.
- Any irregularities during testing.
- Theoretical effective (net) area ratio of tip and friction sleeve.
- water depth to sea floor during test.
- corrections due to tidal variations, if any.
- observed sinking in of the frame.
- the inclination of the cone penetrometer to vertical axis, for a maximum penetration depth spacing of 1 m (for sea-bed testing only).

The measured results (in engineering units) should be presented in digital form, as:

- Depth of penetration (in m).
- Cone tip resistance (in MPa).
- Pore pressure(s) (in MPa).
- Sleeve friction (in kPa).
- Total thrust during test (in kN).

A graphical representation of the results from CPTs in the field (offshore) should be presented. If not otherwise agreed on, the depth scale should be 1 m (field) = 10 mm (plot).

The zero reference for seabed CPTs should be the sea bottom, and for down-hole CPTs, the bottom of the borehole. The selection of the scale for presenting the measured cone resistance, pore pressure, and friction should be reasonable to suit the soil conditions.

The output data are presented in a graph, as shown in Figure 6.3.

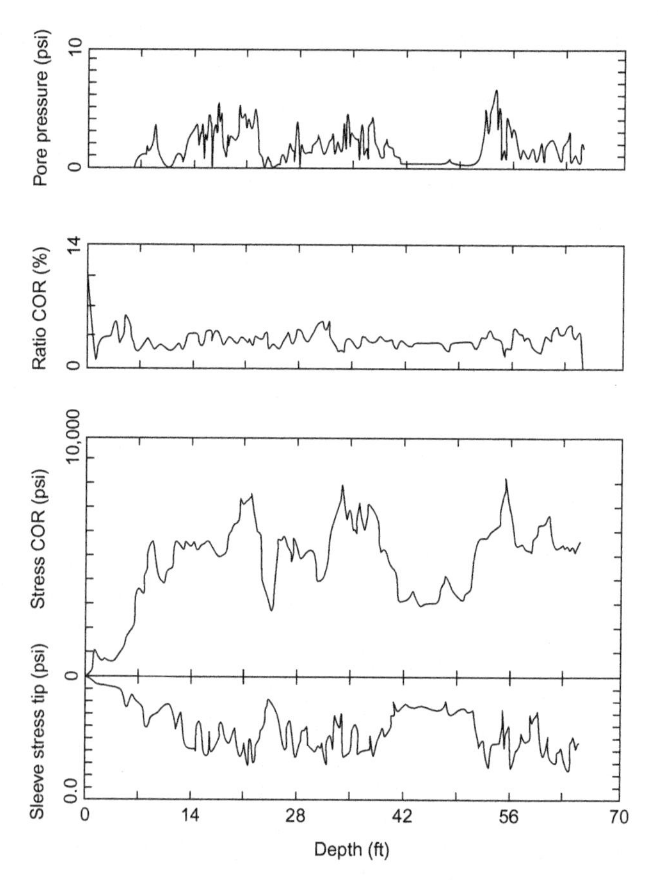

Figure 6.3 Sample of CPT data results.

6.4.2 Field vane test

Vane blades should be rectangular, as defined in ASTM D-2573-01 or BS 5930:1999.

Shear strengths are given in Table 6.2, based on the geometrical dimensions of the vane blades. The vane test in the borehole offshore is presented in Figure 6.4.

As a normal procedure before a vane test is started, the vane blade should be pushed at least 0.5 m below the bottom of the borehole. The pushing rate should be less than 25 mm/s. The time from the instant when the desired test depth has been reached to the beginning of the test (waiting time) should be 2–5 min.

The rotation of the vane should be smooth, and for the initial test (undisturbed), it should be 6–12° per min.

To measure remolded shear strength, the vane should be rotated at least 10 times at a rate ≥4 rpm (rev/min) and until a constant torque over 45° of continuous rotation has been reached. At the end of the rapid rotations, the remolded shear strength should be measured without delay, with a rotation rate equal to that used for intact shear strength.

Table 6.2 **Measuring range with different vane blade dimensions**

Measuring range of s_u	Vane height, mm	Vane width, mm	Vane blade thickness, mm
0–50	130	65	2
30–100	110	55	2
80–250	80	40	2

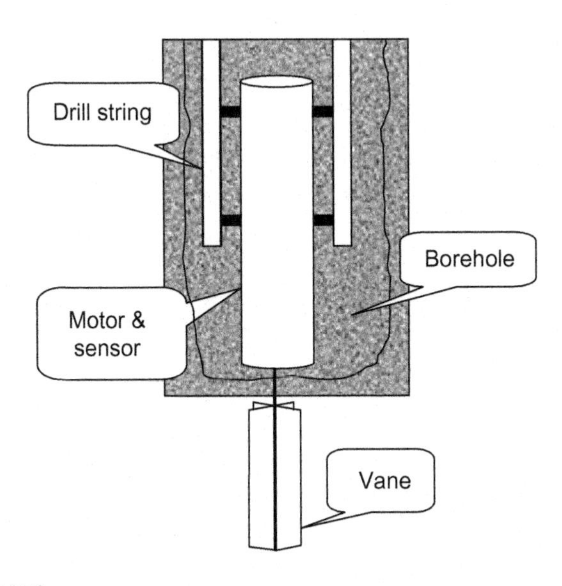

Figure 6.4 Vane test.

It is also possible to do vane tests in the seabed mode. The depth intervals between tests should be at least 0.5 m.

The insertion method and test procedure to be used should be described, giving particular information about the method for insertion and penetration of vanes below the bottom of the borehole, possible rotation rates available, and the method for providing torque and reaction.

The data-acquisition system should be such that the overall accuracy is maintained. Taking into account all sources of error, including the data-acquisition system, the uncertainty in the measured torque should not exceed the smaller of 5% of measured value or 2% of the maximum value of the measured torque of the layer under consideration.

The resolution of the measured result should be within 2% of the measured value.

During testing, the data-acquisition system should allow for real-time inspection of measured results in both digital and graphic form.

It is very important that, at least every year and before each project, the sensor for measuring the torque during vane testing be calibrated. If the sensor is loaded close to its maximum or any damage is suspected, it should be checked and recalibrated. In addition, function checks should be carried out in the field.

For each vane test, the following information should be given:

- Site area.
- Date of test.
- Operator.
- Boring number/test identification.
- Water depth at test location.
- Dimensions of vane.
- Depth below sea bottom to vane tip.
- Depth below bottom of borehole to vane tip.
- Rate of rotation.
- The complete curve of torque versus rotation (degrees).
- Time to failure.
- Any irregularities during testing.
- Formula used to calculate the vane undrained shear strength, s_{uv}, including the assumption made for shear stress distribution on ends of the vane.

6.5 Soil properties

In soil mechanics, the equilibrium and movement of soil bodies is studied, where soil is understood to be the weathered natural material in the upper layers (the upper 20 to 100 m) of the earth's crust. This material may be gravel, sand, clay, peat, or some other soft and loose granular medium. The characteristics of these materials are quite different from those of artificial (human-made) materials, such as steel or concrete. Human-made materials usually are much more consistent than soils and exhibit relatively simple, linear, mechanical behavior, at least if the deformations are not too large.

The mechanical properties of soils are usually strongly nonlinear, with the material exhibiting irreversible plastic deformations when loaded and unloaded, even at low stress levels, and often showing anisotropic behavior, creep, and such typical effects as a volume change during shear. This mechanical behavior of soil is also difficult to predict, because the structure of the soil may be highly inhomogeneous, as a result of its geological history; and it is often not possible to determine the detailed behavior of the soil by tests in the laboratory or in situ.

The behavior of soils may be further complicated by the presence of water in the pores. This relatively stiff fluid in the pores may prevent or retard volume deformations.

For all these reasons, the characterization of the mechanical behavior of soils is often done in a schematic way only, and its form is adapted to the particular type of problem under consideration.

The soil layers have a dominating effect, whereas the settlement of an embankment is mainly governed by the deformation properties of the soil, including creep.

Therefore, in soil mechanics, the range of applicability of a certain parameter is often restricted to a limited class of problems. Many properties cannot be used outside their intended field of application. Nevertheless, various properties may all derive from such common basic phenomena as interparticle friction or the structure of a granular medium, so that good correlations between certain properties may well exist. In this chapter, some of these properties are reviewed and some correlations are discussed. It should be noted that, in engineering practice, nothing can beat the results of experimental determination of the soil parameters in situ or in the laboratory. A correlation may at best give a first estimate of the order of magnitude of a parameter.

6.5.1 Strength

Soils usually cannot transfer stresses beyond a certain limit. This is called the *strength* of the soil. The shear strength of soils is usually expressed by Coulomb's relation between the maximum shear stress and the effective normal stress:

$$\tau_{max} = c + \sigma' \tan \phi \tag{6.1}$$

where c is the cohesion and ϕ is the friction angle. For sand, c is usually negligible, so that ϕ is the only strength parameter.

For clay, it is often most relevant to consider the strength in undrained conditions, during which the effective stress remains constant. The undrained shear strength is usually denoted by s_u, and it is often considered irrelevant to what degree this is to be attributed to cohesion or friction. The shear strength parameters can be determined in the laboratory, for instance, by triaxial testing.

A simple and useful in-situ test is the cone penetration test, in which a cone is pushed into the ground using hydraulic pressure equipment, while recording the stress at the tip of the cone and often also the friction along the lower part of the shaft. The test is used in the Netherlands as a model test for a pile foundation, and the results are used directly to determine the bearing capacity of end-bearing piles, using simple scale rules.

The CPT can also be used to estimate the strength of a soil, however, by using certain correlations. For a penetration test in sand, for instance, the bearing capacity of the cone, q_c, is, according to Brinch Hansen's formula:

$$q_c = s_q N_q \sigma'_v = s_q N_q \gamma' z \tag{6.2}$$

where s_q is a shape factor to express effective weight of the overburden, for which one may use the formula:

$$s_q = 1 + \sin\phi \tag{6.3}$$

The z is the depth, and N_q is a dimensionless constant for which theoretical analysis has given the value as in the following equation:

$$N_q = \frac{1+\sin\phi}{1-\sin\phi}\exp(\pi\tan\phi) \tag{6.4}$$

The predicted cone resistances q_c for various types of sand at a depth of 10 m and 20 m are shown in Table 6.3, assuming that $\gamma = 10\ \text{kN/m}^3$. The values in the table are indeed often observed for sand layers at these depths. They may also be used inversely: if a certain cone resistance is observed, it is indicative for a certain type of material.

For a CPT in clay soil, the Brinch Hansen formula can be used to correlate the CPT result to the undrained shear strength.

The general Brinch Hansen formula is

$$q_c = s_c N_c c + s_q N_q \sigma'_v \tag{6.5}$$

Because the test is performed very quickly, the soil behavior can be considered to be undrained.. The values for the coefficients can then be taken as $N_c = 5.14$, $N_q = 1$, $s_c = 1.3$, $s_q = 1$. Equation (6.5) now reduces to

$$q_c = 6.7 s_u + \sigma'_v \tag{6.6}$$

where the cohesion c has been interpreted as the undrained shear strength s_u.

Table 6.3 Guidance for cone resistance in sand

Soil type	ϕ	N_q	q_c, $z = 10$ m, MPa	q_c, $z = 20$ m; e, MPa
Loose sand	30°	18.4	2.8	5.5
Medium dense sand	35°	33.3	5.2	10.5
Very dense sand	40°	64.2	10.5	21.1

The undrained shear strength of normally consolidated clays depends on the vertical stress σ'_v. A relationship that is often used is the correlation proposed by Ladd,

$$s_u = 0.22\sigma'_v \tag{6.7}$$

Substitution of this result into equation (6.6) gives

$$q_c \approx 11s_u \tag{6.8}$$

For a soft clay, with $s_u = 20$ kPa, the order of magnitude of the cone resistance would be 220 kPa ≈ 0.2 MPa. Such values are indeed often observed. Again, they may also be used to estimate the undrained shear strength from CPT data.

Table 6.4 provides guidance to cohesive soil characteristics based on the rule of thumb.

The relative density for cohesionless soil can be predicted using Table 6.5 as a guideline.

The approximate soil parameters for different types of soil are illustrated in Table 6.6.

Table 6.4 Consistency of cohesive soil

Consistency	Unconfined compressive strength, tons/ft^2	Rule-of-thumb test
Very soft	0–0.25	Core (height twice diameter) sags under own weight
Soft	0.25–0.5	Can be pinched in two by pressing between thumb and finger
Firm	0.5–1.0	Can be imprinted easily with finger
Stiff	1.0–2.0	Can be imprinted with considerable pressure from fingers
Very stiff	2.0–4.0	Barely can be imprinted by pressure from fingers
Hard	>4.0	Cannot be imprinted by fingers

Table 6.5 Degree of compactness for cohesionless soil

Degree of compactness	Relative density, %
Very loose	0–15
Loose	15–35
Medium dense	35–65
Dense	65–85
Very dense	85–100

Table 6.6 Approximate property values for different soil types

Soil type	Undrained shear strength, S_u, in kPa	Effective cohesion, c', in kPa	Friction angle, f, in degrees	Saturated density, D_s, in t/m^3	Voids ratio
Soft to firm clay	10–50	5–10	19–24	1.4–1.8	
Stiff clay	50–100	10–20	22–29	1.8–1.9	
Very stiff to hard clay	100–400	20–50	27–31	1.9–2.2	
Silt	10–50	—	27–35	1.7–2.3	1.1–0.3
Loose sand	—	—	29–30	1.7–1.8	1.1–0.8
Medium dense sand	—	—	30–40	1.8–2.0	0.8–0.5
Dense sand	—	—	35–45	2.0–2.3	0.5–0.2
Gravel	—	—	35–55	1.7–2.4	1.1–2.2

6.5.2 Soil characterization

A key part of developing realistic analytical models to evaluate cyclic loading effects on piles is the characterization of soil-pile interaction behavior. High-quality in-situ, laboratory, and model-prototype pile-load tests are essential in such characterizations. In developing soil-pile interaction characterizations, it is important that pile installation and pile-loading conditions be integrated into the testing programs.

According to McClelland and Ehlers (1986), in-situ tests, such as vane shear, cone penetrometer, and pressure meter tests, can provide important insights into in-place soil behavior and stress-strain properties.

Both low- and high-amplitude stress-strain properties can be developed. Long-term static and creep loadings or short-term dynamic, impulsive, and cyclic loadings sometimes can be simulated with in-situ testing equipment.

Laboratory tests on representative soil samples permit a wide variety of stress-strain conditions to be simulated and evaluated. Soil samples can be modified to simulate pile-installation effects, such as remolding and reconsolidating, to estimate in-situ stresses. The samples can be subjected to different boundary conditions, such as triaxial, simple shear, and interface shear, and to different levels of sustained and cyclic shear time histories to simulate in-place loading conditions.

Another important source of data to develop soil characterizations for cyclic loading analyses are tests on model and prototype piles. Based on Bogard et al. (1985)

and Karlsrud and Haugen (1985), model piles can be highly instrumented, and repeated tests can be performed in soils and for a variety of loadings.

Geometric scale, time scale, and other modeling effects should be carefully considered in applying results from model tests to prototype behavior analyses. As discussed by Pelletier and Doyle (1982) and Arup et al. (1986), the data from load tests on prototype piles are useful for calibrating analytical models.

Such tests, even if not highly instrumented, can provide data to guide development of analytical models. These tests can also provide data to verify results of soil characterizations and analytical models.

Prototype pile-load testing, coupled with in-situ and laboratory soil testing and realistic analytical models, can provide an essential framework for making realistic evaluations of the responses of piles to cyclic axial loadings.

The foundation should be designed to carry static, cyclic, and transient loads without excessive deformation or vibrations in the platform. Special attention should be given to the effects of cyclic and transient loading on the strength of the supporting soils as well as on the structural response of piles.

It is very important to consider the possibility of movement of the sea floor against the foundation members, and the forces caused by such movements, if anticipated, should be considered in the design.

6.6 Pile foundations

Offshore structure platforms commonly use open driven piles. These piles are usually driven into the seabed with impact hammers, which use steam, diesel fuel, or hydraulic power as the source of energy. Therefore, the pile wall thickness should be adequate to resist axial and lateral loads as well as the stresses during pile driving.

According to Smith (1962), it is possible to predict approximately the stresses during pile driving using the principles of one-dimensional elastic stress wave transmission by carefully selecting the parameters that govern the behavior of soil, pile, cushions, cap block, and hammer.

This approach may also be used to optimize the pile hammer cushion and cap block using the computer analysis, commonly known as *wave equation analysis*. The design penetration of driven piles should be determined, rather than correlation of pile capacity with the number of blows required to drive the pile a certain distance into the seabed.

If a pile stops before it reaches design penetration, one or more of the following actions can be taken:

1. Review of all aspects of hammer performance, possibly with the aid of hammer and pile-head instrumentation, which may identify problems that can be solved by improving hammer operation and maintenance or by the use of a more powerful hammer.
2. Reevaluation of design penetration by reconsideration of loads, deformation, and required capacities of both individual piles and the foundation as a whole, which may identify

reserve capacity available. An interpretation of driving records in conjunction with the instrumentation mentioned previously may allow design soil parameters or stratification to be revised and pile capacity to be increased.

3. Modifications to piling procedures (usually, the last course of action), which may include one of the following: The soil plug inside the pile is removed by jetting and air lifting or by drilling to reduce pile-driving resistance. If plug removal results in inadequate pile capacities, the removed soil plug should be replaced by a gravel grout or concrete plug having sufficient load-carrying capacity to replace that of the removed soil plug. Attention should be paid to plug and pile load-transfer characteristics. Note that plug removal may not be effective in some circumstances, particularly in cohesive soils.

Soil below the pile tip is removed by drilling an undersized hole or by lowering jetting equipment through the pile, which acts as the casing pipe for the operation. The effect on pile capacity of drilling an undersized hole is unpredictable unless there has been previous experience under similar conditions.

According to API RP2A (2007), jetting below the pile tip should be avoided because of the unpredictability of the results.

A first stage or outer pile is driven to a predetermined depth, the soil plug is removed, and a second stage or inner pile is driven inside the first-stage pile. In this case, the grouting is inserted in the annulus between the two piles to provide load transfer between the two piles and to develop composite action.

The piles of the offshore structures are exposed to static and cyclic loading, and the loads are axial and lateral. So, all these load effects should be considered in the pile design.

6.6.1 Pile capacity for axial loads

On the basis of API RP2A (2007), the ultimate static axial capacity (Q) of an open-ended pipe pile in compression is given by the equation

$$Q_t = Q_f + Q_s + \{\text{small values from } Q_{\text{fi}} \text{ or } Q_{\text{sp}}\} \tag{6.9}$$

$$Q_t = f \cdot A_s + A_a \cdot Q + A_p \cdot q \tag{6.10}$$

where A_p is the area of pile or

$$Q_t = f \cdot A_s + A_a \cdot Q + f\, A_{\text{si}} \tag{6.11}$$

where

Q_t = the total pile resistance
Q_f = total outside shaft resistance
Q_s = end-bearing capacity of the annulus
Q_{fi} = total inside shaft resistance
Q_{sp} = bearing capacity of the soil beneath the plug
f = unit skin friction capacity (in kPa)
A_s = outside surface area of pile (in m^2)

q = unit end-bearing capacity (in kPa)
A_a = the area of the pile annulus (in m^2)
A_{si} = inside surface area of pile (in m^2)

In computing pile loading and capacity, the weight of the pile-soil plug system and hydrostatic uplift should be considered.

In determining the load capacity of a pile, consideration should be given to the relative deformation between the soil and the pile, as well as the compressibility of the soil-pile system.

Coyle and Reese (1966), Murff (1980), and Randolph (1983) discuss skin friction and assume that the maximum skin friction along the pile and the maximum end bearing are mobilized simultaneously. However, the ultimate skin friction increments along the pile are not necessarily directly additive, nor is the ultimate end bearing necessarily additive with the ultimate skin friction. In some circumstances, this effect may result in the capacity being less than that given by equation (6.11).

In such cases, a more explicit consideration of axial pile performance effects on pile capacity may be warranted. Pile sizing should be based on what experience has shown can be installed consistently, practically, and economically under similar conditions with the installation equipment being used. Alternatives for possible remedial action in the event design objectives cannot be obtained during installation should also be investigated and defined prior to construction.

For the pile system, the pile-capacity factor of safety is defined in Table 6.7, according to API RP2A (2007). The allowable skin friction values on the pile section on the upper surface on the pile should be discounted in computing skin friction resistance, Q_f. The end-bearing area of a pilot hole, if drilled, should be discounted in computing total bearing area.

Table 6.7 Design parameter guide for cohesionless siliceous soil (based on API RP2A)

Soil description	Soil condition	Shaft friction factor	Limited shaft friction values, f, kPa	End-bearing factor, N_q	Limited unit end-bearing values, q, MPa
Sand	Very loose				
Sand	Loose				
Sand-silt	Loose				
Silt	Medium dense				
Silt	Dense				
Sand-silt	Medium dense	0.29	67	12	3
Sand	Medium dense	0.37	81	20	5
Sand-silt	Dense	0.37	81	20	5
Sand	Dense	0.46	96	40	10
Sand-silt	Very dense	0.46	96	40	10
Sand	Very dense	0.56	115	50	12

6.6.1.1 Skin friction and end bearing in cohesive soils

For pipe piles in cohesive soils, the shaft friction, f (in kPa), at any point along the pile may be calculated by

$$f = \alpha c \tag{6.12}$$

where α = a dimensionless factor and c = undrained shear strength of the soil at the point in question.

The factor α can be computed by

$$\begin{aligned} \alpha &= 0.5\psi - 0.5\psi \leq 1.0 \\ \alpha &= 0.5\psi - 0.25\psi > 1.0 \end{aligned} \tag{6.13}$$

with the constraint that $\alpha \leq 1.0$, where $\psi = c/p'$ for the point in question and p' is the effective overburden pressure at the point in question (in kPa). For underconsolidated clays, clays with excess pore pressures undergoing active consolidation, α can usually be taken as 1.0.

A discussion of appropriate methods for determining the undrained shear strength, c, and effective overburden pressure, p', including the effects of various sampling and testing procedures, is important. Due to the lack of pile-load tests in soils having c/p' ratios greater than 3, equation (6.13) should be applied with some engineering judgment for high c/p' values. Similar judgment should be applied for deep-penetrating piles in soils with high undrained shear strength c, where the computed shaft frictions, f, using equation (6.14), are generally higher than previously specified in API RP2A. For very long piles, some reduction in pile capacity would happen, because the shaft friction may reduce to some lesser residual value on continued displacement.

For piles end bearing in cohesive soils, the unit end bearing, q (in kPa), may be computed by

$$q = 9c \tag{6.14}$$

It is obvious that, in open-driven piles, the shaft friction, f, acts on both the inside and outside of the pile. The total resistance is the sum of the external shaft friction, the end bearing on the pile wall annulus, and the total internal shaft friction or the end bearing of the plug, whichever is less.

For piles considered to be plugged, the bearing pressure may be assumed to act over the entire cross section of the pile. For unplugged piles, the bearing pressure acts on only the pile wall annulus. Whether a pile is considered plugged or unplugged may be based on static calculations. For example, a pile could be driven in an unplugged condition but act plugged under static loading.

For piles driven in undersized drilled holes, piles jetted in place, or piles drilled and grouted in place, the selection of shaft friction values should take into account the soil disturbance resulting from installation. In general, f should not exceed values for driven piles; however, in some cases, for drilled and grouted piles in over consolidated clay, f may exceed these values.

In determining f for drilled and grouted piles, the strength of the soil-grout interface, including potential effects of drilling mud, should be considered. A further check should be made of the allowable bond stress between the pile steel and the grout, which was discussed by Kraft and Lyons (1974).

In layered soils, shaft friction values, f, in the cohesive layers should be as given in equation (6.12). End-bearing values for piles tipped in cohesive layers with adjacent weaker layers may be as given in equation (6.14), assuming that the pile achieves penetration of two to three diameters or more into the layer in question and the tip is approximately three diameters above the bottom of the layer, to preclude punch-through.

Where these distances are not achieved, some modification in the end-bearing resistance may be necessary. Where adjacent layers are of comparable strength to the layer of interest, the proximity of the pile tip to the interface is not a concern.

Example 6.1

Calculate the capacity for plug pile for the following data.

Pile diameter = 48 in. = 1.219
Pile thickness = 25.4 mm
Pile length = 40 m
Skin friction $f = 70$ kpa
End bearing, $Q = 2.64$ MPa at the last layer

Solution

Base area = 1.167 m^2
Surface area of the pile, $A_s = 3.83 \times 40 = 153.18$ m^2
Total pile capacity = $153.18 \times 70{,}000 + 1.167 \times 2{,}640{,}000 = 11{,}002{,}680$ N = 1.1 MN

Example 6.2

Calculate the capacity for the pile based on API for the following data.

Pile diameter = 60 in. = 1.524
Pile thickness = 22 mm
Pile length = 40 m
Skin friction, $f = 70$ kPa
End bearing, $Q = 2.64$ MPa at the last layer

Solution

Base area = 1.824 m^2
Area of the annulus = 0.104 m^2
Surface area of the pile, $A_s = 4.788 \times 40 = 191.51$ m^2
Pile internal diameter = 1.48 m
Surface area of pile internal = $4.65 \times 40 = 185.98$ m^2

In case of a plugged pile,

Total pile capacity = $(191.51 \times 70{,}000 + 1.824 \times 2{,}640{,}000) \times 10^{-6} = 18.2$ MN

In case of an unplugged pile,

Total pile capacity = $[(191.51 + 185.98) \times 70{,}000 + 0.104 \times 2{,}640{,}000] \times 10^{-6} = 26.7$ MN
The design pile capacity = 18.2 MN

Example 6.3

Calculate pile capacity at different soil strata using the following input data:

OD = 1.3 m
The pile depth = 33 m
Total base area = 1.327 m^2
Outside shaft area/m' = 4.08 m^2
Cohesive soil. $k_o = 1$
Noncohesive soil. $k_o = 0.5$

The soil consists of nine layers, nc for noncohesive soil, and c for cohesive soil.

Solution Table 6.8 presents the solution. You can get the Excel sheet by contacting the website www.elreedyma.comli.com.

6.6.1.2 Shaft friction and end bearing in cohesionless soils

This section provides a simple method for assessing pile capacity in cohesionless soils. There are reliable methods for predicting pile capacity that are based on direct correlations of pile unit friction and end-bearing data with cone penetration test results. The CPT-based methods are considered fundamentally better, have shown statistically closer predictions of pile-load test results, and although not required, they are, in principle, the preferred methods. CPT-based methods also cover a wider range of cohesionless soils. However, offshore experience with CPT methods is limited, and hence, more experience is needed before they are recommended for routine design. CPT-based methods should be applied only by qualified engineers who are experienced in the interpretation of CPT data and understand the limitations and reliability of the methods. Following installation, pile-driving instrumentation data may be used to give more confidence in predicted capacities.

For pipe piles in cohesionless soils, the unit shaft friction at a given depth, f, may be calculated by

$$f = \beta p' \tag{6.15}$$

where β = a dimensionless shaft friction factor and p' = effective overburden pressure at the depth in question.

Table 6.7 may be used for selection of β values for open-ended pipe piles driven unplugged if other data are not available. Values of β for full-displacement piles (i.e., driven fully plugged or closed ended) may be assumed to be 25% higher than those given in Table 6.11. For long piles, f may not increase linearly with the overburden pressure, as implied by equation (6.15). In such cases, it may be appropriate to limit f to the values given in Table 6.9.

Table 6.8 Pile capacity calculation

Layer number, enter c or nc	Depth, m	Soil p for s'_v, kN/m^3	s'_v, kPa	p'_o, kPa	Cohesive soil			Non cohesive soil				Q shaft for layer, MN*	Q base bottom, MN**	Q total,
					C_u, kPa	Skin friction, f, kPa	End bearing, q, MPa	Soil-pile friction, d, °	N_q	Skin friction, f, kPa	End bearing, q, MPa			
1	0	9.2	0	0	115	0.0	1.035	0	0	NA	NA			
c	5	9.2	46	46	115	45.7	1.035	0	0	NA	NA	0.47	1.37	1.84
2	5	9.2	46	46	112.5	45.0	1.0125	0	0	NA	NA			
c	9	9.2	83	83	112.5	52.1	1.0125	0	0	NA	NA	0.79	1.34	2.60
3	9	9.2	83	83	100	47.7	0.9	0	0	NA	NA			
c	14	9.2	129	129	100	56.7	0.9	0	0	NA	NA	1.07	1.19	3.52
4	14	9.2	129	129	125	63.4	1.125	0	0	NA	NA			
c	21.5	9.2	198	198	125	78.6	1.125	0	0	NA	NA	2.18	1.49	6.00
5	21.5	9.5	198	132	0	NA	NA	30	20	60.9	2.64			
nc	23.5	9.5	217	145	0	NA	NA	30	20	66.8	2.89	0.52	3.84	8.86
6	23.5	9.5	217	145	0	NA	NA	31	20	69.5	2.89			
nc	25	9.5	231	154	0	NA	NA	31	20	74.0	3.08	0.44	4.09	9.55
7	25	9.5	231	231	140	89.9	1.26	0	0	NA	NA			
c	27	9.5	250	250	140	93.6	1.26	0	0	NA	NA	0.75	1.67	7.88
8	27	9.5	250	167	0	NA	NA	32	20	83.3	3.33			
nc	28.5	9.5	264	176	0	NA	NA	32	20	88.1	3.52	0.53	4.68	11.41
9	28.5	9.5	264	264	85	74.9	0.765	0	0	na	na			
c	33	9.5	307	307	85	80.8	0.765	0	0	na	na	1.43	1.02	9.18

*Skin friction by selecting its value from cohesive (c) or non cohesive (nc) soil layer multiply by surface area.

**End bearing calculated by selecting its value from cohesive (c) or non cohesive (nc) soil layer multiply by pile area.

Table 6.9 **Pile-capacity factor of safety in API RP2A (2007)**

Load condition	Factor of safety
Design environmental conditions with appropriate drilling loads	1.5
Operating environmental conditions during drilling operations	2.0
Design environmental conditions with appropriate producing loads	1.5
Operating environmental conditions during producing operations	2. 0
Design environmental conditions with minimum loads (for pullout)	1.5

For pile end bearings in cohesionless soils, the unit end bearing q may be calculated by

$$q = Nq\,p' \tag{6.16}$$

where Nq = a dimensionless bearing capacity factor and p' = effective overburden pressure at the depth in question. Recommended Nq values are also presented in Table 6.7.

For long piles, q may not increase linearly with the overburden pressure, as implied by equation (6.16). In such cases, it may be appropriate to limit q to the values given in Table 6.9. For plugged piles, the unit end bearing q acts over the entire cross section of the pile. For unplugged piles, q acts on the pile annulus only. In this case, additional resistance is offered by friction between the soil plug and the inner pile wall.

Whether a pile is considered to be plugged or unplugged may be based on static calculations using a unit skin friction on the soil plug equal to the outer skin friction. It is noted that a pile could be driven in an unplugged condition but can act plugged under static loading.

The design parameters in Table 6.9 are just a guide from API RP2A, and detailed information must be obtained from the CPT results, strength tests, and other soil and pile response tests.

Olson (1987) compared the load test data for piles in sand (obtained by measuring the axial load capacities for open steel piles) and the calculated capacity from API RP2A. Studies done by Lehane et al. (2005a, 2005b) indicate that variability in capacity predictions using the API calculation method may exceed those for piles in clay. These researchers also indicated that the calculation method is conservative for short offshore piles (short = piles less than 45 m, or 150 ft, long) in dense to very dense sands loaded in compression and may not be conservative in all other conditions. In unfamiliar situations, the designer may want to account for this uncertainty through a selection of conservative design parameters or by going toward higher factors of safety.

For soils that do not fall within the ranges of soil density and descriptions given in Table 6.7 or for materials with unusually weak grains or compressible structure, Table 6.7 may not be appropriate for selection of design parameters. For example, very loose silts or soils containing large amounts of mica or volcanic grains may require special laboratory or field tests for selection of design parameters. Of particular importance are sands containing calcium carbonate, which are found extensively in many ocean areas. Experience suggests that driven piles in these soils may have substantially lower design strength parameters rather than those described in Table 6.9.

Drilled and grouted piles in carbonate sand may have significantly higher capacities than driven piles and have been used successfully in many areas with carbonate soils. The characteristics of carbonate sands are highly variable and local experience should dictate the design parameters selected. For example, experience suggests that capacity is improved in carbonate soils of high densities and higher quartz contents.

Cementation may increase end-bearing capacity but result in a loss of lateral pressure and a corresponding decrease in frictional capacity.

For piles driven in undersized drilled or jetted holes in cohesionless soils, the values of f and q should be determined by some reliable method that accounts for the amount of soil disturbance due to installation, but they should not exceed values for driven piles. Except in unusual soil types, such as described previously, the f and q values given in Table 6.7 may be used for drilled and grouted piles, with consideration given to the strength of the soil-grout interface.

In layered soils, unit shaft friction values in cohesionless layers and the end-bearing values for piles tipped in cohesionless layers with adjacent layers of lower strength may also be taken from Table 6.7, provided that the pile achieves penetration of two to three diameters or more into the cohesionless layer, and the tip is at least three diameters above the bottom of the layer, to preclude punch-through. Where these pile tip penetrations are not achieved, some modification in the tabulated values may be necessary. Where adjacent layers are of comparable strength to the layer of interest, the proximity of the pile tip to the layer interface is not a concern.

6.6.2 Foundation size

In most cases, in the FEED engineering phase, the pile configuration is defined based on past experience. During selection of the size of the pile foundation, the following items should be considered: diameter, penetration, wall thickness, type of tip, spacing, number of piles, geometry, location, mudline restraint, material strength, installation method, and other parameters as may be considered appropriate.

A number of different analysis procedures may be utilized to determine the requirements of a foundation. At a minimum, the procedure used should properly simulate the nonlinear response behavior of the soil and ensure load-deflection compatibility between the structure and the pile-soil system.

For deflections and rotations of individual piles, the total foundation system should be checked at all critical locations, which may include pile tops, points of contra lecture, mudline, and so forth. Deflections and rotations should not exceed serviceability limits, which would render the structure inadequate for its intended function.

6.6.2.1 Pile penetration

The design pile penetration should be sufficient to develop adequate capacity to resist the maximum computed axial bearing and pullout loads with an appropriate factor of safety.

The ultimate pile capacities can be computed in accordance with previous sections or by other methods that are supported by reliable comprehensive data. API RP2A (2007) defined the minimum factor of safety by dividing the ultimate pile capacity into the actual load, as shown in Table 6.9.

Two safety factors in API RP2A depend on the desaign considering environmental conditions with a 100-year storm wave effect and the operating environmental conditions, which includes the maximum wave height per year.

The provisions of API RP2A (2007) for sizing the foundation pile are based on an allowable stress (working stress) method, except for pile penetration. In this method, the foundation piles should conform to the requirements of specification and design. Any alternative method supported by sound engineering methods and empirical evidence may also be utilized. Such alternative methods include the limit state design approach or ultimate strength design of the total foundation system.

6.6.3 Axial pile performance

6.6.3.1 Static load-deflection behavior

The static pile axial deflection should be compatible with the structural forces and deflection, so it should be within the service limit. An analytical method for determining axial pile performance is provided in Meyer, Holmquist, and Matlock (1975).

This method uses axial pile shear transition versus local pile deflection (*t-z*) curves to model the axial support provided by the soil along the sides of the pile. An additional (*Q-z*) curve is used to model the tip and bearing versus the deflection response. (Methods for constructing *t-z* and *Q-z* curves are given later.) Pile response is affected by load directions, load types, load rates, loading sequence, installation technique, soil type, axial pile stiffness, and other parameters. Some of these effects for cohesive soils have been observed in both laboratory and field tests.

In some circumstances, as for soils that exhibit strain-softening behavior and where the piles are axially flexible, the actual capacity of the pile may be less than that given by equation (6.9). In these cases, an explicit consideration of the effects on ultimate axial capacity may be warranted. Note that other factors, such as increased axial capacity under loading rates associated with storm waves, may counteract these effects, as discussed by Dunnavant et al. (1990).

6.6.3.2 Cyclic response

Unusual pile loading conditions or limitations on design pile penetrations may warrant detailed consideration of cyclic loading effects.

Cyclic loading, which includes inertial loadings developed by environmental conditions such as storm waves and earthquakes, can have two potentially counteractive effects on the static axial capacity. Repetitive loadings can cause a temporary or permanent decrease in load-carrying resistance or an accumulation of deformation. Rapidly applied loadings can cause an increase in load-carrying resistance or stiffness of the pile. Very slowly applied loadings can cause a decrease in load-carrying resistance or stiffness of the pile. The resultant influence of cyclic loadings are a function of the combined effects of the magnitudes, cycles, and rates of applied pile loads; the structural characteristics of the pile; the types of soils; and the factors of safety used in design of the piles. The design pile penetration should be sufficient to develop an effective pile capacity to resist the design static and cyclic loadings.

The design pile penetration can be confirmed by performing pile response analyses of the pile-soil system subjected to static and cyclic loadings. The pile-soil resistance-displacement *t-z* and *Q-z* characterizations are discussed next.

When any of the preceding effects are explicitly considered in pile-response analysis, the design static and cyclic loadings should be imposed on the pile top and the resistance-displacements of the pile determined. At the completion of the design loadings, the maximum pile resistance and displacement should be determined. Pile deformations should meet structure serviceability requirements.

6.6.3.3 Axial load-deflection (t-z and Q-z) data

The pile foundation should be designed to resist static and cyclic axial loads. The axial resistance of the soil is provided by a combination of axial soil-pile adhesion or load transfer along the sides of the pile and end-bearing resistance at the pile tip. The plotted relationship between mobilized soil-pile shear transfer and local pile deflection at any depth is described using a *t-z* curve. Similarly, the relationship between mobilized end-bearing resistance and axial tip deflection is described using a *Q-z* curve.

Axial deformation of piles may be modeled in a way similar to the lateral case, to permit stress transfer to be computed and axial pile stiffness to be assessed. For axial loading, *t-z* curves are used to represent the resistance along the pile shaft, and *Q-z* curves are introduced to model end bearing. Characteristic shapes of the curves for sand and clay are shown in Figures 6.5 and 6.6 for clay and sand soil, respectively.

Axial support to a pile is provided by surrounding soil, and axial pile deformation at the end of pile depth may be considered to consist of four components: elastic pile deformation, elastic soil deformation, plastic soil deformation, and plastic soil-pile slip deformation. The purpose of *t -z* curves is to model the latter three components. *Q-z* curves model elastic and plastic soil deformation around the pile tip. Elastic pile deformation is not directly related to soil characteristics and is modeled in the beam-column representation of the pile.

In the past, *t-z* curves were based directly on experimental evidence from Coyle and Reese (1966)., This led to an adopted standard that peak shaft resistance was

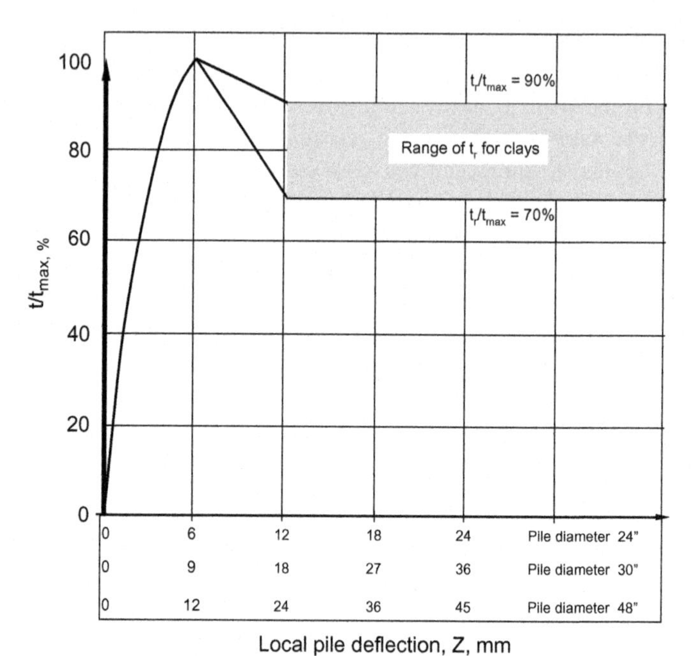

Figure 6.5 The shape of the t-z curve for clay soil for different pile diameters.

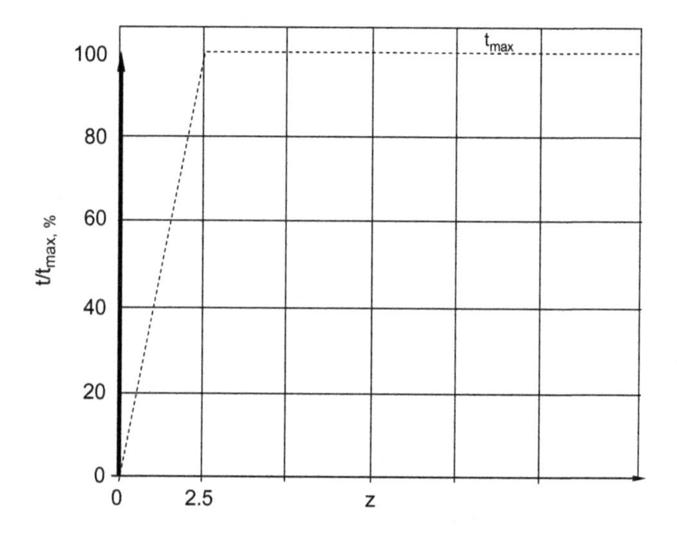

Figure 6.6 The shape of the t-z curve for sand soil for different pile diameters.

mobilized at a vertical relative pile-soil movement (Z_c) of 2.54 mm (0.1 in.) in sand. For clay, the value is 1% of pile diameter. It has been shown theoretically and experimentally that the form of the *t-z* curve is a function of the pile length and diameter, soil stiffness, and shaft resistance, as discussed by Kraft et al. (1981a, 1981b). However, to account for these characteristics, the average shear modulus of the soil must be known. In sand, the appropriate strain-level shear modulus is known only within an order of magnitude, and hence *t-z* curves generated by this method will contain considerable uncertainty. However, a parametric study by Meyer et al. (1975) showed that a sixfold variation in soil yield displacement had only a small effect on the predicted pile head displacement.

The data include a peak-residual behavior for *t-z* curves in clay, the governing parameter of which is the ratio of peak to residual unit skin friction. The recommended range for this parameter is 0.7 to 0.9. Vijayvergiya (1977) indicates that this parameter decreases with an increasing overconsolidation ratio. The peak residual behavior in sand has been adopted as per Wiltsie et al. (1982).

For some projects, in the absence of more definitive criteria, the recommended *t-z* curves for noncarbonate soils, according to API RP2A, are shown in Figure 6.6.

Table 6.10 presents the relation between the vertical displacement of the pile to the pile diameter and the adhesion between the pile and the soil as a percentage of the total friction capacity .diameter The shape of the *t-z* curve at displacements greater than z_{max}, as shown in Figure 6.7, should be carefully considered.

The ratio between the residual adhesion to the maximum adhesion between pile and soil, t_r/t_{max}, at the axial pile displacement at which it occurs (z_r) is affected by

Table 6.10 Relation between the ratio of pile deflection to the diameter and the skin friction capacity

Soil type	z/D	Pile axial deflection for different pile diameters, mm			t/t_{max}
		24″	36″	48″	
Clays	0.0016	0.98	1.46	2.0	0.30
	0.0031	1.89	2.83	3.8	0.50
	0.0057	3.5	5.21	7.0	0.75
	0.0080	4.9	7.32	9.8	0.90
	0.0100	6.1	9.14	12.2	1.00
	0.0200	12.2	18.3	24.4	0.70–0.90
	∞				0.70–0.90
Sands	**z (mm)**				t/t_{max}
	0.000	0	0	0	0.00
	2.5	2.5	2.5	2.5	1.00
	∞				1.00

Note: z = local pile deflection (in mm), D = pile diameter (in mm), t = mobilized soil pile adhesion (in lb/ft^2 or kPa), and t_{max} = maximum soil pile adhesion or unit skin friction capacity computed (in lb/ft^2 or kPa).

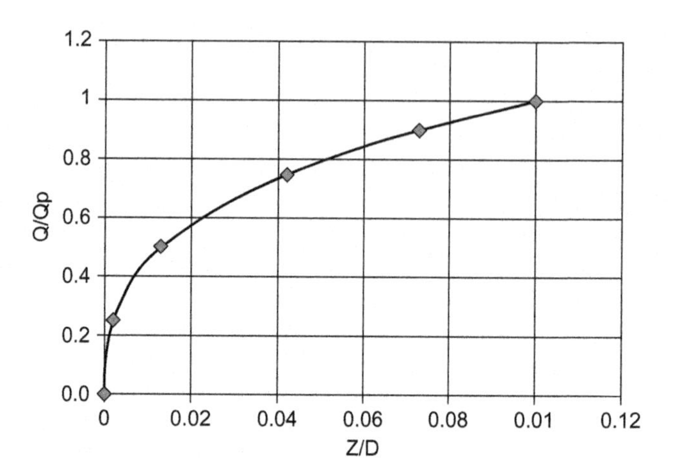

Figure 6.7 Pile tip load displacement (Q-z) curve.

the soil stress-strain behavior, stress history, pipe installation method, pile load sequence, and other factors.

The value of t_r/t_{max} can range from 70% to 90%. Laboratory, in-situ or model pile tests can provide valuable information for determining values of t_r/t_{max} and z_r for various soils.

The end-bearing or tip-load capacity should be determined. However, relatively large pile tip movements are required to mobilize the full end-bearing resistance. A pile-tip displacement up to 10% of the pile diameter may be required for full mobilization in both sand and clay soils. In the absence of more definitive criteria, the curve in Figure 6.7 is recommended for both sands and clays.

Table 6.11 presents the relation between the axial displacement of the pile relative to the pile diameter and the end-bearing capacity as a percentage for the total end bearing capacity.

6.6.3.4 *Axial pile capacity*

A number of studies were performed that were aimed at collecting and comparing axial capacities from relevant pile load tests to those predicted by traditional offshore pile design procedures. Studies like these can be very useful in tempering one's judgment in the design process. It is clear, for example, that there is considerable scatter in the various plots of measured versus predicted capacities. The designer should be aware of the many limitations of such comparisons when making use of the results. Limitations of particular importance include

1. There is considerable uncertainty in the determination of both predicted capacities and measured capacities. For example, determination of the predicted capacities is very sensitive to the selection of the undrained shear strength profile, which itself is subject to

Table 6.11 Relation between the axial deflection to pile diameter ratio and percentage of end-bearing capacity

z/D	Pile axial deflection for different pile diameters, mm			Q/Q_p
	24″	36″	48″	
0.002	1.2	1.8	2.4	0.25
0.013	7.9	11.9	15.8	0.50
0.042	25.6	38.4	51.2	0.75
0.073	44.5	66.8	89	0.90
0.100	61	91.4	121.9	1.00

Note: z = axial tip deflection (in mm), D = pile diameter (in mm), Q = mobilized end-bearing capacity (in lb or KN) and Q_p = total end bearing (in lb or KN).

considerable uncertainty. The measured capacities are also subject to interpretation as well as possible measurement errors.

2. Conditions under which pile load tests are conducted generally vary significantly from design loads and field conditions. One clear limitation is the limited number of tests on deeply embedded, large-diameter, high-capacity piles. Generally, pile load tests have capacities that are 10% or less of the prototype capacities. Briaud et al. (1984) mentioned that the rate of loading and the cyclic load history are usually not well represented in load tests. According to Clarke (1993), pile load tests are often conducted before full setup occurs, for practical reasons. Furthermore, pile-tip conditions (closed versus open ended) may differ from offshore piles.
3. In most of the studies, an attempt has been made to eliminate the factors thought to be significantly affected by extraneous conditions in load testing, such as protrusions on the exterior of the pile shaft (weld beads, cover plates, etc.), installation effects (jetting, drilled-out plugs, etc.), and artesian conditions, but it is not possible to be absolutely certain in all cases.

The tests most relevant to offshore applications have all been conducted in the United States or Europe. As regional geology and particularly operating experience are considered very important in foundation design, care should be exercised in applying these results to other regions of the world. In addition, the designer should note that certain important tests in silty clays of low plasticity, indicate overprediction of frictional resistance by equations (6.12) and (6.13). The reason for the overprediction is not well understood and has been an area of active research. The designer is therefore cautioned that pile design for soils of this type should be given special consideration. Additional considerations that apply to drilled and grouted piles are discussed by Kraft and Lyons (1974) and O'Neill and Hassan (1994).

Pile load tests are commonly used as basis for determining pile load-movement characteristics. As discussed by El-Reedy (2011), in clay, the ultimate capacity of the pile reaches a maximum value at some movement, beyond which there is a gradual drop to a residual value.

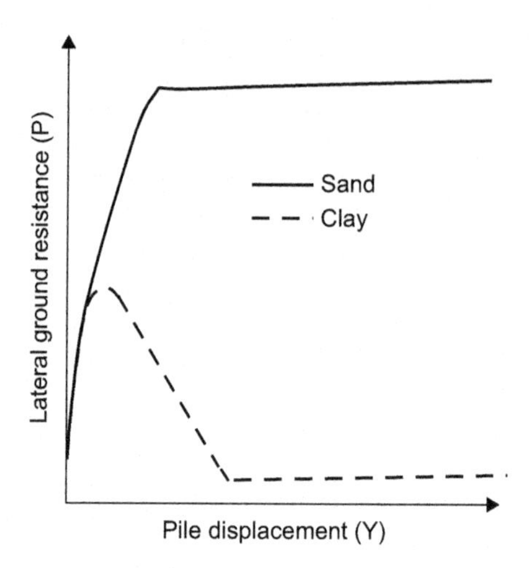

Figure 6.8 Example of typical *p-y* curve for 36″ pile.

The frictional resistance increases rapidly and reaches a maximum value at a very small displacement, referred to as the *critical movement*. However, the point resistance continues to increase beyond this critical movement and tends to reach a maximum value at a relatively larger movement. This maximum value is referred to as the *end-bearing capacity*.

In sand, the ultimate capacity seems to increase and reach a constant value. The point resistance in sand continues to increase gradually. This is probably why a pile in sand does not usually reach a plunging failure during a load test.

The relation between the lateral resistance and displacement for a 36-in. diameter pile in clay and sand is shown in Figure 6.8.

6.6.3.5 Laterally loaded piles reaction

The pile foundation should be designed to sustain lateral loads, whether static or cyclic. Additionally, the designer should consider overload cases in which the design lateral loads on the platform foundation are increased by an appropriate safety factor.

The designer should be satisfied that the overall structural foundation system will not fail under overloads. The lateral resistance of the soil near the surface is significant to pile design, and the effects of scour and soil disturbance on this resistance during pile installation should be considered. Generally, under lateral loading, clay soils behave as a plastic material, which makes it necessary to relate pile-soil deformation to soil resistance. To facilitate this procedure, lateral soil resistance deflection (*p-y*) curves should be constructed using stress-strain data from laboratory soil samples. The ordinate for these curves is soil resistance, *p*, and the abscissa is soil deflection, *y*. By iterative procedures, a compatible set of load-deflection values for the pile-soil system can be developed.

Matlock (1970) performed a comprehensive study of the design of laterally loaded piles in soft clay, and Reese and Cox (1975) performed a study of laterally loaded piles in stiff clay.

It is noted that these *p-y* curves are recommended for estimating pile bending moment, displacement, and rotation profiles for various (static or cyclic) loads. Different criteria may be applicable for fatigue analysis of a pile that has previously been subjected to loads larger than those used in the fatigue analysis and that resulted in "gapping" around the top of the pile.

6.6.3.6 *Lateral bearing capacity for soft clay*

According to API RP2KA (2007), for static lateral loads, the ultimate unit lateral bearing capacity of soft clay p_u has been found to vary between $8c$ and $12c$, except at shallow depths, where failure occurs in a different mode due to minimum overburden pressure. Cyclic loads cause deterioration of lateral bearing capacity below that for static loads.

In the absence of more definitive criteria, the following is recommended: p_u increases from $3c$ to $9c$ as X increases from 0 to X_R according to

$$p_u = 3c + \gamma X + J(cX/D) \tag{6.17}$$

and

$$p_u = 9c, \quad \text{for } X \geq X_R \tag{6.18}$$

For a condition of constant strength with depth, equations (6.17) and (6.18) are solved simultaneously to give the following equation:

$$X_R = 6D/[(\gamma D/c) + J] \tag{6.19}$$

where p_u = ultimate resistance (in kPa), c = undrained shear strength for undisturbed clay soil samples (in kPa), D = pile diameter (in mm), γ = effective unit weight of soil (in MN/m^3), J = a dimensionless empirical constant with values ranging from 0.25 to 0.5 (having been determined by field testing; a value of 0.5 is appropriate for Gulf of Mexico clays, so it should be checked in another areas), X = depth below soil surface (in mm), and X_R = depth below soil surface to bottom of reduced resistance zone (in mm).

Where the strength varies with depth, equations (6.19) and (6.20) may be solved by plotting the two equations, that is, p_u versus depth. The point of first intersection of the two equations is taken to be X_R.

These empirical relationships may not apply where strength variations are erratic. In general, minimum values of X_R should be about 2.5 pile diameters.

On the other hand, in soft clay, the load-deflection (*p-y*) curves for lateral soil resistance-deflection relationships for piles are generally nonlinear. The *p-y* curves

Table 6.12 Relation between pile lateral load and lateral deflection

p/p_u	y/y_c
0.00	0.00
0.23	0.1
0.33	0.3
0.5	1.0
0.72	3.0
1.00	8.00
1.00	∞

Table 6.13 Relation between pile load and lateral displacement

$X > X_R$		$X < X_R$	
p/p_u	y/y_c	p/p_u	y/y_c
0.00	0.00	0.00	0.0
0.23	0.1	0.23	0.1
0.33	0.3	0.33	0.3
0.50	1.0	0.50	1.0
0.72	3.0	0.72	3.1
0.72	∞	0.72 X/X_R	15.0
		0.72 X/X_R	∞

for the short-term static load case may be generated from Table 6.12, where p = actual lateral resistance (in kPa), y = actual lateral deflection (in m),

$$y_c = 2.5\varepsilon_c D \tag{6.20}$$

and ε_c is the strain that occurs at one half the maximum stress on laboratory unconsolidated, undrained, compression tests of undisturbed soil samples.

For the case where equilibrium has been reached under cyclic loading, the *p-y* curves may be generated from Table 6.13.

6.6.3.7 Lateral bearing capacity for stiff clay

For static lateral loads, the ultimate bearing capacity p_u of stiff clay ($c > r$ 96 kPa), as for soft clay, would vary between $8c$ and $12c$. Due to rapid deterioration under cyclic loading, the ultimate resistance is reduced to something considerably less and should be so considered in cyclic design. Furthermore, because stiff clays also have nonlinear stress-strain relationships, they are generally more brittle than soft clays. In developing stress-strain curves and subsequent *p-y* curves for cyclic loads, good judgment should reflect the rapid deterioration of load capacity at large deflections for stiff clays.

In general, under lateral loads, clay soils behave as a plastic material, which makes it necessary to relate pile-soil deformation to soil resistance.

For a more detailed study of the construction of *p-y* curves, see Matlock (1970) for soft clay, Reese and Cox (1975) for stiff clay, O'Neill and Murchison (1983) for sand, and Georgiadis (1983) for layered soils.

6.6.3.8 Lateral bearing capacity for sand

A series of studies verified the theoretical studies with the field-test results during lateral loading of a 24-in. diameter test pile installed at sites with clean, fine sand, and silty sand. (The studies were funded by Amoco's production company, Chevron oil field research, Esso's production research company, Mobil Oil Corporation, and Shell's development company.) The results suggest a shape for the *p-y* curve as shown in Figure 6.8 where the initial part is a straight line representing the elastic behavior and the horizontal straight line represents the plastic behavior, with the straight lines connected by a parabola.

The values of P_m and p_c are a function of the ultimate soil resistance. A difference between the ultimate resistance from theory and that from experiments was observed, which was covered by empirical factors. Another study, by O'Neill and Murchison (1983), evaluated *p-y* relationships in sand. API RP2A (2007) recommends that the *p-y* curve be calculated using the information from that study.

The ultimate lateral bearing capacity for sand has been found to vary from a value at shallow depths determined by equation (6.21) to a value at deep depths determined by equation (6.22). At a given depth, the equation giving the smallest value of p_u should be used as the ultimate bearing capacity.

$$p_{us} = (C_1 H + C_2 D)\gamma H \tag{6.21}$$

$$p_{ud} = C_3 D \gamma H \tag{6.22}$$

where p_u = ultimate resistance (force/unit length) (in kN/m) (s = shallow, d = deep), γ = effective soil weight (in KN/m^3), H = depth (in m), ϕ' = angle of internal friction of sand (in degrees), C_1, C_2, C_3 = coefficients determined from Table 6.14 as function of ϕ', and D = average pile diameter from surface to depth (in m).

The relationship between lateral soil resistance and deflection (*p-y* curve) for sand is nonlinear. If no definitive information is available, the curve may be approximated at any specific depth H, according to API RP2A, by the following equations:

$$P = Ap_u \tan\text{h}\left[\frac{kH}{Ap_u}y\right] \tag{6.23}$$

where A is a factor to account for the cyclic or static loading condition, evaluated by $A = 0.9$ for cyclic loading and by $A \geq 0.9$ for static loading, so that

Table 6.14 **Coefficients C_1, C_2, C_3**

Angle of internal friction, ϕ	C_1	C_2	C_3
20	0.6	1.5	8.5
21	0.7	1.6	9.6
22	0.8	1.7	10.8
23	0.9	1.8	12.2
24	1.0	1.9	13.8
25	1.1	2.0	15.6
26	1.2	2.1	17.6
27	1.3	2.2	19.9
28	1.4	2.3	22.5
29	1.6	2.5	25.4
30	1.7	2.6	28.7
31	1.9	2.7	32.4
32	2.1	2.9	36.6
33	2.3	3.0	41.4
34	2.5	3.2	46.7
35	2.8	3.4	52.8
36	3.1	3.6	59.6
37	3.4	3.8	67.4
38	3.8	4.0	76.1
39	4.2	4.2	86.0
40	4.6	4.4	101.5

$$A = [3.0 - 0.8(H/D)] \tag{6.24}$$

where p_u = ultimate bearing capacity at depth H (in kN/m); k = initial modulus of subgrade reaction (in kN/m^3), as determined from Table 6.15 as function of the angle of internal friction, ϕ', and the relative density for sand under the water table. y = lateral deflection (in inches or m); and H = depth (in m).

6.6.3.9 *Changes in axial capacity in clay with time*

The pile capacity calculated from the previous equation does not consider the effect of time on the pile capacity. Note that, in the old platform constructed 40 years ago and more, if you review the calculation, you find that its factor of safety does not follow that proposed by the API and the environment over time will surely affect the pile capacity. Normal phenomena after longtime there will be a good bond between the pile and the surrounding soil and this additional adhesion is not considered in the calculation. Therefore, a studies were performed to define the behavior of the axial capacity in clay soil with time. Clarke (1993) and Bogard and Matlock (1990)) conducted field measurements studies in which it was shown that the time

Table 6.15 Relation between subgrade reaction, angle of internal friction, and relative density for sand below the water table

Soil type	Angle of internal friction, ϕ	Relative density, %	Subgrade reaction, K, t/m^3
Very loose	<29	20	265.7
Loose sand	29–30	25	426.3
		30	553.6
		35	744.6
		40	996.4
Medium dense	30–36	45	1356.3
		50	1716.1
		55	2026.1
		60	2491.1
Dense	36–40	65	2850.9
		70	3293.8
		75	3792.0
		80	4262.6

required for driven piles to reach ultimate capacity in a cohesive soil can be relatively long, as much as 2–3 years.

It is worth mentioning that the rate of strength gain is highest immediately after driving, and the rate decreases during the dissipation process. Therefore, a significant strength increase can occur in a relatively short time.

During pile driving in normally to lightly overconsolidated clay, the soil surrounding a pile is significantly disturbed, the stress state is altered, and large excess pore pressures can be generated. After installation, these excess pore pressures begin to dissipate, which means that the soil mass around the piles begins to consolidate, so the pile capacity increases with time. This process is usually referred to as *setup*. The rate of excess pore pressure dissipation is a function of the coefficient of radial consolidation, pile radius, plug characteristics, and soil layering.

In the most popular case, where the driven pipe piles supporting a structure have design loads applied to the piles shortly after installation, the time-consolidation characteristics should be considered in pile design. Noting that, in traditional, fixed offshore structure installations, the time between the pile installation and the platform being total loaded is in the range of 1–3 months. In some cases, the commissioning and startup come early; in this case, such information should be transferred to the engineering office, as the expected increase in capacity with time is an important design variable that can affect the safety of the foundation system during early stages of the consolidation process.

The relation between the pore pressure and load test data at different times after pile driving is expressed by an empirical correlation: the degree of consolidation, degree of plugging, and pile shaft shear transfer capacity. The test results for closed-ended steel piles in heavily overconsolidated clay indicate no significant change in capacity with time. This is contrary to tests on 0.273 m (10.75 in.) diameter closed-ended steel piles in overconsolidated Beaumont clay, where considerable and rapid setup in 4 days was found, so the pile capacity at the end of installation was never fully recovered.

So, it is very important to highlight that the axial capacity of the pile over time is under research and development, and there is no solid formula or equation to follow. Focus is on the research done on the specific site locations and depends on the previous history of this location.

6.6.4 Pile capacity calculation methods

API RP2A (2007) presents new methods for calculating pile capacity based on the cone penetration test.

As it is presented previously, a simple method for assessing pile capacity in cohesionless soils, this method is recommended in previous editions of API RP2A-WSD. Changes were made to remove potential nonconservatism in previous editions. There are reliable CPT-based methods for predicting pile capacity. All these methods are based on direct correlations of pile unit friction and end-bearing data with cone tip resistance (q_c) values from CPT. These CPT-based methods cover a wider range of cohesionless soils, are considered fundamentally better, and have shown statistically closer predictions of pile load test results.

The new CPT-based methods for assessing pile capacity in sand are preferred to the previous method. However, more experience is required with all these new methods before any single one can be recommended for routine design over previously presented methods. API states clearly that the new CPT-based methods should be used only by qualified engineers who are experienced in interpreting CPT data and who understand the limitations and reliability of the CPT-based methods.

The assumption is made that friction and end-bearing components are uncoupled. Hence, for all methods, the ultimate bearing capacity in compression (Q_d) and tensile capacity (Q_t) of plugged open-ended piles are determined by

$$Q_t = Q_f + Q_p = P_o \int f_{c,z} dz + A_q q_p \quad (6.25)$$

$$Q_f = P_o \int f_{t,z} dz \quad (6.26)$$

Note that, since the friction component, Q_f, involves numerical integration, results are sensitive to the depth increment used, particularly for CPT-based methods. As guidance, depth increments for CPT-based methods should be on the order of {1/100} of the pile length (or smaller). In any case, the depth increment should not exceed 0.3 m.

Four recommended CPT-based methods are mentioned in API RP2A:

1. Simplified ICP-05.
2. Offshore UWA-05 (Lehane et al., 2005a,b).
3. Fugro-05 (Lehane et al., 2005a; Kolk et al., 2005).
4. NGI-05 (Lehane et al., 2005a; Clausen et al., 2005).

The first method is a simplified version of the design method recommended by Jardine et al. (2005), whereas the second is a simplified version of the UWA-05 method applicable to offshore pipe piles. Methods 2, 3 and 4 are summarized by Lehane et al. (2005a). Friction and end-bearing components should not be taken from different methods.

The unit skin friction formulas for open-ended steel pipe piles for the first three recommended CPT-based methods (Simplified ICP-05, Offshore UWA-05, and Fugro-05) can be considered as special cases of the general formula

$$f_z = u \cdot q_{\int cz}\left(\frac{\sigma'_{vo}}{P_a}\right)A_r^b\left[\max\left(\frac{L-z}{D},v\right)\right]^{-c}[\tan\delta_{cv}]^d\left[\min\left(\frac{L-z}{D}\frac{1}{v},1\right)\right] \quad (6.27)$$

where f_z is the unit skin friction, δ_{cv} is pile–soil constant-volume interface friction angle, L *is* the pile length underneath the seabed, $A_r = 1-(D_i/D)^2$, D_i is the pile inner diameter ($D_i = D - 2t$), z is the depth under the seabed, q_{cz} is the CPT tip resistance at depth z, D is the outer diameter, t is the wall thickness, and P_a is the atmospheric pressure equal to 100 kPa.

Table 6.16 provides the recommended values for parameters *a, b, c, d, e, u*, and *v* for compression and tension, which are the unit skin friction parameter values for driven open-ended steel piles for the Simplified ICP-05, Offshore UWA-05, and Fugro-05 methods.

Additional recommendations for computing unit friction and end bearing for all four CPT-based methods are presented in the following subsections.

Table 6.16 Unit skin friction parameter values for driven open-ended steel pipes (simplified ICP-05, Offshore UWA-05, and Fugro-05 methods)

Method	Load direction	*a*	*b*	*c*	*d*	*e*	*u*	*v*
Simplified ICP-05	Compression	0.1	0.2	0.4	1	0	0.023	$4\sqrt{Ar}$
	Tension	0.1	0.2	0.4	1	0	0.016	$4\sqrt{Ar}$
Offshore UWA-05	Compression	0	0.3	0.5	1	0	0.030	2
	Tension	0	0.3	0.5	1	0	0.022	2
Fugro-05	Compression	0.05	0.45	0.90	0	1	0.043	$2\sqrt{Ar}$
	Tension	0.15	0.42	0.85	0	0	0.025	$2\sqrt{Ar}$

6.6.4.1 Application of CPT

By using CPT measurement as the basis for the previous methods to calculate the unit skin friction and end bearing for the pile, some precautions and information should be considered as in obtaining *t-z* data for an axial load-deformation response. The peak unit skin friction in compression and tension at a given depth, f_{cz} and f_{tz}, are not unique and both depend on the pile geometry. In general, the axial load and deformation response are affected by the pile penetration depth, the pile diameter, and its wall thickness. Note that an increased pile penetration decreases these ultimate values at a given depth.

In doing the test to obtain the *q-z* data for an axial load-deformation response, the end bearing (Q_p) is assumed to be fully mobilized at a pile tip-displacement value of $0.1D$.

Soil types such as carbonate sands, micaceous sands, glauconitic sands, volcanic sands, silts, and clayey sands have unusually weak structures with compressible grains. These types require special consideration in-situ and laboratory tests for selection of an appropriate design method and design parameters, according to Thompson and Jardine (1998) and Kolk (2000) for pile design in carbonate sand and to Jardine et al. (2005).

It is worth to mention that using CPT in cohesionless soil, such as gravels when particle sizes are in excess of 10% of the CPT cone diameter, is misleading, and one possible approach could be to use the lower-bound q_c profile. In this case, one can estimate the end-bearing capacity profile from the adjacent sand layers.

In using CPT in weaker clay layers near the pile tip, it is recommend that q_c data averaged between $1.5D$ above the pile tip to $1.5D$ below the pile tip level should generally be satisfactory, provided q_c does not vary significantly. The UWA method should be used if significant q_c variations occur to compute $q_{c,av}$.

Using CPT data on a thin clay layer, which is less than around 0.1 D thick, is a problem especially when CPT data are discontinuous vertically or not all pile locations have been investigated. From practical point of view, the offshore piles usually develop only a small percentage of q_p under extreme loading conditions. So, the finite element method can be used in calculating the pile capacity, and settlement of a pile tip on sand containing weaker layers may be considered to assessthe axial pile response under such conditions.

It is recommended that the end-bearing component be reduced if the pile tip is within a zone up to $\pm 3D$ from such layers. When q_c data averaging is also applied to this $\pm 3D$ zone, the combined effects may be unduly cautious, and such results should be critically reviewed. This rule applies also in cases of large pile diameter ($D > 2$ m).

6.7 Pile wall thickness

The wall thickness of the pile may vary along its length and may be controlled at a particular point by any one of several loading conditions or requirements. The minimum wall thickness is shown in Table 6.17.

Table 6.17 Minimum pile wall thickness

Pile diameter, in mm (in.)	Nominal wall thickness, in mm (in.)
610 (24)	13 (½″)
762 (30)	14 (9/16″)
914 (36)	16 (⅝″)
1067 (42)	17 (11/16″)
1219 (48)	19 (¾″)
1524 (60)	22 (⅞″)
1829 (72)	25 (1″)
2134 (84)	28 (1⅛″)
2438 (96)	31(1¼″)
2743 (108)	34 (1⅜″)
3048 (120)	37 (1½″)

The allowable pile stresses should be the same as those permitted by the American Institute of Steel Construction (AISC) specification for a compact hot-rolled sections. A rational analysis considering the restraints placed on the pile by the structure and the soil should be used to determine the allowable stresses for the portion of the pile that is not laterally restrained by the soil. General column buckling of the portion of pile below the mudline need not be considered unless the pile is believed to be laterally unsupported because of extremely low soil shear strengths, large computed lateral deflections, or some other reason.

6.7.1 Design pile stresses

The pile wall thickness in the vicinity of the mudline, and possibly at other points, is normally controlled by the combined axial load and bending moment that result from the design loading conditions for the platform.

The moment curve for the pile may be computed with soil reactions determined taking into consideration possible soil removal by scour. It may be assumed that the axial load is removed from the pile by the soil at a rate equal to the ultimate soil-pile adhesion divided by the appropriate pile safety factor, as specified in Table 6.9. When lateral deflections associated with cyclic loads at or near the mudline are relatively large, exceeding y_c for soft clay, consideration should be given to reducing or neglecting the soil-pile adhesion through this zone.

6.7.2 Stresses due to the weight of the hammer during hammer placement

Each pile or conductor section on which a pile hammer will be placed should be checked for stresses due to placing the equipment. These loads may be the limiting factors in establishing the maximum length of add-on sections. This is particularly true in cases where piling will be driven or drilled on a batter. The most frequent effects include static bending, axial loads, and arresting lateral loads generated during initial hammer placement.

Experience indicates that reasonable protection from failure of the pile wall due to these loads is provided if the static stresses are calculated as follows:

1. The pile projecting section should be considered a freestanding column with a minimum effective length factor, K, of 2.1 and a minimum reduction factor, C_m, of 1.0.
2. Bending moments and axial loads should be calculated using the full weight of the pile hammer, cap, and leads acting through the center of gravity of their combined masses, and the weight of the pile add-on section, with due consideration to pile batter eccentricities.

The bending moment so determined should not be less than that corresponding to a load equal to 2% of the combined weight of the hammer, cap, and leads applied at the pile head and perpendicular to its centerline.

Allowable stresses in the pile should be calculated with the allowable stress design by AISC. Note that the one-third increase in stress should not be allowed.

Consideration should also be given to the stresses that occur in the freestanding pile section during driving, as shown in Figures 6.9 and 6.10. Generally, stresses

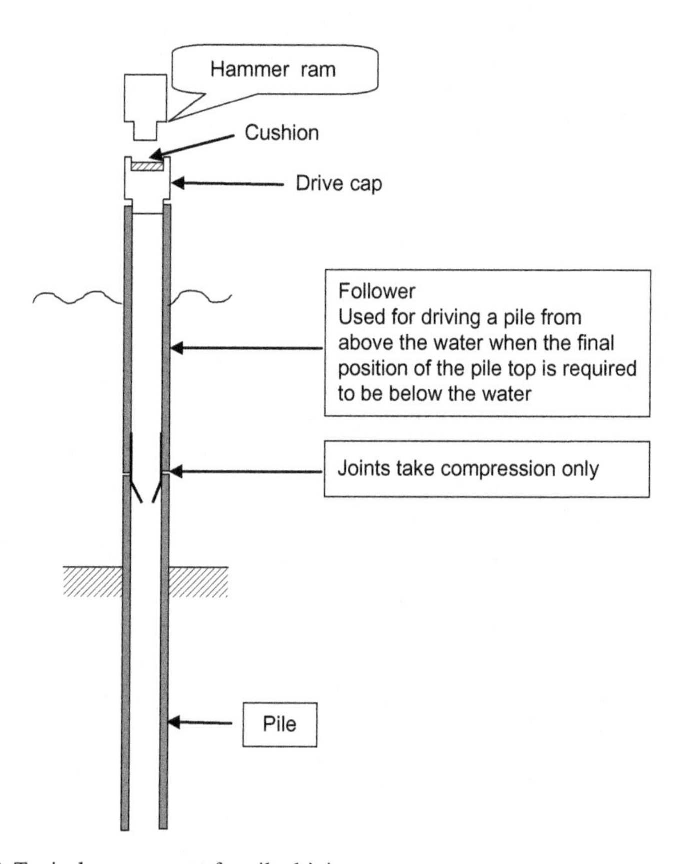

Figure 6.9 Typical arrangement for pile driving.

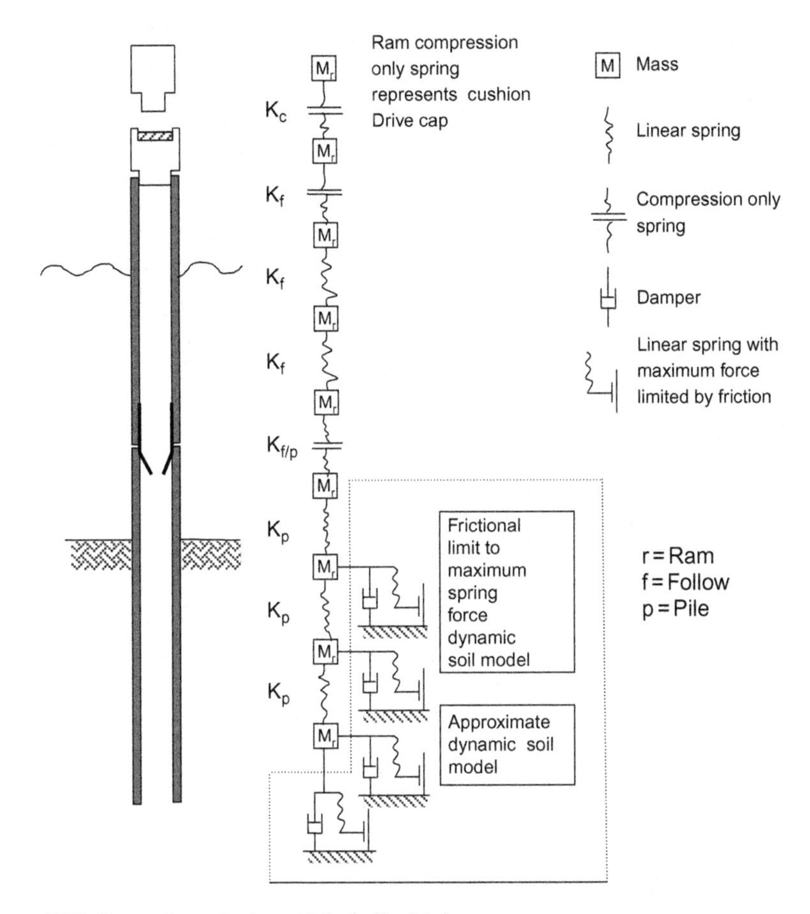

Figure 6.10 Dynamic analysis model of pile driving.

are checked based on the conservative criterion that the sum of the stresses due to the dynamic stresses caused by the impact from the hammer and the stresses due to axial load and bending (the static stresses) should not exceed the minimum yield stress of the steel.

Less conservative criteria are permitted, provided that they are supported by sound engineering analysis and empirical evidence. A method of analysis based on wave propagation theory should be used to determine the dynamic stresses.

In general, it may be assumed that column buckling will not occur as a result of the dynamic portion of the driving stresses. The dynamic stresses should be less than 80–90% of yield, depending on specific circumstances, such as the location of the maximum stresses down the length of pile, the number of blows, previous experience with the pile-hammer combination, and the confidence level in the analyses. Separate considerations apply when significant driving stresses may be transmitted into the structure and damage to appurtenances must be avoided.

The static stress during driving may be taken to be the stress resulting from the weight of the pile above the point of evaluation plus the pile hammer components actually supported by the pile during the hammer blows, including any bending stresses resulting from these. With hydraulic hammers, it is possible that the driving energy may exceed the rated energy, and this should be considered in the analyses. Also, the static stresses induced by hydraulic hammers need to be computed with special care, due to the possible variations in driving configurations, for example, when driving vertical piles without lateral restraint and exposed to environmental forces. The pile hammers evaluated for use during driving should be noted by the designer on the installation drawings or specifications.

In the past, several case histories were reported that describe some of the unusual characteristics of foundations on carbonate soils and their often poor performance. It has been shown from numerous pile load tests that piles driven into weakly cemented and compressible carbonate soils mobilize only a fraction of the capacity, as low as 15%, predicted by conventional design/prediction methods for siliceous material of the type generally encountered in the Gulf of Mexico. On the other hand, dense, strongly cemented carbonate deposits can be very competent foundation material. Unfortunately, the difficulty in obtaining high-quality samples and the lack of generalized design methods sometimes make it difficult to predict where problems may occur.

The energy is determined primarily by the mass of the ram and its impact velocity:

$$E = 0.5\, mV^2 \tag{6.28}$$

Note that only 60–70% of the energy is typically transferred to the drive cap from the ram. It is obvious that the greater the energy, the greater will be the penetration; on the other hand, the greater will be the risk of damaging the pile. The maximum stress in the stress wave is largely determined by the velocity of the ram. It is worth mentioning that, for easy driving conditions, long duration and a low stress waveforms are the best and this could be achieved by a heavier and slower ram and a soft cushion.

Using the finite difference method, the ram is represented by a concentrated mass, and the required information about the ram is available from the hammer manufacturer (Table 6.18). The efficiency of the hammer depends on the conditions as well as the operating procedures, as shown in Table 6.19 for various hammer types.

6.7.3 Minimum wall thickness

API RP2A defines the minimum wall thickness of the pile based on the *D/t* ratio of the entire length of a pile should be small enough to preclude local buckling at stresses up to the yield strength of the pile material. Consideration should be given to the different loading situations occurring during the installation and service life of a piling. For in-service conditions and those installations where normal pile driving is anticipated or where piling installation will be by means other than driving, the limitations should be considered to be the minimum requirements.

Table 6.18 Pile hammering data

Contract no.:	Project:	Structure name:	
Pile driving contractor			
Hammer	Manufacturer:	Model No.	
	Hammer type:	Serial no.	
	Manufacturers maximum rated energy, m-KN	Stroke at maximum rated energy, m	Range in operating energy to m-kN
	Range in operating stroke to m-kN	Rame weight, kips	
Strike plate	Weight, kg	Diameter mm	Thickness mm
Hammer cushion	Area, mm^2	Plate thickness, mm	
	No. of plates, mm	Total thickness of hammer cushion, mm	
Helmet (drive head)	Weight, kg		
Pile cushion	Materials:	Area, mm^2	
	Sheet thickness, mm	No. of sheet:	
	Total thickness of pile cushion, mm		
Pile	Pile type:	Wall thickness, mm	
	Cross-sectional area: mm^2	Weight/m:	
	Ordered length: m	Design load, ton	
	Ultimate pile capacity, ton		

Table 6.19 Efficiency for different hammer types

Hammer	Efficiency
Single-acting steam or air	0.75–0.85
Double-acting steam or air	0.70–0.80
Diesel	0.85–1.0
Hydraulic	0.85–0.95

For piles that are to be installed by driving where sustained hard driving (250 blows per foot, 820 blows per meter) with the largest size hammer is anticipated to be used, the minimum piling wall thickness used should be more than

$$t = 6.35 + D/100 \tag{6.29}$$

where t = wall thickness (in in. or mm) and D = diameter (in in. or mm).

The preceding requirement for a smaller *D/t* ratio when hard driving is expected may be relaxed when it can be shown by past experience or by detailed analysis that the pile will not be damaged during its installation.

6.7.4 Driving shoe and head

The purpose of driving shoes is to assist piles to penetrate through hard layers or to reduce driving resistances, allowing greater penetrations to be achieved than would otherwise be the case. Different design considerations apply for each use.

If an internal driving shoe is provided to drive through a hard layer, it should be designed to ensure that unacceptably high driving stresses do not occur at and above the transition point between the normal and the thickened section at the pile tip. Also, it is important to check that the shoe does not reduce the end-bearing capacity of the soil plug below the value assumed in the design. External shoes are not normally used, as they tend to reduce the skin friction along the length of pile above them.

The installation contractor is responsible for designing the driving head at the top of the pile, and it should be designed to ensure that it is fully compatible with the proposed installation procedures and equipment.

6.7.5 Pile section lengths

In defining the pile section lengths, consideration should be given to the following:

1. The capability of the lift equipment to raise, lower, and stab the sections.
2. The capability of the lift equipment to place the pile-driving hammer on the sections to be driven.
3. The possibility of a large amount of downward pile movement immediately following the penetration of a jacket leg closure.
4. Stresses developed in the pile section while lifting.
5. The wall thickness and material properties at field welds.
6. Avoiding interference with the planned concurrent driving of neighboring piles.
7. The type of soil in which the pile tip is positioned during driving interruptions for field welding to attach additional sections. In addition, static and dynamic stresses due to the hammer weight and operation should be considered. Each pile section on which driving is required should contain a cutoff allowance to permit the removal of material damaged by the impact of the pile-driving hammer. The normal allowance is 2 to 5 ft (0.5 to 1.5 m) per section. Where possible, the cut for the removal of the cutoff allowance should be made at a conveniently accessible elevation.

6.8 Pile drivability analysis

The pile drivability analysis has three stages:

1. Evaluation of soil resistance to driving.
2. Wave equation analysis.
3. Estimate of blow count versus pile penetration.

The procedures used to evaluate the soil resistance to driving are empirical and have been developed from the back-analysis of pile-driving records. Their use is therefore limited to pile drivability assessment by wave equation analysis and they are not intended to provide an estimate of the ultimate axial capacity of foundation piles.

In the present case, as is often true, some of the parameters required for the wave equation analysis (step 2) depend on the maximum achievable penetration, which is calculated in step 3. Therefore, to ensure consistency between the steps, an iterative analysis has been carried out.

6.8.1 Evaluation of soil resistance drive

Different procedures are used for evaluating the soil resistance drive (SRD) in cohesionless and cohesive soils. They are discussed next. As is the case for static pile capacity analysis, the components of shaft resistance and end bearing in SRD are evaluated separately, then combined to give total driving resistance (Toolan and Fox, 1977).

The variability of the soil conditions across the site and some anticipated variation in hammer performance are likely to influence the apparent driving resistance. Furthermore, the driving resistance during continuous driving is known to be considerably lower than when driving is restarted after an interruption, long enough to allow soil setup. To account for these factors, upper-bound and lower-bound SRD profiles have been formulated for a given design soil profile, based on the recommendations of Stevens et al. (1982).

6.8.2 Unit shaft resistance and unit end bearing for uncemented materials

In cohesive soil, the unit skin friction has been assessed based on the method proposed by Semple and Gemeinhardt (1981). This method was developed from back-analysis of pile installations in normally consolidated to heavily overconsolidated clays from many areas.

The unit shaft resistance component of SRD is derived using the API RP (1984) procedure for static pile capacity, modified by a pile capacity factor F_p, which is a function of the overconsolidation ratio (OCR), as follows:

$$F_p = 0.5(\mathrm{OCR})^{0.3} \tag{6.30}$$

The unit end-bearing component of SRD is taken as $9S_u$.

The OCR is defined as the ratio of the maximum past effective consolidation stress and the present effective overburden stress. OCR is a function for undrain shear strength ratio (S_u/p') which is equal to 0.22 in normally consolidated clay with shear stress angle equal to 26°. In general, OCR is obtained by CPT.

In cohesionless (granular) soil, shaft resistance and end-bearing components of SRD can be derived using the API (1984) procedure for static pile capacity together with the soil parameters specified by Stevens et al. (1982) for the particular soil type. A limiting skin friction of 15 kPa has been taken for the calcareous and carbonate sand layers.

6.8.3 Upper- and lower-bound SRD

Based on Stevens et al. (1982), four cases have been assessed to obtain upper- and lower-bound SRD values in uncemented and weakly cemented soil layers:

1. SRD for lower-bound, coring pile is the outside shaft friction in addition to the inside shaft friction as half the outside. End bearing is unit end bearing multiplied by the steel annulus area.
2. SRD for upper-bound, coring pile is similar to the previous case but with the full shaft friction on the inside of the pile.
3. SRD for lower-bound, plugged pile is the outside shaft friction in addition to the end bearing multiplied by full cross-section area.
4. SRD for upper-bound, plugged pile for cohesionless soil layers is with outside shaft friction from case 3 increased by 30% and end bearing from case 3 increased by 50%. For cohesive layers, the outside shaft friction is as in case 3 and end bearing from case 3 is increased by 67%.

For the sandstone and limestone layers, plugged conditions are unrealistic, and only the coring cases are considered.

The software GRLEAP input data and the method of obtaining the compressive strength on the pile and the number of blows every 250 mm is presented in Figure 6.11.

SRD calculations have been made to cover the range in unit values just set out. Sample results are shown in Figure 6.12. As shown in the figure, a range in SRD is provided, reflecting the various combinations in unit values given previously. The lower bounds are likely to be applicable to the minimum resistances during continuous driving, with upper bounds indicative of local variations in soil conditions and resistances expected immediately on restarting a drive (i.e., soil setup condition).

Example 6.4

Figure 6.11 presents the output from the GRLWEAP software by choosing the hammer type and pile length and its penetration. From the output curve, you have the load in the pile about 1480 KN.

Solution As per the chart of the analysis for hammer type Delmag D12-42,

Efficency = 0.80
Helmet = 7.60 kN
Hammer cushion = 10,535 kN/mm

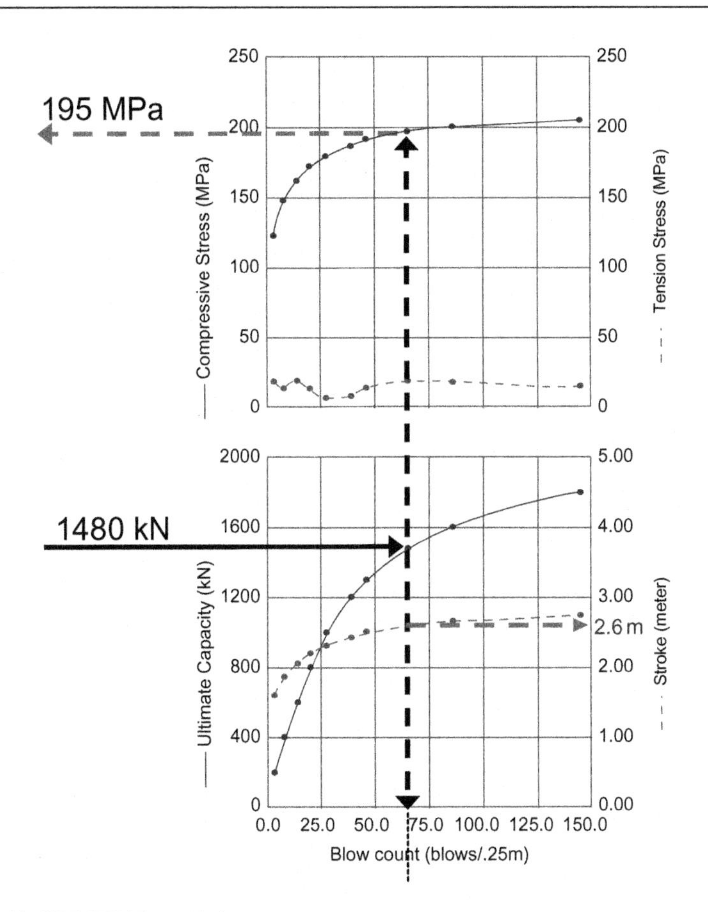

Figure 6.11 GRLWEAP example.

Based on the number of counts is around 60 counts for 250 mm depth and the compressive stress on the pile is 195 MPa, which should be check against the pile capacity.

Toe damping = 0.5 s/m
Skin quake = 2.5 mm
Toe quake = 3 mm
Skin damping = 016 s/m
Pile length = 25 m

6.8.4 Results of wave equation analysis

Blow count versus SRD curves have been developed for the hammers listed in Table 6.20, using a software program.

Achievable penetration is based on refusal criteria of 15 blows/0.25 m.

The input parameters for the wave equation analysis differed for each hammer, pile, and SRD bound, primarily in terms of the percentage of SRD that was due to

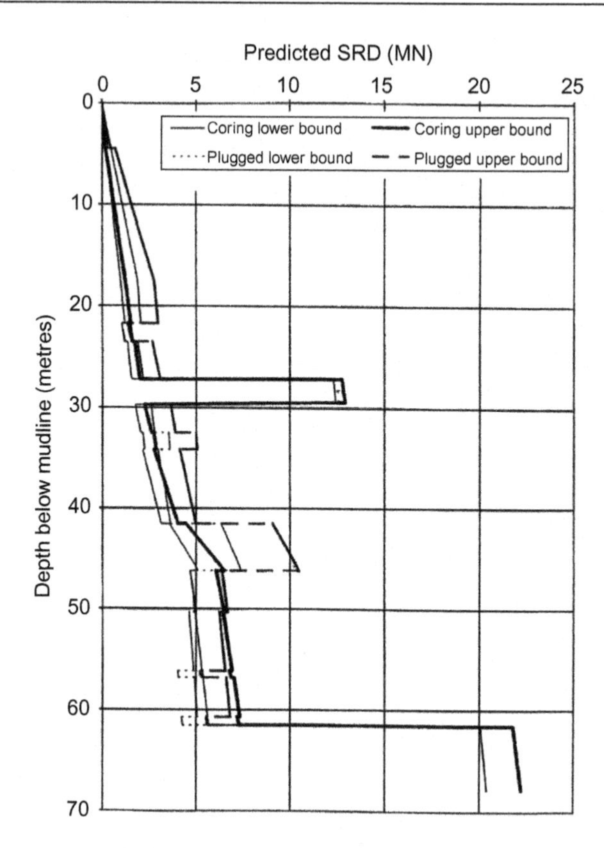

Figure 6.12 Present example for soil resistance to driving fora pipe pile 30″ with 1¼″ thickness.

Table 6.20 **Summary of drivability analysis for pile 30 in. in diameter**

Hammer	Achievable penetration	Maximum driving stresses
Delmag D-80	27.2	202
MRBS 3000	27.2	220
MRBS 3900	61.5	252

skin friction. This has an effect on the amount of energy lost through damping, because of differences in damping values for skin friction and end bearing. The following damping values were used:

Skin damping for purely cohesive soil, 0.65 s/m.
Skin damping for cohesionless soil, 0.15 s/m.
Tip damping, all soils, 0.50 s/m.

Skin damping for a combination of soil types was assessed by taking account of the fraction of total skin resistance due to each type and linearly interpolating

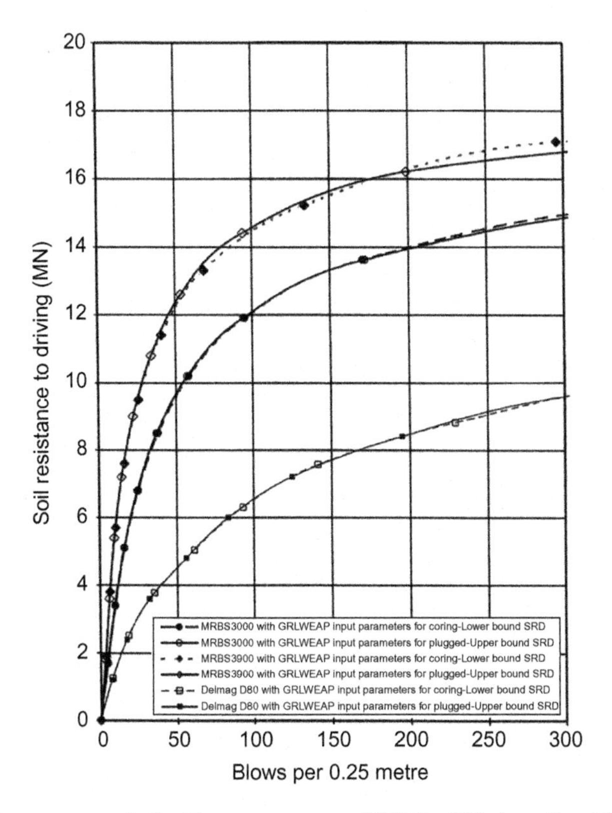

Figure 6.13 Present example for blow count versus SRD for 30″ pipe pile with 1¼″ wall thickness with three different hammers, 3000, 3900, and D80, Delmag.

between the purely cohesive and purely cohesionless values. Quakes of 0.1 in. were taken in all calculations.

The results are shown in Figure 6.13. There are two curves for each hammer analyzed, corresponding to input parameters associated with the relevant lower- and upper-bound SRD cases.

6.8.5 Results of drivability calculations

By combining the SRD versus depth curves and the wave equation results, one obtains the predicted blow count versus depth. The results are shown in Figure 6.14 and are summarized in Table 6.20.

According to a case study in the Red Sea, refusal will occur at 27.2 m penetration if the Delmag 080 hammer is used. For the Menck MRBS 3000, high blow counts will be experienced during penetration and refusal may occur, for example, if the thickness or strength of cemented material at 27.2 m increases significantly from that identified in the borehole. For the Menck MRBS 3900, and indeed for all the hammers analyzed, refusal is indicated at 61.5 m penetration.

6.8.6 Recommendations for pile installation

The lower-bound curve of blow counts versus depth in Figure 6.14 applies if there is no interruption during driving. The upper-bound curves represent estimates of effects of delays during driving. To ensure an efficient offshore pile-driving operation, it is recommended that delays during driving be avoided if possible. Attention should be paid to ensuring that the blow count does not become excessive and that no pile-tip damage occurs, for all hammers analyzed. Depending on the pile-driving plant finally selected, it may be advisable to have equipment readily available during piling operations in case refusal or particularly high blow counts occur at 27.2 m below the mudline. If refusal occurs, an assessment may be needed of the effect on ultimate pile capacity and therefore target penetration.

The pile analysis usually assumes uniform wall thickness over the entire pile length. Should the wall thicknesses of the piles finally selected differ from this, then the drivability analysis should be repeated to assess the effect on blow counts and achievable penetration. It is recommended that consideration be given to

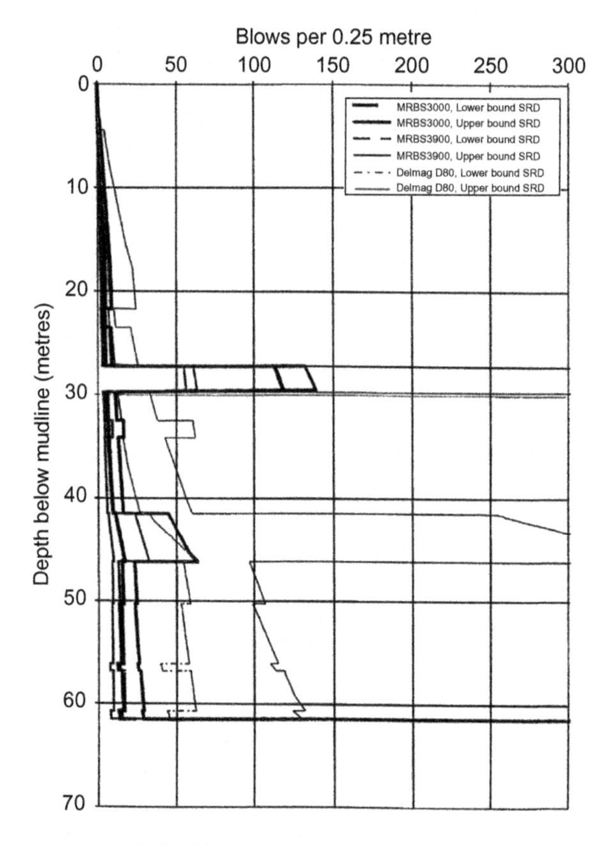

Figure 6.14 Presents example for blow count versus depth for three type of hammers for pile 30″ with 1¼″ wall thickness.

incorporation of a pile shoe to prevent undue distress to the pile during driving through the cemented layers.

The shoe should be externally flush, should have an appropriate outer bevel profile at the tip, and should have an increased wall thickness over a suitable length. As with an overall increase in pile wall thickness requiring some additional drivability assessment, the incorporation of a pile shoe also needs to be readdressed with regard to its effect on pile-driving behavior.

6.9 Soil investigation report

After all soil investigation tests on site and in the laboratory are completed, the following curves, which assist in pile design, should be presented in the soil investigation report: the curve for the relationship between the depth below the mudline and the unit skin friction (Figure 6.15) and the curve for the relationship between the pile depth below the mudline and the unit of end bearing (Figure 6.16).

The relation between the tension and compression pile capacity in different depth is as shown in Figure 6.17.

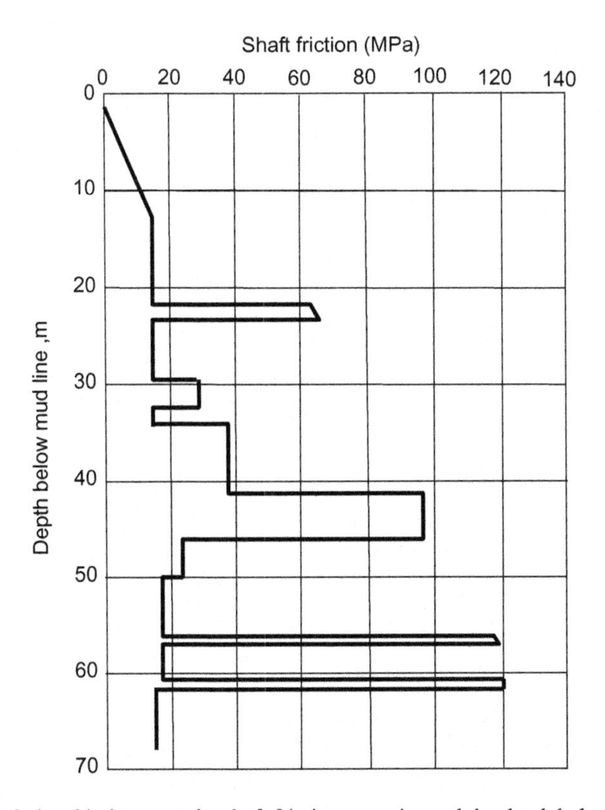

Figure 6.15 Relationship between the shaft friction capacity and the depth below the mudline.

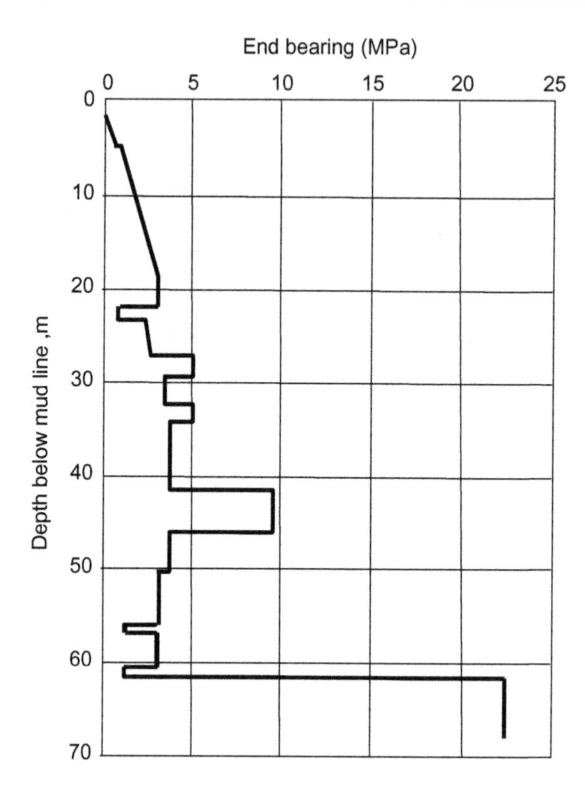

Figure 6.16 Relationship between the end-bearing capacity and the depth below the mudline.

If the pile diameter is not identified, the study should include the pile capacity in compression for different pile diameters, as presented in Figure 6.18, and the pile capacity in tension for different pile diameters, as shown in Figure 6.19.

Example 6.5

For a pile with a diameter of 1.22 m (48 in) and a depth of 80 m, the structure analysis by SACS present the compression axial force is equal to 8.2 MN in case of operating condition and 11 MN in case of extreme condition and the maximum tension force is 9 MN.

Solution As in Figure 6.18, the compression capacity os 19.6 MN and the tension capacity, as shown in Figure 6.19, is 13.8 MN.

To calculate the factor of safety under extreme conditions = 19.6/11 = 1.78 > 1.5 OK.

To calculate factor of safety under operating conditions = 1 9.6/8.2 = 2.39 > 2.0 OK.

To calculate thc factor of safety under pile pullout = 13.8/9 = 1.53 > 1.5 OK.

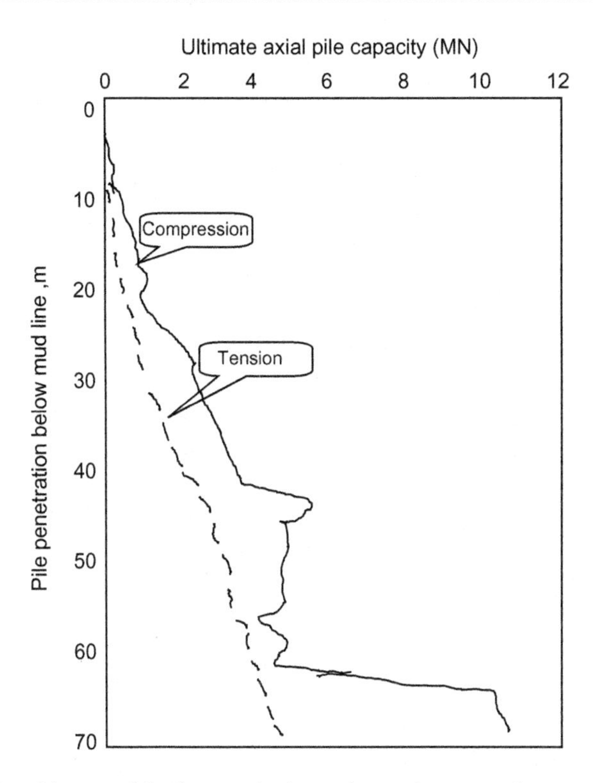

Figure 6.17 The ultimate axial pile capacity in tension and compression.

Example 6.6

For a pile with a diameter of 1.22 m (48 in) and a depth 80 m, the structure analysis by SACS presents the compression axial force as equal to 8.2 MN under operating conditions, 11 MN under extreme conditions, and the maximum tension force as 9 MN.

Solution As in Figure 6.18, the compression capacity is 19.6 MN and the tension capacity, as shown in Figure 6.19, is 13.8 MN.

To calculate the factor of safety under extreme conditions = 19.6/11 = 1.78 > 1.5 OK.

To calculate factor of safety under operating conditions = 19.6/8.2 = 2.39 > 2.0 OK.

To calculate the factor of safety under pile pullout = 13.8/9 = 1.53 > 1.5 OK.

6.10 Composite pile

Grouting of piles platform loads may be transferred to steel piles by grouting the annulus between the jacket leg (or sleeve) and the pile. The load is transferred to the pile from the structure across the grout. Experimental work indicates that the mechanism of load transfer is a combination of bond and confinement friction between the grout, the steel surfaces, and the bearing of the grout against

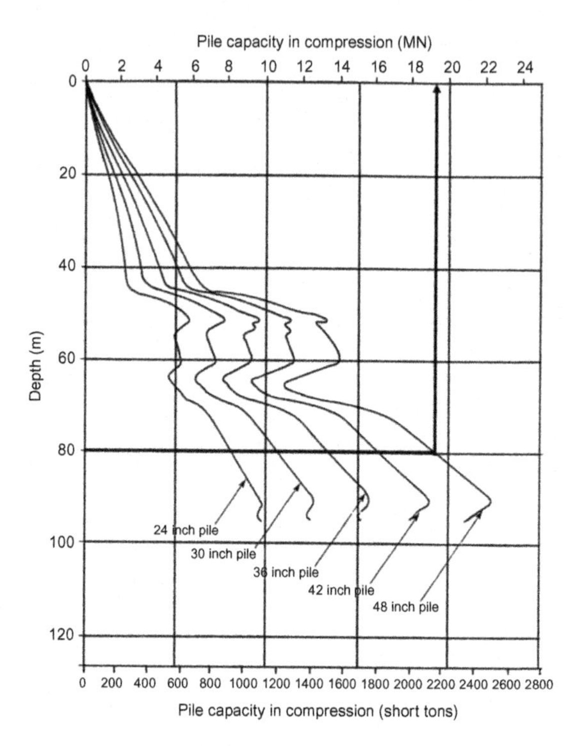

Figure 6.18 Relationship between the pile capacity in compression and depth by different pile diameters.

mechanical aids, such as shear keys. Centralizers should be used to maintain a uniform annulus or space between the pile and the surrounding structure. A minimum annulus width of 1½ in. (38 mm) should be provided where grout is the only means of load transfer. Adequate clearance between pile and sleeve should be provided, taking into account the shear keys' outer dimension. Note that packers should be used as necessary to confine the grout. Proper means for the introduction of grout into the annulus should be provided, to minimize the possibility of dilution of the grout or formation of voids in the grout. The use of wipers or other means of minimizing mud intrusion into the spaces to be occupied by piles should be considered at sites having soft mud bottoms.

Grout is extensively used to "cement" the annulus between pile leg and jacket sleeve. An annular gap of 50–100 mm is usually selected. The grout should flow from the bottom up.

The mix is generally cement plus water. Fly ash may be used to replace part of the cement in order to reduce heat of hydration. Silica fume may be added to promote thixotropic behavior, increase strength, and reduce bleed. Admixtures may be used to provide water reduction, retardation, and expansion characteristics. It is important that trial batches be made to ensure that the grout has the proper flow characteristics as well as strength. Flow rate should be kept low to avoid entrapment of voids. Grout

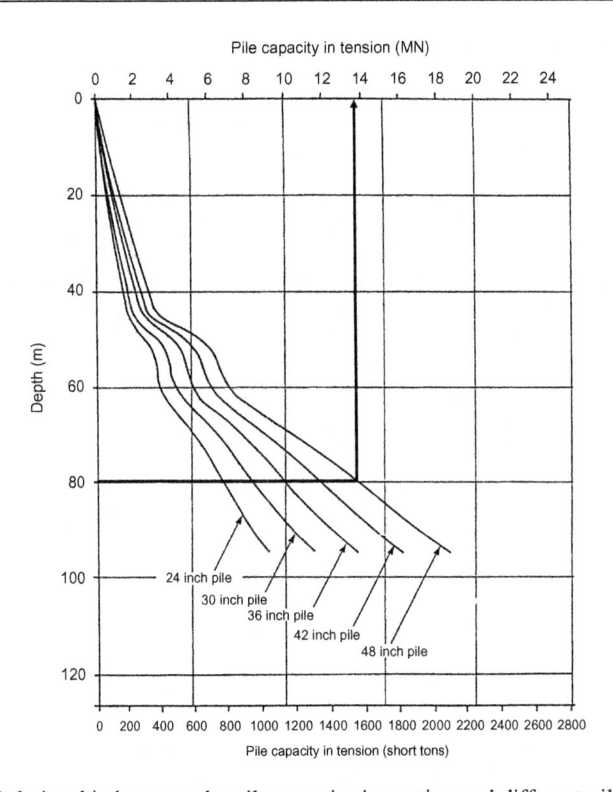

Figure 6.19 Relationship between the pile capacity in tension and different pile diameters.

should be overflowed to ensure that the initial mixture of cement and seawater is cleared. Pressures should be carefully controlled to prevent forcing the grout out from under the jacket sleeve; usually, this exit is restricted by a grout retainer, but many times the grout retainer will have been damaged during pile driving. Therefore, a second entry grouting pipe is often provided, to permit the first grout to set and form a plug; then the main grouting is carried out through the upper entry port.

6.10.1.1 *Computation of allowable axial force*

Until now, little research has been done into the transfer of axial force from the pile to the leg through the grouting. The transfer of axial force from the pile to the leg is accomplished by the bond strength between the pile and the leg through the grouting. The allowable axial load transfer should be calculated as the smaller value from pile or leg of the force, calculated by a multiplication of the contact area between the grout and steel surfaces, and the allowable bond strength due to axial load f_{ba}. The value of the allowable bond strength due to axial load, f_{ba}, should be taken as 0.138 MPa for API loading conditions 1 and 2 and 26.7 psi (0.184 MPa) for loading conditions 3 and 4, where the API loading conditions are

1. $E_o + DL + L_{max}$
2. $E_o + DL + L_{min}$

3. ED + DL + L_{max}
4. ED + DL + L_{min}

where E_o is the operating environmental condition; DL is the dead load; L_{max} and L_{min} are the maximum and minimum live loads, respectively, and in both cases, the live load is appropriate to the normal operating condition of the platform; ED is the design environmental condition in extreme environmental conditions.

In the case of higher values of axial load, shear keys are required at the interface between the steel and the grout, and the value of the nominal allowable bond strength for axial load transfer f_{ba}, for loading conditions 1 and 2 should be taken as

$$f_{ba} = 0.138 + 0.5 f_{cu} \cdot \frac{h}{s}, \text{MPa} \tag{6.31}$$

For loading conditions 3 and 4, f_{ba} should be taken as

$$f_{ba} = 0.184 + 0.67 f_{cu} \cdot \frac{h}{s}, \text{MPa} \tag{6.32}$$

where f_{cu} = unconfined grout compressive strength (in MPa), h = shear key outer dimension (in mm), and s = shear key spacing (in mm). Shear keys designed according to equations (6.31) and (6.32) should be detailed in accordance with the following requirements:

1. Shear keys may be circular hoops at spacing s or a continuous helix with a pitch of S_s, as in Figure 6.20.
2. Shear keys should be one of the types indicated in Figure 6.21.
3. For driven piles, shear keys on the pile should be applied to sufficient length to ensure that, after driving, the length of the pile in contact with the grout has the required number of shear keys.
4. Each shear key cross section and weld should be designed to transmit the part of the connection capacity that is attributable to the shear key for loading conditions 1 and 2.

The shear key and weld should be designed at basic allowable steel and weld stresses to transmit an average force equal to the shear key bearing area multiplied by 1.7 f_{cu}, except for a distance of two pile diameters from the top and the bottom end of the connections, where 2.5 f_{cu} should be used. The following limitations should be observed when designing a connection:

- 17.25 MPa $\leq f_{cu} \leq$ 110 MPa
- Sleeve geometry, $D_s/t_s \leq 80$
- Pile geometry, $D_p/t_p \leq 40$
- Grout annulus geometry, $7 \leq D_g/t_g \leq 45$
- Shear key spacing ratio, $2.5 \leq D_p/s \leq 8$
- Shear key ratio, $h/s \leq 0.10$
- Shear key shape factor, $1.5 \leq w/h \leq 3$
- f_{cu} $(h/s) \leq 5.5$ MPa.

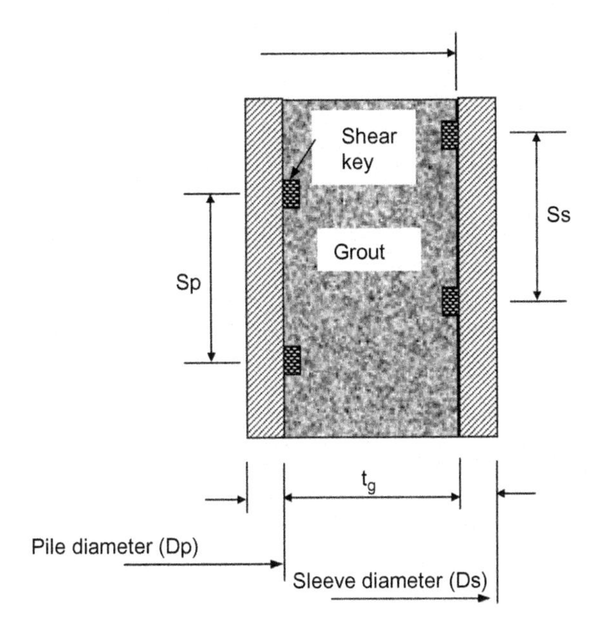

Figure 6.20 Grouting the annulus between piles and legs.

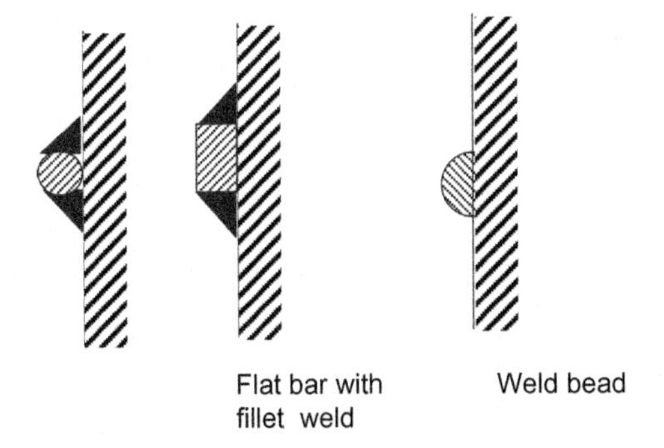

Figure 6.21 Recommended API shear key detail.

Example 6.7

Calculate the composite pipe pile of Figure 6.22, where the actual moment of inertia is described as follows:

$D_{o1} = 1800$ mm
$D_{i1} = 1730$ mm
$D_{o2} = 1600$ mm
$D_{i2} = 1540$ mm
$I_1 = \pi[(D_{o1})^4 - (D_{i1})^4]/64 = 7{,}5601{,}916{,}209$ mm^4
$I_2 = 45{,}607{,}749{,}335$

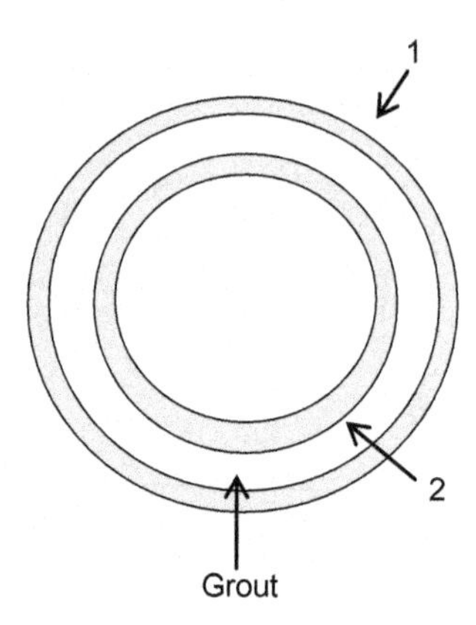

Figure 6.22 Composite pipe pile of Example 6.7.

$I_t = I_1 + I_2 = 1.2121 \times 10^{11}$ mm^4
$A_1 = \pi[(D_{o1})^2 - (D_{i1})^2]/4 = 194{,}071.8862$ mm^2
$A_2 = 147{,}969.014$ mm^2
$A_G = 339{,}998.8649$ mm^2

The equivalent composite pile is

$I = 1.2121 \times 10^{11}$ mm^4
$D_o = 1800$ mm
$D_i = \{[I\,(64)/\pi] - (D_o)^4\}^{0.25} = 1683.28$ mm
$t = D_o - D_i = 58.36$ mm
Area $= \pi[(D_{o1})^2 - (D_{i1})^2]/4 = 319{,}317.75$ mm^2
Elastic modulus for steel, $E_s = 210{,}000$ MPa
Elastic modulus for grout, $E_G = 30{,}000$ MPa
$E_t = \{E_s\,[(D_{o1})^2 - (D_{i1})^2] + E_s\,(D_{o2})^2 - (D_{i2})^2\} + E_G\,[(D_{i1})^2 - (D_{o2})^2]/D_o = 32{,}235$ MPa

6.11 Mud mat design

The geotechnical investigation report should include the bearing capacity analysis at the mudline. In most cases, it will be done for circular, triangular, square, and rectangular mud mats (with ratios $L/B = 1.5$ and $L/B = 2$) and define the where the lengths of the sides range. In addition, the pure vertical bearing capacity analysis for mud mats with skirt length of 1 m and 2 m also are performed by considering the skirt depth as the foundation base level.

Temporary on-bottom stability of the platform jacket following touchdown on the seabed and prior to piling is provided by mud mats. The ultimate bearing capacity of a mud mat is calculated as

$$Q_u = q_u A$$

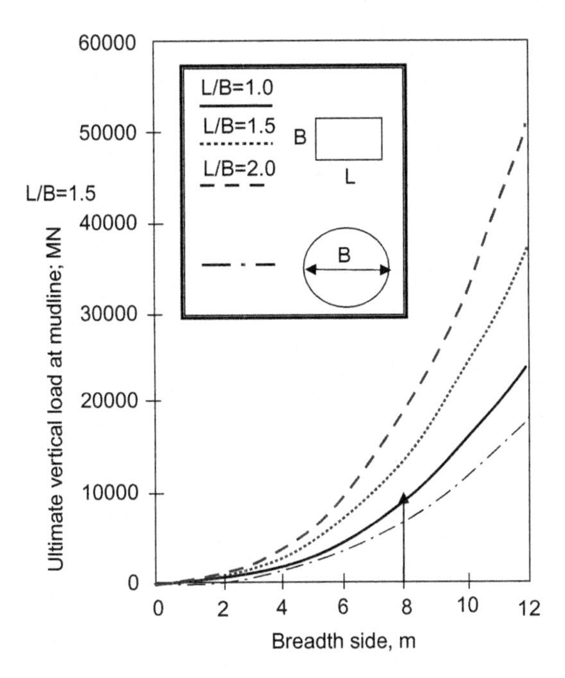

Figure 6.23 Relationship between foundation size and ultimate strength.

where

Q_u = ultimate bearing capacity
q_u = ultimate unit bearing capacity
A = mud mat area

ISO 19901-4(2003) provides a further analysis of bearing capacity for combined loading performed according to mud mat size. Figure 6.23 presents 10.0 × 10.0 m, 10.0 × 7.6 m, and 8.0 × 12.0 m mud mats equipped with 0 m, 1 m, and 2,m long skirts. For mud mats without skirts, full contact between mud mats and seabed has been assumed.

The ultimate bearing capacities discussed in the examples do not include a factor of safety. A factor of 2.0 is recommended by API RP2A-WSD (2000) for pure vertical loading.

The relation between the vertical load on the mud mat and the horizontal load and depend on whether there is a skirt pile or not and the depth of the skirt pile. Figures 6.24 through 6.26 present cases of no skirt pile and skirts with a depth 1 and 2 m.

Example 6.8

Based on Figure 6.23, if the mud mat is square with area 8 m × 8 m, what will be the maximum vertical load?

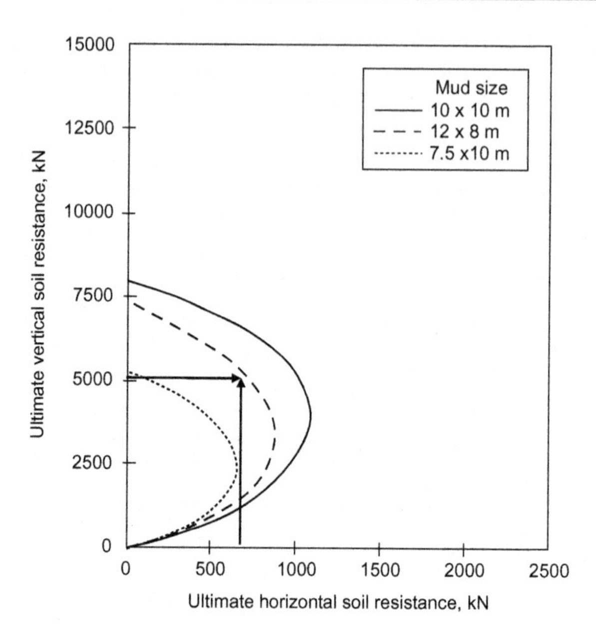

Figure 6.24 Relationship between ultimate vertical and horizontal soil resistance with no skirt.

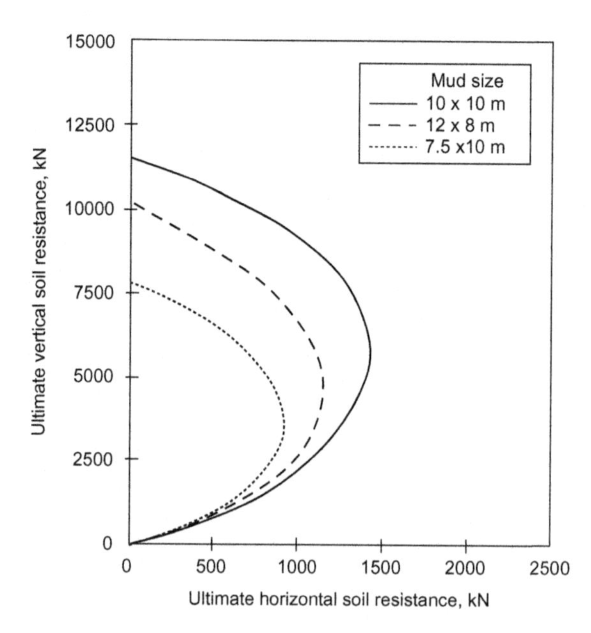

Figure 6.25 Relationship between ultimate vertical and horizontal soil resistance with skirt 1 m.

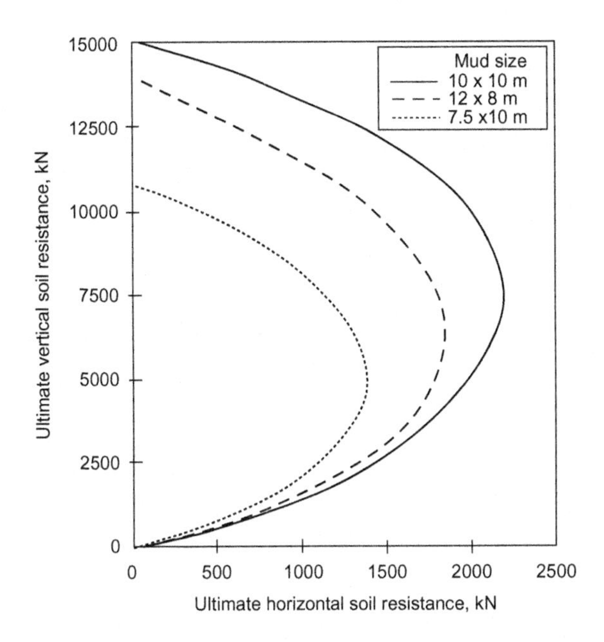

Figure 6.26 Relationship between ultimate vertical and horizontal soil resistance with skirt 2 m.

Solution As in Figure 6.23, the maximum ultimate strength is 9000 KN, around 900 ton

Example 6.9

Select the better dimensions of a mud mat with no skirt:

1. The vertical load is 5000 KN and the horizontal load is 1000 kN.
2. The vertical load 5000 KN and the horizontal load is 600 KN.

Solution As per Figure 6.24,

1. In this case, 10×10 m is the better solution.
2. In this case, 8×12 m is the better solution.

Example 6.10

Choose the better dimensions of a mud mat with a skirt of 1 m:

1. The vertical load is 5000 KN and the horizontal load is 1000 kN.
2. The vertical load 5000 KN and the horizontal load is 600 KN.

Solution As per Figure 6.25,

1. In this case, 8.0×12 m is the better solution.
2. In this case, 7.6×12 m is the better solution.

Further reading

API RP2A, 2007. Recommended Practice for Planning, Designing, and Constructing Fixed Offshore Platforms. fifteenth ed. American Petroleum Institute, Washington, DC.

American Society for Testing and Materials, 2004. Standard Method of Deep Quasi-Static Cone and Friction-Cone Penetration Tests of Soil. ASTM International, West Conshohocken, PA, ASTM Standard D 3441.

Arup, O., et al., 1986. Research on the behavior of piles as anchors for buoyant structures—Summary report. Department of Energy, London, Offshore Technology Report OTH 86 215.

Begemann, H.K.S. 1965. The friction jacket cone as an aid in determining the soil profile. In: Proceedings of the Sixth ICSMFE, vol. I. Montreal, pp. 17–20.

Bogard, J.D., Matlock, H., 1990. Applications of model pile tests to axial pile design. *Proceedings of the 22nd Annual Offshore Technology Conference, Houston Texas, May 1990*. Department of Energy, London.

Bogard, J.D., Matlock, H., Audibert, J.M.E., Bamford, S.R., 1985. Three years' experience with model pile segment tool tests. *Proceedings of the Offshore Technology Conference, Houston Texas, May 6–9, 1985*. Department of Energy, London.

Briaud, J.L., Garland, E.E., Felio, G.Y., 1984. Loading rate parameters for piles in clay. In: Proceedings of the Offshore Technology Conference, Houston, TX.

Clarke, J. (Ed.), 1993. Large-Scale Pile Tests in Clay. Thomas Telford, London.

Clausen, C.J.F., Aas, P.M., Karlsrud, K., 2005. Bearing capacity of driven piles in sand, the NGI approach. Proceedings of the International Symposium on Frontiers in Offshore Geotechnics, Perth, September. A.A. Balkema, Rotterdam, the Netherlands, pp. 677–682.

Coyle, H.M., Reese, L.C., 1966. Load transfer for axially loaded piles in clay. ASCE J. Soil Mech. Found. Div. 92, 1052.

De Reister, J., 1971. Electric penetrometer for site investigations. ASCE J. SMFE Div. 97 (SM-2), 457–472.

Dunnavant, T.W., Clukey, E.C., Murff, J.D., 1990. Effects of Cyclic Loading and Pile Flexibility on Axial Pile Capacities in Clay. OTC, Houston, TX, OTC 6374.

El-Reedy, M.A., 2011. Offshore Structure Design Construction and Maintenance. Elsevier, Waltham, MA.

Georgiadis, M., 1983. Development of p-y curves for layered soils. Proceedings of the Conference on Geotechnical Practice in Offshore Engineering, Austin Texas. American Society of Civil Engineers, New York, pp. 536–545.

IRTP, International Society of Soil Mechanics and Geotechnical Engineering, 1999. International reference test procedure for the cone penetration test (CPT) and the cone penetration test with pore pressure (CPTU). Report of the ISSMGE Technical Committee 16 on ground property characterisation from in-situ testing. Proceedings of the Twelfth European Conference on Soil Mechanics and Geotechnical Engineering. Balkema, Amsterdam.

ISO (International Organization for Standardization), 2005. Geotechnical Investigation and Testing—Field Testing—Electrical Cone and Piezocone Penetration Tests. ISOInternational Organization for Standardization), 2005, Amsterdam, International Standard ISO 22476-1 (DIS 2005).

Jardine, R., Chow, F., Overy, R., Standing, J., 2005. ICP Design Methods for Driven Piles in Sands and Clays. Thomas Telford, London.

Karlsrud, K., Haugen, T., 1985. Behavior of piles in clay under cyclic axial loading—Results of field model tests. In: Proceedings of the Fourth International Conference on Behavior of Offshore Structures, Delft, the Netherlands, McGraw Hill publisher. July 1–5, 1985.
Kolk, H.J., 2000. Deep foundations in calcareous sediments. In: Al-Shafei, K.A. (Ed.), Engineering for Calcareous Sediments: Proceedings of the Second International Conference on Engineering for Calcareous Sediments, Bahrain, February 21–24, 1999, vol. 2. A.A. Balkema, Rotterdam, the Netherlands, pp. 313–344.
Kolk, H.J., Baaijens, A.E., Senders, M., 2005. Design criteria for pipe piles in silica sands. In: Gourvenec, S., Cassidy, M. (Eds.), Frontiers in Offshore Geotechnics ISFOG 2005: Proceedings of the First International Symposium on Frontiers in Offshore Geotechnics, University of Western Australia, Perth, September 19–21, 2005. Taylor & Francis6, London, pp. 711–716.
Kraft, L.M., Lyons, C.G. 1974. State of the art: Ultimate axial capacity of grouted piles. In: Proceedings of the Annual Offshore Technology Conference, Houston Texas, May 1996. Department of Energy, London.
Kraft Jr., L.M., Cox, W.R., Verner, E.A., 1981a. Pile load tests: Cyclic loads and varying load rates. ASCE J. Geotech. Eng. Div. 107 (GT1), .
Kraft, L.M., Focht, J.A., Amarasinghe, S.F., 1981b. Friction capacity of piles driven into clay. ASCE J. Geotech. Eng. Div. 107 (GT11), 1521–1541.
Lehane, B.M., Schneider, J.A., Xu, X., 2005a. A Review of Design Methods for Offshore Driven Piles in Siliceous Sand. University of Western Australia, Perth, Geomechanics Group, Report No. GEO: 05358.
Lehane, B.M., Schneider, J.A., Xu, X., 2005b. The UWA-05 method for prediction of axial capacity of driven piles in sand. In: Gourvenec, S., Cassidy, M. (Eds.), Frontiers in Offshore Geotechnics ISFOG 2005: Proceedings of the First International Symposium on Frontiers in Offshore Geotechnics, University of Western Australia, Perth, 19–21 September 2005. Taylor & Francis, London, pp. 683–689.
Matlock, H. 1970. Correlations for design of laterally loaded piles in soft clay. In: Proceedings of the Second Annual Offshore Technology Conference, Houston, Texas, SPE publisher. April 1970.
McClelland, B., Ehlers, C.J., 1986. Offshore geotechnical site investigations. In: McClelland, B., Reifel, M.D. (Eds.), Planning and Design of Fixed Offshore Platforms. Van Nostrand Reinhold, New York.
Meyer, P.C., Holmquist, D.V., Matlock, H.T.I., 1975. Computer predictions of axially loaded piles with non-linear supports. In: Proceeding of the Annual Offshore Technology Conference, OTC, Houston Texas. Department of Energy, London.
Murff, J.D., 1980. Pile capacity in a softening soil. Int. J. Numer. Anal. Methods Geomech. 4 (2), 185–189.
NORSOK Standard G-001 Rev. 2, 2004. Marine soil investigations.
O'Neill, M.W., Hassan, K.M., 1994. Drilled shafts: Effects of construction on performance and design criteria, Proceedings of the International Conference on Design and Construction of Deep Foundations, vol. 1. U.S. Federal Highway Administration, Washingt5on, DC, pp. 137–187.
O'Neill, M.W., Murchison, J.M., 1983. An evaluation of *p-y* relationships in sands. University of Houston, Houston, TX, Prepared for the American Petroleum Institute Report PRAC 82-41-1.
Pelletier, J.H., Doyle, E.H., 1982. Tension capacity in silty clays—Beta pile test. In: Proceedings of the Second International Conference on Numerical Methods of Offshore Piling, university of Texas at Austin, Austin Texas. April 29–30, 1982.

Randolph, M.F., 1983. Design considerations for offshore piles. Proceedings of the Conference on Geotechnical Practice in Offshore Engineering, Austin, Texas. American Society of Civil Engineers, New York, pp. 422–439.

Reese, L.C., Cox, W.R., 1975. Field testing and analysis of laterally loaded piles in stiff clay. In: Proceedings of the Fifth Annual Offshore Technology Conference, Houston, Texas, SPE publisher. April 1975.

Semple, R.M., Gemeinhardt, J.P., 1981. Stress history approach to analysis of soil resistance to pile driving. Offshore Technology Conference, Houston, Texas, SPE publisher. vol. 1, pp. 165–172.

Smith, E.A.L., 1962. Pile driving analysis by the wave equation, Part 1, Paper No. 3306. Trans. ASCE. 127, 1145–1193

Stevens, R.S. Wiltsie, E.A., Turton, H., 1982. Evaluating pile drivability for hard clay, very dense sand and rock. In: Proceedings of the 14th Offshore Technology Conference, Houston, Texas, SPE publisher. vol.1, pp. 465–482.

Thompson, G.W.L., Jardine, R.J., 1998. The applicability of the new Imperial College design method to calcareous sands. Proceedings of the Conference on Offshore Site Investigations and Foundation Behavior. Society for Underwater Technology, London, pp. 383–400.

Toolan, F.E., Fox, D.A., 1977. Geotechnical planning of piled foundations for offshore well-head jacket centers. Proc. Inst. Civ. Eng. 62 (Part 1), 221–244.

Vijayvergiya, V.N., 1977. Load movement characteristics of piles. In: Proceedings of the Ports '77 Conference, vol. 2, pp. 269–284. New York: American Society of Civil Engineers.

Construction and installation lifting analysis

7

7.1 Introduction

The lifting process exists during the construction and installation phases. So, it is important to understand the lifting calculation as most marine work depending on lifting calculation.

The construction of the fixed offshore platform is very specific for this type of structure. Therefore, the contractor company who is responsible for fabrication and installation should be a specialist in this type of structures and have a competent staff and reasonable cranes and facilities to do this construction project. The design of this platform should be checked against transportation , lifting, and installation. In this phase, the major interface management is between the engineering and construction. For example, in some cases, the sea fastening or lifting is done by the contractor but checked by the engineering company. Note that the launching and lifting of the platform component is designed by the engineering company but the input data for the analysis is delivered to the engineering company based on the available barge, cranes, and other equipment and capability for launching, transportation, and installation.

7.2 Construction procedure

The engineering firm performing the design of a jacket should take into consideration its lifting, launching, or self-floating, which depends primarily on the available offshore installation equipment and the water depth. In general, the preference is to lift the jacket in place. The size of such jackets has been increasing as offshore lifting capacity has grown. Nowadays, the lifting capacity can be reach to 14,000 tons.

For jackets in shallow water, in most cases, the height of the jacket is on the same order as the plan dimensions, so the erection is usually carried out vertically in the same direction as the final installation. Therefore, in this case, the jacket may be lifted or skidded onto the barge.

Jackets designed for deeper water are usually erected on their side. Such jackets are loaded by skidding out on a barge. Historically, most large jackets have been barge launched. This method of construction usually involves additional flotation tanks and extensive pipework and valves to enable the legs to be flooded in order

Marine Structural Design Calculations. DOI: http://dx.doi.org/10.1016/B978-0-08-099987-6.00007-6

to ballast the jacket into the vertical position on site. This method of construction is currently applicable for jackets up to 25,000 tons. Very large jackets, in excess of this, have been constructed as self-floaters in a graving dock and towed offshore subsequent to flooding the dock.

The construction of offshore structure jackets has a series of very distinct stages, from fabrication until loadout. These stages start by obtaining the steel sections, which will be delivered under full quality control from the manufacturer, then the fabrication is done manually under the responsible quality control team on site.

7.3 Engineering the execution

For practical purposes, the engineering firm performing the design should follow the execution during each phase to ensure that the design requirements are fulfilled and to be on line in case of a request for change or clarification. A general method of execution is envisaged at the jacket design stage, since the shape of the jacket, its form, and properties require quite specific methods of loadout, offshore transportation, and installation. Note that this phase is the contractor's responsibility and involves considerable interface with the engineering firm. In the earlier phases, that is, procurement through assembly and erection, the contractor, while limited by design specification requirements, has freedom of choice with regard to the exact method of execution. However, in all phases, the contractor is required to demonstrate that the methods adopted are compatible with the specification requirements and do not affect the integrity of the structure.

Each phase of execution has its own specific engineering requirements, which are determined by the processes executed during that phase. These processes range from those that are largely repetitive early in execution to one-off activities in the latter phases. Accordingly, the engineering that supports procurement and shop fabrication is voluminous but repetitive, such as material takeoffs, shop drawings, and cutting plans. The assembly and erection phases are supported by a mix of repetitive engineering, such as scaffolding, and specific studies for a limited series of activities.

The volume of construction work for a large jacket is usually around (130,000–150,000) hours. When designing larger components, consideration must be given to their subdivision into elements that will not distort when fabricated and can be relatively easily assembled without welding or dimensional problems.

7.4 Construction process

All fabrication work should be performed based on safe practices, using equipment that complies with all applicable regulatory requirements, local standards, and any appropriate regional requirements. Choosing the construction staff to be competent and qualified to perform their task, all fabrication, assembly, and erection should be in accordance with approved procedures.

Unless detailed otherwise on the design drawings, the "default" layout rules and fabrication are based on API, ISO, or another standard mention in the project specification.

To have good quality in construction, it is recommended that the subassembly and welding be at the ground level or undercover. When such subassemblies are moved and installed into larger subassemblies or within the structure, care should be exercised to ensure that no members or joints within the subassemblies are overstressed or distorted.

Suitable supports may include webbing slings, wire slings when the components are protected from damage, and temporary lifting attachments welded to the subassemblies.

Subassemblies may be rolled into the correct orientation in the structure or to allow lifting at the correct orientation into the structure. Rollup saddles should be designed and positioned to ensure that the structure is not overstressed during rollup and is well greased or otherwise lubricated.

ISO contains the fabrication requirements and tolerances for fixed steel offshore structures.

The following sections present the tolerance allowed during construction that will affect the weight of the structure. Such weight is overcome by accounting for a contingency factor in the lifting process analysis, and this is the main reason that the weighing be done before lifting the jacket.

7.4.1 Fabrication tolerances

The contractor should provide a qualified persons approved by the owner's representative and deliver good equipment and the necessary instruments to perform, monitor, and control the dimensions within the allowable tolerance. Tolerances should be checked at each stage in accordance with the fabrication procedures, and the final survey should meet the defined tolerances.

Based on ISO, personnel responsible for the final survey should be either qualified surveyors or have had at least five years' experience with similar work. The instruments used should be accurate and have current valid calibration certificates.

Based on ISO 19902, the tolerances are guided by the following:

- Local tolerances for structural components and subassemblies should be controlled so that the accumulation of such tolerances does not affect the specified global tolerances. The specified tolerances apply at all stages of fabrication and assembly.
- Allowance should be made for weld gap tolerances and weld shrinkage in all component, subassembly, and global tolerance calculations.
- Where tolerances have to be derived from a formula, such as tolerances expressed in terms of the component dimension (e.g., wall thickness), the results should be taken to the nearest mm (0.04″).
- Tolerances should be based on theoretical setting-out points and centerlines of the structure referenced to permanent approved datum points (e.g., coordinated survey stations) and corrected to a temperature of +20°C (68°F).

- Compensation should be made for significant deflection differences between temporary support and final conditions.
- Fabrication and yard assembly supports should be set to within ±5 mm (¼″) of the appropriate position shown on the approved workshop drawings. Where no such drawings exist, fabrication should be carried out from a level plane to within ±5 mm (¼″).
- The dimensional tolerance of launchway centerlines should be within ±20 mm (¾″) of the theoretical position and within ±6 mm (¼″) of its reference elevation. The variation in elevation between any two points on a launchway should not exceed 3 mm (⅛″) within any 3 m (10′).

7.4.1.1 Leg spacing tolerances

The global tolerances for leg spacing at the plan bracing levels are as shown in Figure 7.1 and detailed next:

- The horizontal center-to-center distance between adjacent legs at the top of a structure where a deck or other structure is to be placed (stab-in nodes) should be within 10 mm (⅜″) tolerance from the construction drawings.
- The horizontal center-to-center distance between legs at other locations should be within 20 mm (¾″) of the design values.
- The horizontal center-to-center diagonal distances between legs at the top of a structure where a deck or other structure is to be placed (stab-in nodes) should be within 10 mm (⅜″) of the construction drawings.
- The horizontal center-to-center diagonal distance between legs at other locations should be within 20 mm (¾″) of the design values.

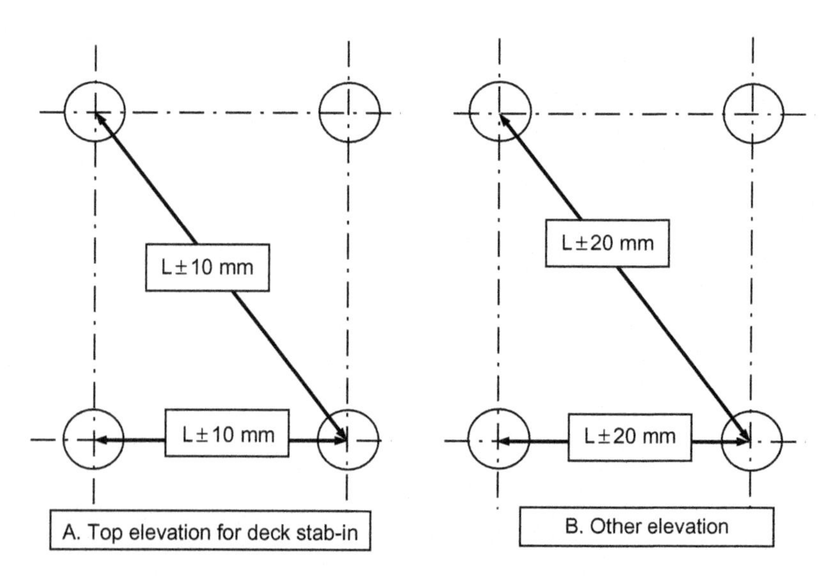

Figure 7.1 Horizontal tolerance.

7.4.1.2 Vertical level tolerances

The global tolerances for vertical levels of plan bracing are as shown in Figure 7.2 and detailed next:

- The elevation of the plan bracing levels should be within ±13 mm (±½″) of the construction drawings.
- The vertical level of braces within a horizontal plane should be within ±13 mm (±½″) of the construction drawings.
- The vertical distance between plan bracing elevations should within ±13 mm (±½″) of the construction drawings.

7.4.1.3 Tubular member tolerances

For tubular members with thicknesses of 50 mm (2″) or less, the difference between the major and the minor outside diameters (the out-of-roundness) at any point of a tubular member should not exceed the smaller of 1% of the diameter or 6 mm (¼″).

For tubular members with thicknesses of more than 50 mm (2″), the difference between the major and the minor outside diameters at any point of a tubular member should not exceed 12.5% of the wall thickness.

For tubular members with a diameter of 1200 mm or more and a thickness of 100 mm or less, the difference between the major and the minor outside diameters

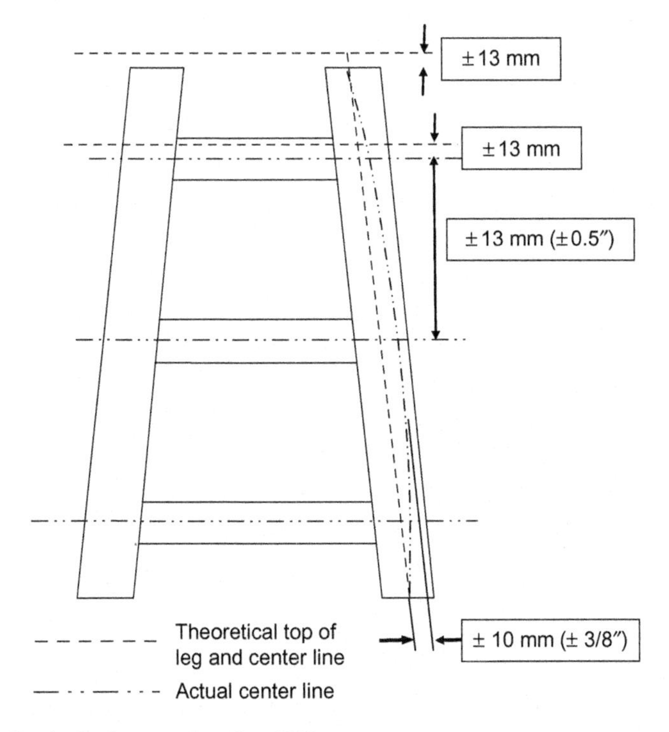

Figure 7.2 Vertical tolerances based on ISO.

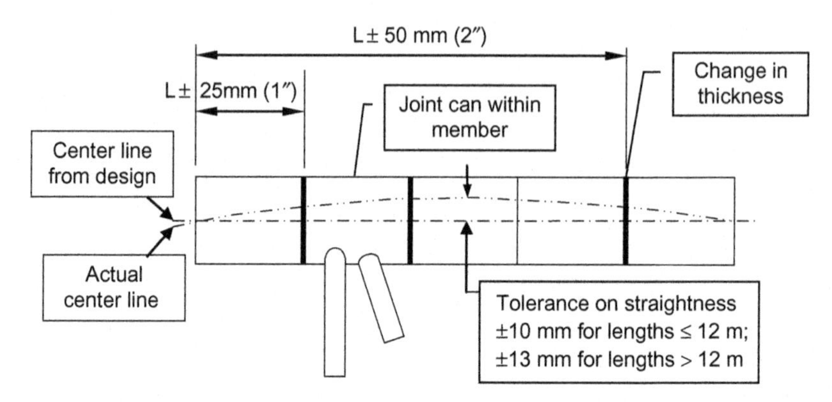

Figure 7.3 Tolerances on positioning of cans within members and straightness of members.

at any point of a tubular member may increase to 13 mm (½″) provided the tolerance on the circumference is less than 6 mm (¼″).

The difference between the actual and nominal outside circumferences at any point of a tubular member should not exceed the smaller of 1% of the nominal circumference or 13 mm (½″).

Figure 7.3 presents the tolerances for the straightness of tubular members and beams, based on ISO, as follows:

- Straight to be within ±10 mm (±⅜″) for length ≤12 m.
- Straight to be within ±13 mm (±½″) for length >12 m.

Additionally the tolerances for the straightness of tubular members and beams when assembled into the structure are

- Between the ends of the node stubs, braces should be straight to within 0.12% of the length.
- Between the ends of the node chords, braces should be straight to within 0.10% of the length.

The tolerances for the location of cans with different wall thickness within a tubular member are as shown in Figure 7.4 and as detailed next:

- For joint cans, within 25 mm (1″) from the construction drawings.
- For other changes of wall thickness, within 50 mm (2″) of the construction drawings.

7.4.1.4 Tolerances of leg alignment and straightness

In addition to the tolerances for all tubular members, the tolerances for the alignment and straightness of legs are as shown in Figure 7.4 and detailed next:

- For structures with more than 4 legs, at each plan bracing level the legs will be aligned within 10 mm (⅜3/8″) of a straight line.
- Between horizontal bracing plans legs should be straight to within 10 mm (⅜″).
- Between nodes legs should be straight to within 10 mm (⅜″).

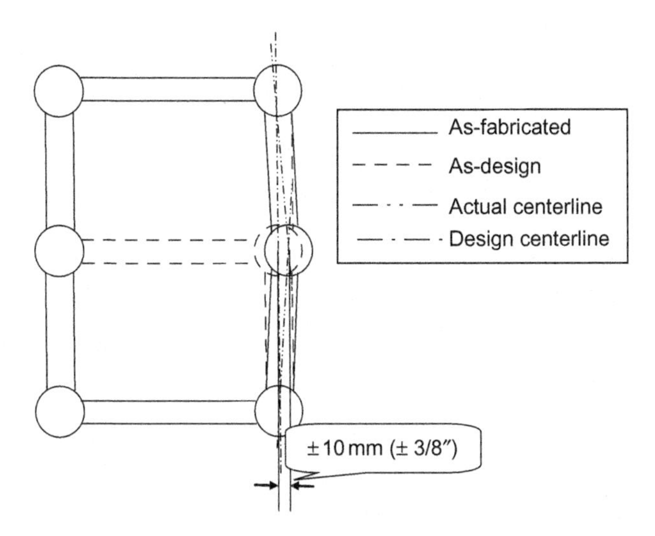

Figure 7.4 Allowable tolerances in leg alignment.

7.4.1.5 *Tubular joint tolerances*

For tubular joint tolerances, a best-fit work point should be determined taking into account all the design and as-built dimensions of the complete tubular joint. The best-fit work point should be within 15 mm (⅝″) of the design position.

The alignment of a brace stub best-fit centerline or, for point-to-point construction, the brace centerline, should be within 13 mm (½″) of the design work point, as in Figure 7.5.

The lengths of cans within the chord of a tubular joint should be within the range of design length +10 mm to design length −5 mm and the location of the circumferential weld between chord cans should be within 5 mm of design location.

The lengths of the brace stub of a tubular joint should be within the range of design length +50 mm to design length −5 mm.

The tolerances on the positioning and alignment of stubs on tubular joints are as shown in Figure 7.6.

- The centerline of the brace at the intersection with the chord wall should be within 5 mm of the construction drawing in the case of a chord diameter <3.5 m.
- The centerline of the brace at the intersection with the chord wall should be within 10 mm of the design position where a chord diameter >3.5 m.
- The angular orientation of the brace stub should be within 10′ of the design orientation.
- The position of the centerline of the brace end of the stub should be within 5 mm of the design position.

Where plates carrying in-plane faces are arranged to form a cruciform, the misalignment should not exceed the lesser of 50% of the thickness of the thinnest noncontinuous member or 10 mm, whichever is smaller.

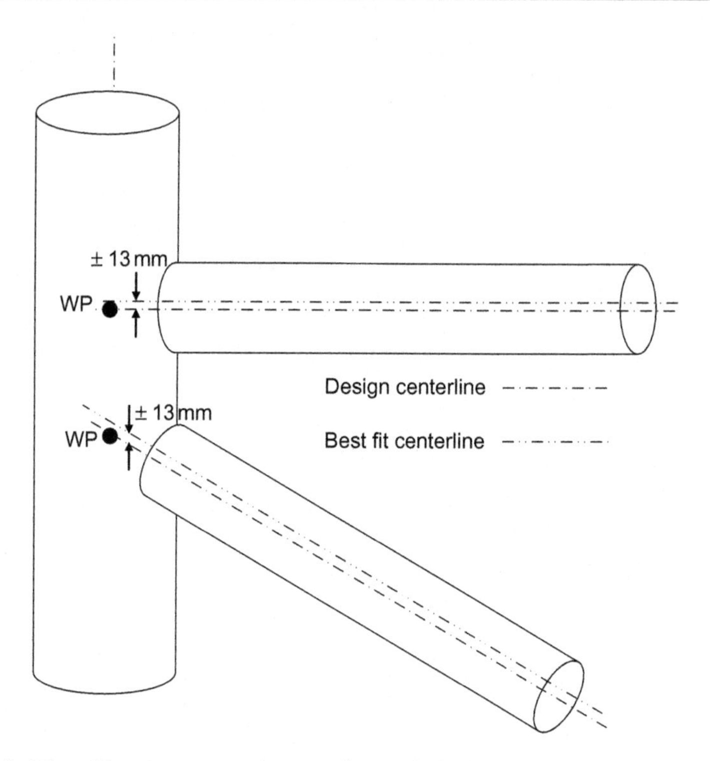

Figure 7.5 Allowable tolerances on brace stubs at tubular joints and best-fit centerline.

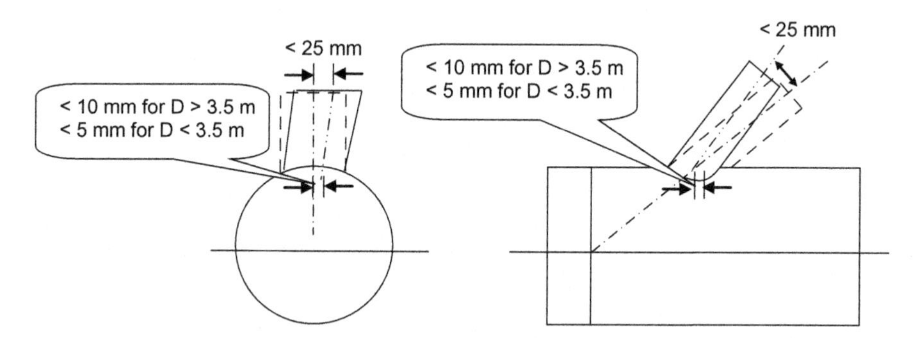

Figure 7.6 Allowable tolerances for tubular joints by ISO.

7.4.2 Stiffener tolerances

The stiffener tolerances are as follows:

- At or within 150 mm of a conical transition, to within 3 mm (⅛″) of the construction drawing.
- In launch legs, to within 3 mm (⅛″) of the construction drawing.

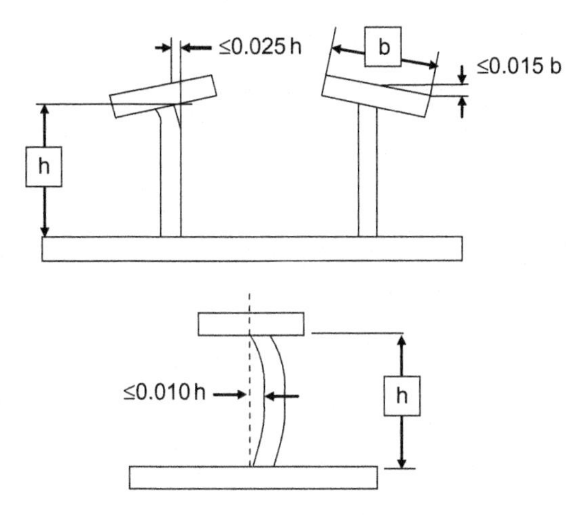

Figure 7.7 Tolerances for ring stiffener cross sections based on ISO.

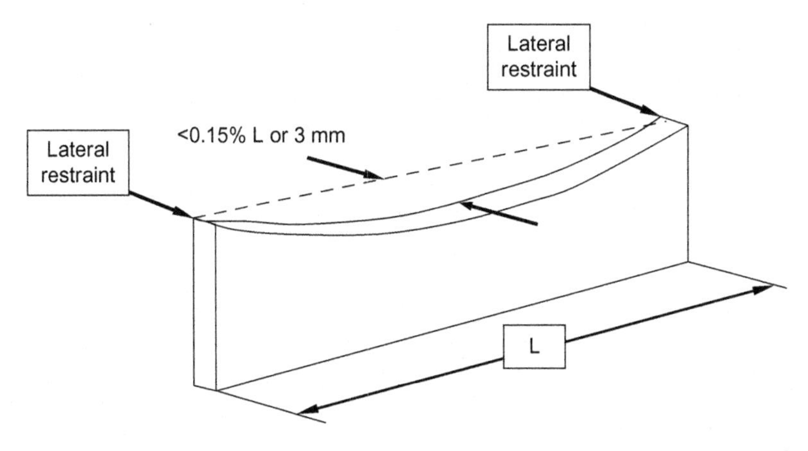

Figure 7.8 Tolerances for ring stiffener straightness.

- In tubular joints other than at conical transitions of launch leg nodes, to within 5 mm (¼″) of the design position.
- At all other locations, to within 10 mm (⅜″) of the construction drawing.

The tolerances on the web of ring stiffener cross sections should be perpendicular to the centerline of the tubular member and within 2.5% of the web height, as shown in Figure 7.7. On the other hand, the flange of the stiffener should be parallel to the centerline of the tubular member to within 1.5% of the flange widt, and the web of the stiffener should be flat over its height to within 1.0% of the web height.

Out-of-straightness for longitudinal or diaphragm stiffeners in tubular member as should be limited to 0.15% L or 3 mm, whichever is larger, as shown in Figure 7.8.

7.4.3 Conductor guides and piles tolerances

The tolerances on conductors are the same as those on pile guides, sleeves, and appurtenance supports, related to a best-fit line through the centers of the guides, sleeves, and supports, as in Figure 7.9. The tolerance at the center of each guide or sleeve and the best-fit line should be less than ±10 mm (±⅜″) at the top guide and where the vertical spacing between guides is less than 12 m (40 ft). The tolerance at the center of each guide and the construction drawing line should be less than 13 mm (½″) in other cases. For pile sleeves, the tolerance should be checked at the mid-height of each set of centralizers.

The tolerance at the center of each appurtenance support and the best-fit line should be less than 10 mm (⅜″) at the top and less than ±25 mm (±1″) elsewhere.

In addition to the tolerances for other dimensions for tubular members, the tolerances of the straightness of a pile should be as follows:

- In any 3 m (10 ft) length, piles should be straight to within 3 mm (⅛″).
- In any 12 m (40 ft) length, piles should be straight to within 10 mm (⅜″).
- In any length over 12 m (40 ft), piles should be straight to within 0.1% of the length considered.

The design of anodes is discussed in ElReedy MA (2012), but during construction, the tolerance of the locations of anodes should be less than 300 mm (12″) and less than 10° circumferentially of the design position.

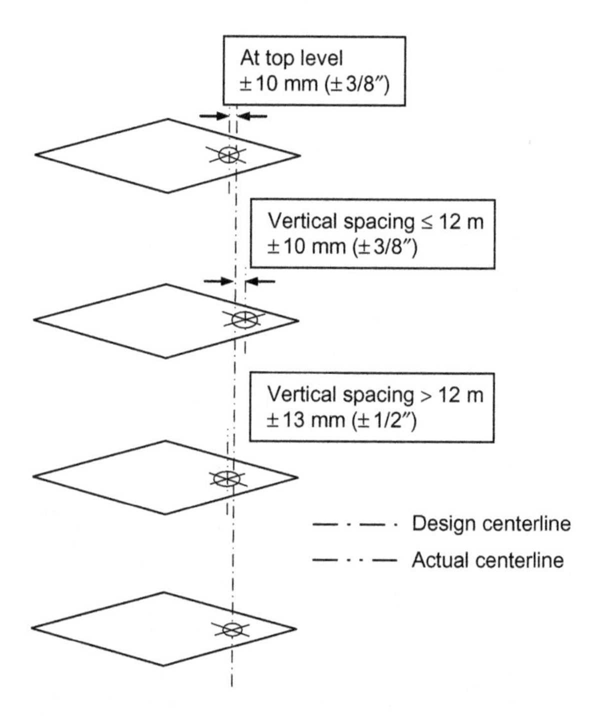

Figure 7.9 Tolerances for conductor guide alignment.

7.4.4 Dimensional control

Quality control should be within the overall quality plan that was established before the start of the fabrication. The owner should verify that all persons and required equipment to control the quality are available.

All the areas of quality control (QC) that require attention, such as that of dimensional control, are emphasized in the code and specifications, especially for offshore structures. However, it is clear that attention must be paid to the dimensions that have structural significance, such as the straightness of elements, ovality of tubular members, and eccentricities at node joints. It is also clear that on a jacket, the global alignment and verticality of items such as pile sleeves, conductor guides, and launch runners are also important.

The principal reason for requiring such accurate dimensional control of joints and tubular members during fabrication is not because of the structural consequences of out of tolerance but rather because the parts may not fit together in the yard. In a tubular steel jacket, the theoretical tolerances on node stub eccentricity are matched from the structural viewpoint, while the actual tolerances are very tight, because of considerations regarding the fitting together of components during the later phases of construction.

Tolerances are listed on the approved construction drawings, so during fabrication, any measurement verification must be within the allowable tolerance based on ISO, API, or AISC or any other standard within the project specification document.

It is worth mentioning that fabrication for all jackets and topside structures should be checked for loads applied during fabrication. These loads are based on the proposed fabrication methodology. Consideration should be given to structure support points used for weighing and loadout. Site wind loads are included as part of this load condition. These checks are carried out during detailed design and not basic design.

7.4.5 Jacket assembly and erection

Subassembly can be considered an intermediate stage between standard shop fabrication of parts, such as joints, tubular members, beams, and assembly or erection. The work should be managed so that the maximum number of welds is done in the shop. Shops have a highest weld quality, since many node and tubular welds can be double-sided or automatic when performed in a fabrication shop.

The main factors that should be considered for the subassemblies are as follows:

- Dimension, size, and weight are the main factors governing the available ways of transportation.
- Subassemblies should not need a difficult welding sequence, which can induces stresses during the subassembly welding or erection, or avoid distorting any member.
- Certain subassemblies may have specific construction difficulties associated with them, as short, large-diameter infillings are difficult to erect vertically and best included in subassemblies, if possible.

As usual, the assembly of a jacket frame, often having a spread at the base of 50 m or more, places severe demands on field layout and survey and may require temporary support and adjustment bracing. Such large dimensions mean that the thermal changes can be significant. Temperature differences may be as great as 30°C between dawn and afternoon and as much as 15°C between various parts of the structure, resulting in several centimeters distortion. High temperatures will tend to induce residual stresses on the structure. Because of the difficulty associated with thermal distortion, it is normal to "correct" all measurements to a standard temperature, at 20°C.

Note that, for joints, elastic deflections are also a source of difficulty in maintaining tolerances. Foundation settlement or displacements under the skid beams and temporary erection skids must be carefully calculated and monitored.

In the construction plan, each assembly should be completed before starting the lifting process, so it is required to define the location, orientation, and if the face-up or face-down of each assembly to match it with the lifting procedure.

Central coordinates for each assembly is usually shown in the layout drawings. The central coordinates are then used as local benchmarks for erecting the assembly, the subassemblies, loose items, appurtenances, and temporary attachments that comprise field welds, overall dimensions, weight, reference drawings, and others.

It is worth mentioning that the quality control check on dimensions should be performed before and after welding to verify the measurements.

The measurements should take into consideration that the assembly is located in position according to theoretical dimensions using allowable positive tolerances to compensate for weld shrinkage. Perhaps the most fundamental rule in fitting is the avoidance of excess force on the member during fitting prior to welding or stresses on unwelded members through the welding sequence, since such conditions cannot have been foreseen by the designer and account for more stress not considering in design calculation.

In this phase assembled, subassembled and fabricated structures, together with loose items, are incorporated into the final structure according to the erection sequence shown in Figure 7.10.

Jacket frames are typically laid out flat then lifted upward by more than one crawler crane. Coordinating such a rigging and lifting operation requires thoroughly developed three-dimensional layouts, firm and level foundations for the cranes, and competent operators.

Structural analysis should be performed to check the structure members' stresses during the erection process for a given assembly, and usually, it is sone using a computer model with all relevant structural characteristics. The assembly is analyzed for a number of load cases that correspond approximately to the support conditions of the assembly with the assumption of the locations of the cranes, saddles, and others. The structural analysis for lift and transportation identifies the worst cases from structural stresses perspective. These cases are then analyzed to determine the maximum stresses and displacements. The calculations should show that global and local stresses are within allowable limits according to API and AISC codes.

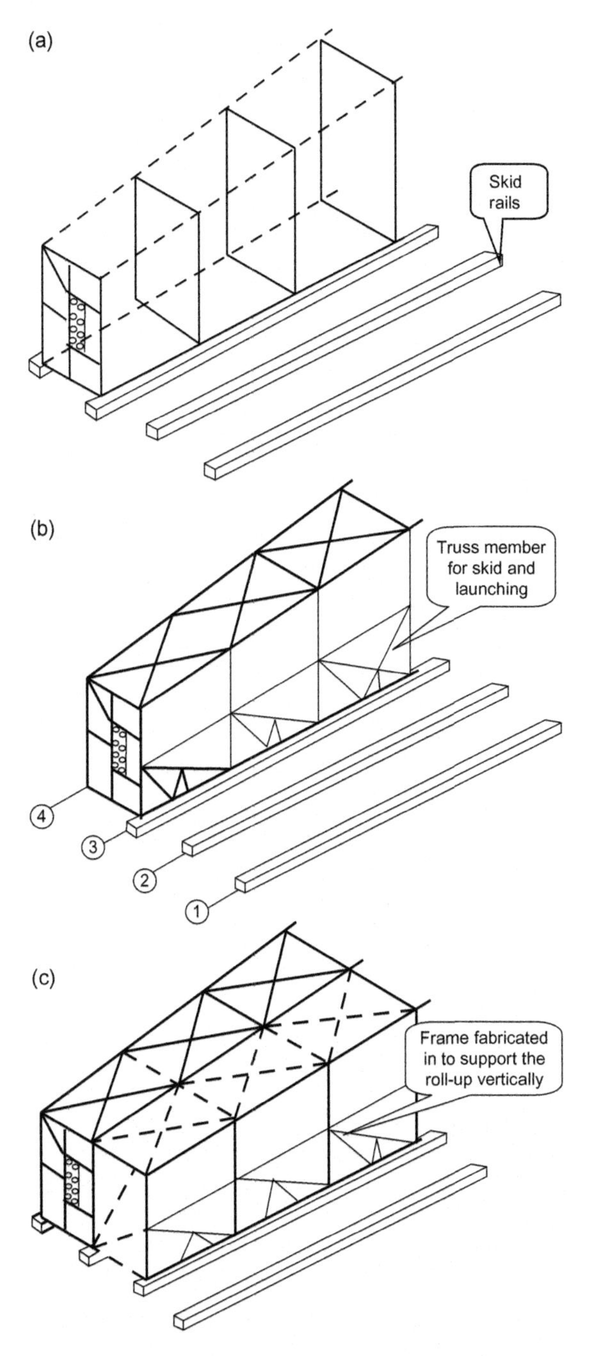

Figure 7.10 (a) Step 1, erection of plan frames; (b) Step 2, erection of bay 1; (c) Step 3, frame roll up; (d) Step 4, rolling up the last frame and filling in between; (e) Step 5, the jacket is ready for loadout and launching.

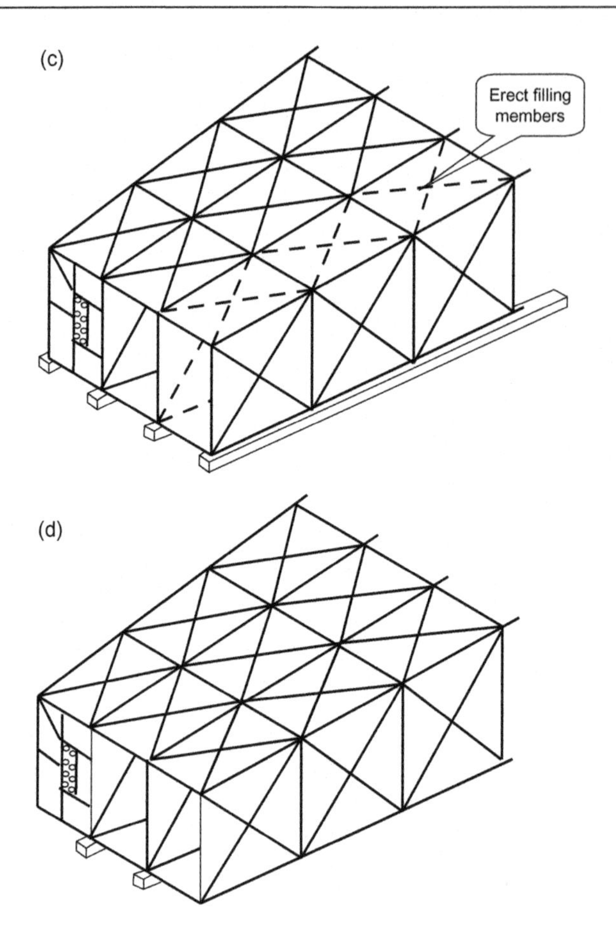

Figure 7.10 (Continued).

7.5 Installation process

The installation process for fixed offshore structure is very complicated, as it depends on the interface between the construction yard, the installation company, and the engineering firm. Mainly, it depends on lifting process and the lifting capability, and the structure weight governs the design and the process of installation. So, the capacity of the barge crane should be identified after the FEED process, and the lifting analysis for the jacket upending and installation should be considered in design. The topside lifting from the construction yard to the materials barge and from materials barge to installation over the pile should be calculated in these two steps. Also, the sea fastening analysis during the jacket and topside transportation, in addition to the load out of the jacket and the topside, should be considered in design, as presented in Chapter 4 and in Chapter 8 using the software.

7.5.1 Loadout process

After finishing the erection of the jacket and the topside, we start the process of the loadout, which applies loadout forces to the structure to move it from the fabrication yard to the barge.

The loadout forces are generated when the jacket is loaded from the fabrication yard onto the barge. If the loadout is carried out by direct lifting. Then, unless the lifting arrangement is different from that to be used for installation, lifting forces need not be computed, because lifting in the open sea creates a more severe loading condition, which requires higher dynamic load factors. If loadout is done by skidding the structure onto the barge, a number of static loading conditions must be considered, with the jacket structure supported on its side. In the case of the loadout jacket, the loading conditions affect the structure from a different positions of the jacket during the loadout phases, as shown in Figure 7.11, from movement of the barge due to tidal fluctuations, marine traffic, or change of draft and from possible support settlements. Since movement of the jacket is slow, all loading conditions can be taken as static. Typical values of friction coefficients for calculation of skidding forces are the following:

- Steel on steel without lubrication, 0.25
- Steel on steel with lubrication, 0.15
- Steel on teflon, 0.10
- Teflon on teflon, 0.08

All structures should be checked for the loads applied during loadout. The proposed method of loadout should be defined explicitly by the contractor, and the structure could be skidded, trolleyed, or lifted. The following should be considered:

- Only dry loads should be used, together with weights for lifting gear, sea-fastenings, and others. The loads should be based on the weight control report.
- For a horizontal load, 15% of the vertical reaction on one skid rail should be applied.
- The total loss of vertical support at one gridline with the structure being supported by the remaining gridlines only should be considered.
- The supports are assumed to be hinged supports in the calculation, and a friction coefficient 0.2–0.3 is considered.
- Wind loads for a return period of one year should be included with this load condition in the structure analysis check.

The direction of the pull out force is shown in Figure 7.12. The structure analysis for this type of loading will be done by SACS software in Chapter 8.

7.5.2 Transportation process

During transportation of the deck or jacket structure from the fabrication yard to its location, some forces will affect the structure. These forces depend on the weight, geometry, and support conditions of the structure (by barge or by buoyancy) and also on the environmental conditions that are predicted during the transportation period.

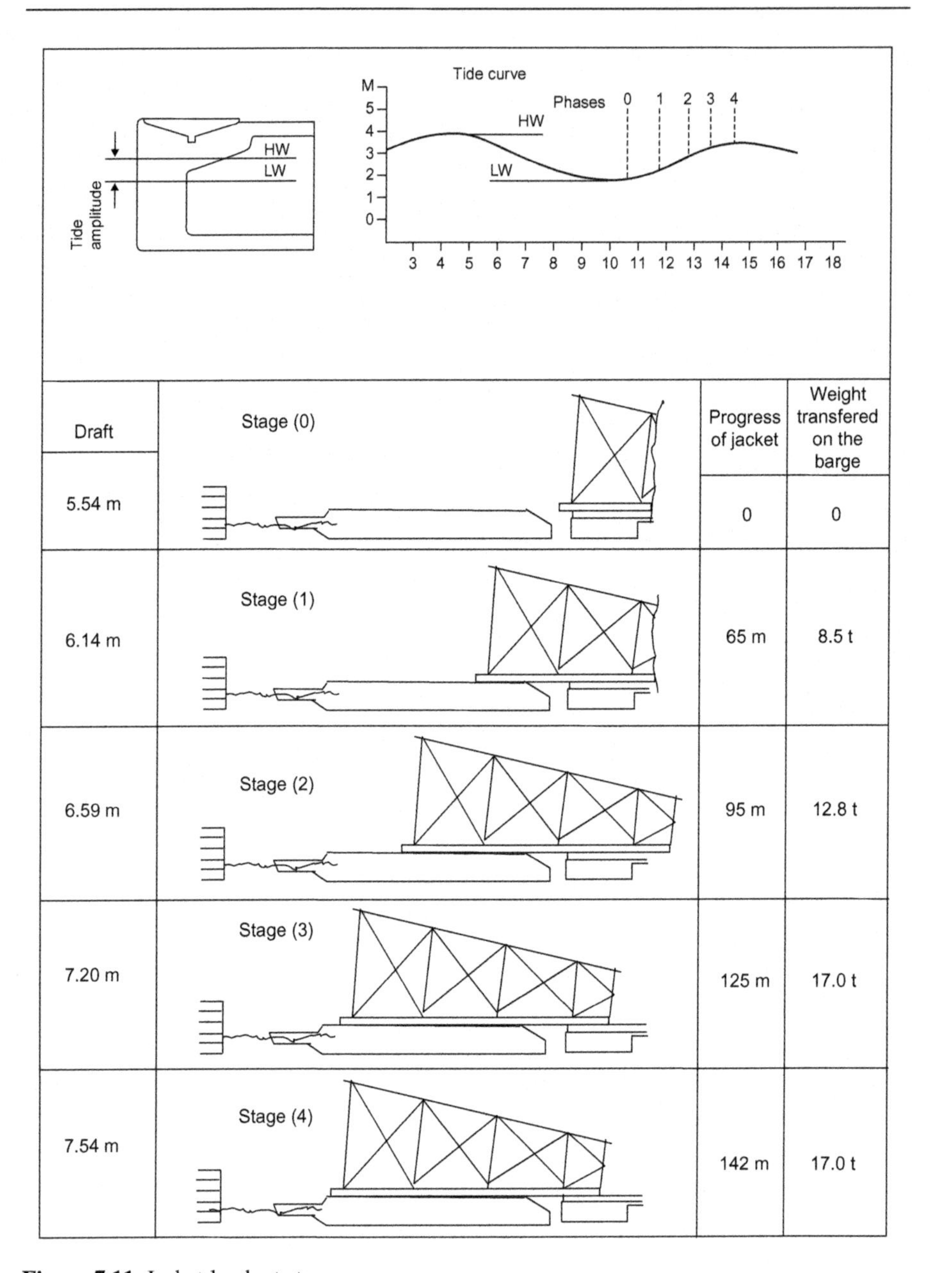

Figure 7.11 Jacket loadout stages.

Figure 7.12 Loadout to the topside, another view.

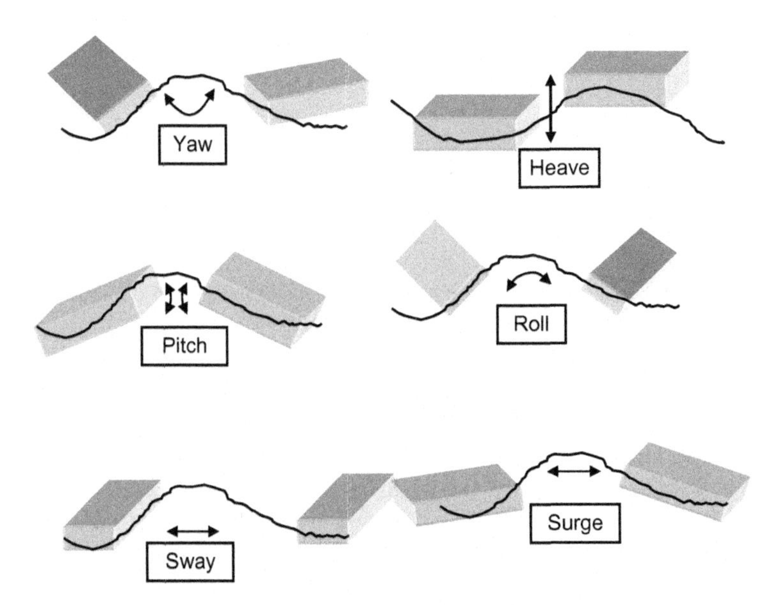

Figure 7.13 Types of motion for any boat.

The main part in the offshore structure project during construction and the operation as well is the vessel that transfers people and equipment from onshore to the platform and performs the construction. Figure 7.13 presents the types of motion that affect the floating structure.

To minimize the associated risks and secure safe transport from the fabrication yard to the platform site, it is important to plan the operation carefully by follow these recommendation from API-RP2A (2007):

- Have previous experience along the tow route.
- Consider exposure time and reliability of predicted "weather windows" during transportation.
- Consider the accessibility of safe havens
- Seasonal weather system
- Consider an appropriate return period for determining and designing for wind, wave, and current conditions, taking into account the tow's characteristics, such as size, structure, sensitivity, and cost.

Transportation forces are generated by the motion of the tow as well as the structure and supporting barge. These loads will be a result of the designed-for winds, waves, and currents. If the structure is self-floating, the loads can be calculated directly. According to API-RP2A, towing analyses must be based on the results of model basin tests or appropriate analytical methods and must consider wind and wave directions parallel, perpendicular, and at 45° to the tow axis. Inertial loads may be computed from a rigid body analysis of the tow by combining roll and pitch with heave motions, when the size of the tow, magnitude of the sea state, and experience make such assumptions reasonable. For open sea conditions, the following may be considered typical design values:

- Structure self weight.
- Equipment and bulk self weight.
- Transportation inertia loads.
- Roll: 20 deg; Period: 10 s.
- Pitch: 12.5 deg; Period: 10 s.
- Heave: ± 0.2 g.
- Center of rotation is 60% above barge keel at longitudinal midship of the transport barge.
- The transportation inertia loads are combined as roll $\pm$ heave and pitch $\pm$ heave.
- Wind loads for a return period of 10 years (1 minute mean) are included with this load condition.
- The support points reflect the support points adopted during loadout.

When transporting a large jacket by barge, stability against sizing is a primary design consideration because of the high center of gravity (CoG) of the jacket. Moreover, the relative stiffness of jacket and barge may need to be taken into account together with the wave forces that could result during a heavy roll motion of the tow, as in Figure 7.14, when structural analyses are carried out for designing the tie-down braces and the jacket members affected by the induced loads. Special computer programs or a module in the special software for structure analysis is available to compute the transportation loads in the structure-barge system and the resulting stresses for any specified environmental condition.

The metacentric height (GM; Figure 7.15) is the distance between the center of gravity of a ship and its metacenter. The GM is used to calculate the stability of a ship, and this must be done before it proceeds out to sea. The GM must equal or

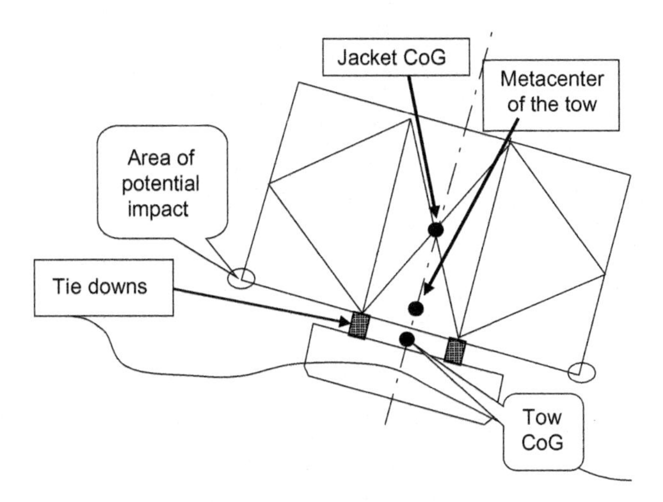

Figure 7.14 Center of gravity for the barge launching a jacket.

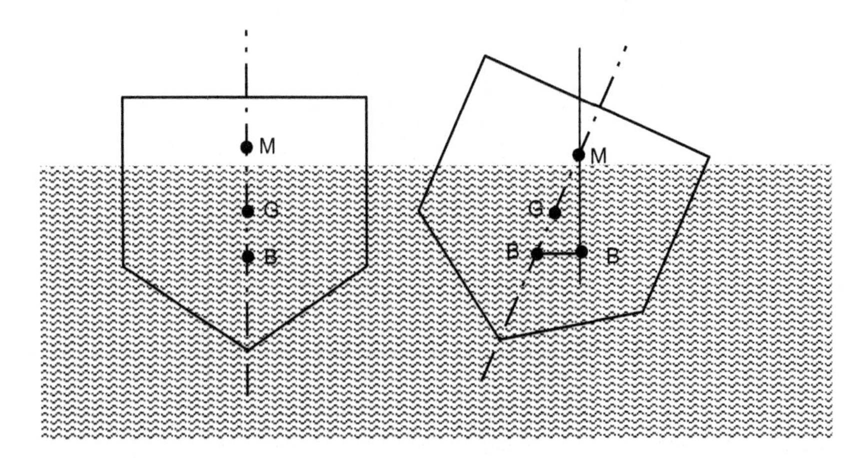

Figure 7.15 The metacentric height.

exceed the minimum required GM for that ship for the duration of the forthcoming voyage. This is to ensure that the ship has adequate stability.

When a ship is heeled, the center of buoyancy of the ship moves laterally. The point at which a vertical line through the heeled center of buoyancy crosses the line through the original, vertical center of buoyancy is the metacenter.

So the cooperation between the installation company and the engineering company should be started early to avoid any change on structure configuration.

The skew load factor is to account for the sling fabrication tolerance or any other inaccuracy in the sling length. The skew load factor is calculated based on the Det Norske Veritas, DNV (2008) rules. In the absence of exact information, this factor is set to 1.25 for a typical indeterminate four-point single hook lift.

As an alternative to the skew load factor, the lift weight (hook weight) might be distributed on a 75–25% split between each pair of slings in turn. All structural members, padeyes, shackles, and rigging components should be designed or checked for both load distributions.

For the padeye design, an additional lateral force equivalent to 5% of the sling force is applied to the padeye. The force should be applied at the eye of the padeye in conjunction with the design sling load.

The criterion of 7–25% split or skew load factor of 1.25 is based on the variation in fabrication tolerance of $\pm 0.25\%$. If, for any reason, this cannot be achieved, then the skew load factor must be modified.

Example 7.1

This example calculates the forces in the sea fastening members due to transportation for a deck weight of 50 tons. The input data are as follows:

Steel deck weight, $W = 50$ tons
Length of the CoG from the middle of the ship, LCG = 8.4 m
Distance between the CoG of the deck and the vessel center line in transverse direction, TCG = 0.0 m
Vertical distance from the deck CoG to the barge keel, VCG = 5.0 m
Barge draft = 3.65 m
Area of the deck in longitudinal direction = 10.0 m^2
Area of the deck in transverse direction = 4.2 m^2
Maximum rolling angle, $\phi = 20°$; Time, $t = 10$ s
Maximum pitch angle, $\psi = 12.5°$; Time, $t = 10$ s
Heave acceleration = ± 0.2 g
Radius in rolling, $R_r = 5.0 - 3.65 = 1.35$ m
Rolling acceleration, $a_r = R_r(\phi\pi/180)(2\pi/t)^2 = 0.186$ m/s^2
$\beta = 0.0°$
Pitching acceleration, $a_p = R_p(\psi\pi/180)(2\pi/t)^2 = 0.82$ m/s^2
$\delta = 80.91°$
Heaving acceleration, $a_z = 0.2g = 0.2 \times 9.81 = 1.962$ m/s^2
In the case of rolling, $K_r = 1$ $k_p = 0.6$ $k_z = 0.8$
In the case of pitching and heaving, $K_r = 0.6$ $k_p = 1$ $k_z = 1$

Calculate the rolling force due to transportation vessel motion (Figure 7.16) as

$F_r = (W/9.81)[k_r a_r \cos\beta + \sin\phi(9.81 + k_p a_p \sin\delta + k_z a_z)] = 21.53$ tons
$F_w = 15$ tons due to wind perpendicular to the longitudinal direction of the vessel
Total load = 21.53 + 15 = 36.53 tons

Calculate the pitching force due to transportation vessel motion (Figure 7.16) as

$F_p = (W/9.81)[\mathrm{k_{pap}} \cos\delta + \sin\phi(9.81 + k_r a_r \sin\beta + k_z a_z)] = 13.57$ tons
$F_w = 11$ tons due to wind perpendicular to the transverse direction of the vessel
Total load in transverse direction = 24.57 tons

Calculate the heaving force due to transportation vessel motion as

$F_z = (W/9.81)[k_r a_r \sin\beta + \sin\phi(9.81 + k_p a_p \sin\delta + k_z a_z)] = 59.82$ tons

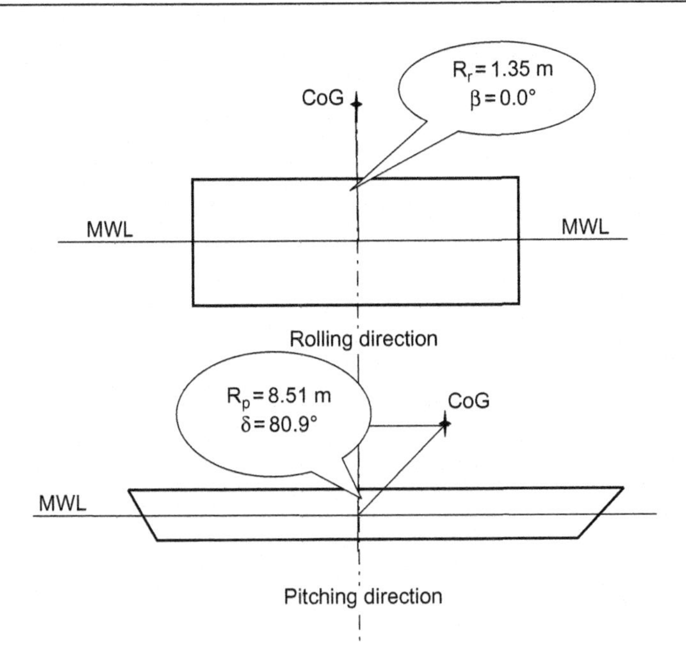

Figure 7.16 Example 7.1.

You have the forces applied on the steel deck CoG, so you can calculate the forces in the supported element.

7.5.3 Barges

A barge is consider a floating workshop. The offshore construction barge must be long enough to have minimal pitch and surge response to the waves in which it normally works, wide enough in the beam to have minimum roll, and deep enough to have adequate bending strength against hog, sag, and torsion as well as an adequate freeboard. The plate of the deck should be continuous enough to enable it to resist the membrane compression, tension, and torsion introduced by wave loading. Side plates must carry high shear, so the sides usually have stiffeners to resist buckling.

Impact loadings can come from wave slams on the bow, from ice, and from boats and other barges hitting the sides. Unequal loads may be incurred in the bending of the bottom hull plates during intentional or accidental grounding and the deck plates due to cargo loads. Note that corrosion may reduce the thickness of hull plates.

Typical offshore barges run from 80 to 160 m in length. The barge width is ($\frac{1}{3}$–$\frac{1}{5}$) the barge length, and the depth typically runs from $\frac{1}{15}$ the length. From a practical point of view, barges have been found to give a reasonably balanced structural performance under wave loadings. Inland barges, subjected to minimal wave loadings and required for operations in shallow water, may have depths as low as $\frac{1}{20}$ of the length. They may be stiffened by external trussing.

Offshore barges typically have natural periods in roll of 5–7 s. This is unfortunately the typical period of wind waves; hence, resonant response does occur. Fortunately, damping is very high, so that, while motion in a beam sea is significant, it reaches a situation of dynamic stability.

Consideration must be given to the need to temporarily weld padeyes to the deck in order to lift the cargo for sea. These padeyes must distribute their load into the hull. They will be subjected to fatigue and impact loads in both tension and shear; therefore, a better design often is to have special doubler plates fixed over the internal bulkheads so that padeyes may be attached along them.

Thus, sea fastenings are designed to resist the static and dynamic forces developed under any combination of the six fundamental barge motions (roll, pitch, heave, yaw, sway, and surge). The dynamic component is due to the inertial forces that develop due to acceleration as the direction of motion changes.

Roll accelerations are directly proportional to the transverse stiffness of the barge, which is measured by its metacentric height where the mean center height is the distance between the center of gravity of the ship and its metacenter.

Since barges typically have large metacentric heights, accelerations are severe. Conversely, if, due to high cargo, the metacentric height is low, the period and amplitude of roll and the static force as a result from the load are greater, but the dynamic component may be less.

These loads are cyclic. Sea fastenings tend to work loose as the wire rope stretches and wedges and blocking fall out. Under repeated loads, fatigue may occur, especially at welds. Welds made at sea may be especially vulnerable because the surfaces may be wet or cold. Using low hydrogen electrodes in welding will help in this case. Chains are a preferred method for securing cargo for sea, since chains do not stretch. If structural posts are used, they should be run through the deck to be welded in shear to the internal bulkheads. The slot through the deck should then be seal welded to prevent water in-leakage.

The effect of the accelerations is to increase the lateral loading exerted by the cargo due to the inclination of the barge by a factor of 2 or more. Flexing the barge can also have a significant effect on support forces and the sea fastenings. Therefore, deeper, and hence stiffer, barges experience a smaller range of loads than shallow, less stiff barges.

In case of using a jack-up rig for construction or in drilling activity, a general rule of thumb is to plot the proposed leg positions and space the new leg locations four to five diameters away. In clay soil, where jack-ups have previously worked around the site, holes will have been left that now may be partially empty or filled with loose sediment. If a leg is now seated adjacent to such a hole, it may kick over into it, losing both vertical and lateral support and bending the leg. This, of course, is another advantage of the mat-supported jack-up legs: The mats can span local anomalies.

If the sea conditions are highly variable, the large jack-up construction rigs are suitable to use with frequent periods of calm, so that the rig has a convenient times to move.

Based on Gerwick (2007), the statistical studies covering jack-up drilling rigs and jack-up construction rigs show that they have been six times more likely to suffer serious damage or loss during relocation and transit than when on location. This is primarily due to the barge having its legs fully raised, thus creating a very high center of gravity. Some jack-ups have telescoping legs.

7.5.4 Launching and upending forces

The launch is a very critical process in constructing the platform, as in this process, the jacket is affected by different stresses due to transfer from the barge to the sea and during the subsequent upending into its proper vertical position to rest on the seabed. A schematic view of these operations can be seen in Figure 7.17.

There are six stages in a launch-upending operation:

1. Jacket is in a stability position on the barge.
2. Jacket slides along the skid beams.
3. Jacket rotates on the rocker arms.
4. Jacket rotates and slides simultaneously.
5. Jacket detaches completely and comes to its floating equilibrium position.
6. Jacket is upended by a combination of controlled flooding and simultaneous lifting by a derrick barge.

The loads induced, static as well as dynamic, can be evaluated by appropriate analyses, which also consider the action of wind, waves, and currents expected during the operation.

To start the launch, the barge must be ballasted to an appropriate draft and trim angle; subsequently, the jacket must be pulled toward the stern by a winch. The sliding of the jacket starts as soon as the downward force from gravity and the winch pull exceeds the friction force. As the jacket slides, its weight is supported on the two legs that are part of the launch trusses. The support length keeps decreasing and reaches a minimum, equal to the length of the rocker beams, when rotation starts. It is generally at this instant that the most severe launching forces develop, as reactions to the weight of the jacket. During the last two stages, variable hydrostatic forces arise, which have to be considered at all members affected.

Figure 7.17 Launching and installing the jacket.

Buoyancy calculations are required for every stage of the operation to ensure fully controlled, stable motion. Computer programs are available to perform the stress analyses required for launching and upending and to portray the whole operation graphically, as presented in Figure 7.18.

The typical launch barge is very large and strongly built, long and wide, subdivided internally into numerous ballast compartments. Since it must support a progressively moving jacket weighing thousands of tons, heavy runner beams or skid beams extend the length of the barge. These girders distribute the jacket's load to the barge structure. The stern end of the barge, over which the jacket will rotate and slide into the water, requires special construction.

First, for a short period of time the stern will have to support the full weight of the jacket. Second, since this reaction force has to be transmitted into the jacket, it must distribute the reaction over as long a length as feasible to avoid a point reaction. The jacket will slide on its specially reinforced runners; even so, they need a distributed rather than a point reaction. Hence, the stern of the barge is fitted with a rocker section that rotates with the jacket as it slides off.

For loading out the jacket at the fabrication yard, the usual method is to ground the launch barge on a screeded sand pad at the appropriate depth so that the barge deck matches the yard level. Then, the jacket can be skidded out onto the barge with no change in relative elevation. This means that the hull bottom must withstand high local pressures from irregularities in the prepared sand bed. Not only must the bottom be of heavy plate, but the stiffeners must be adequate to prevent buckling as the jacket is moved onto the barge. When the loadout is performed with the barge afloat, the ballast must be rapidly adjusted to maintain the relative elevation at the barge deck as the load of the jacket comes on.

Step-by-step adjustments or computer control are used to adjust deck elevation and trim. A launch barge is also fitted with heavy winches or linear jacks on the bow to pull the jacket onto the barge and later, by rerigging through sheaves on the stern, to pull the jacket off the barge during launching. The beam width of a launch barge is often less than the base width of the jacket. The base of a deepwater jacket may be 60 m wide, overhanging the sides of the barge significantly. Several large launch barges can carry and launch a jacket of 40,000 tons. A 300-m-long launch barge was fabricated in Japan to transport the 55,000-ton, 415-m-long Bullwinkle jacket. During transport, the barge must have enough freeboard to prevent the outside legs of the jacket from dipping into the waves as the barge rolls. The beam width of the barge is designed to give stability transversely during launching. This is often the critical condition during launch; if the barge lists and the jacket rolls sidewise, it may buckle a jacket leg.

The topside and jacket structures should be checked for impact loads during the float over installation. Mating loads are determined by motions analysis, by naval architects. The analyses consider the selected barge, stiffness of the structures and leg mating units, and the full range of installation scenarios from initial docking through to final load transfer. Vertical loads are based on the dry weight of the structure obtained from the weight control report.

After launching the jacket, the crane barge is used to uplift the jacket, as shown in Figure 7.19, to its location and, after that, start driving the piles into the legs.

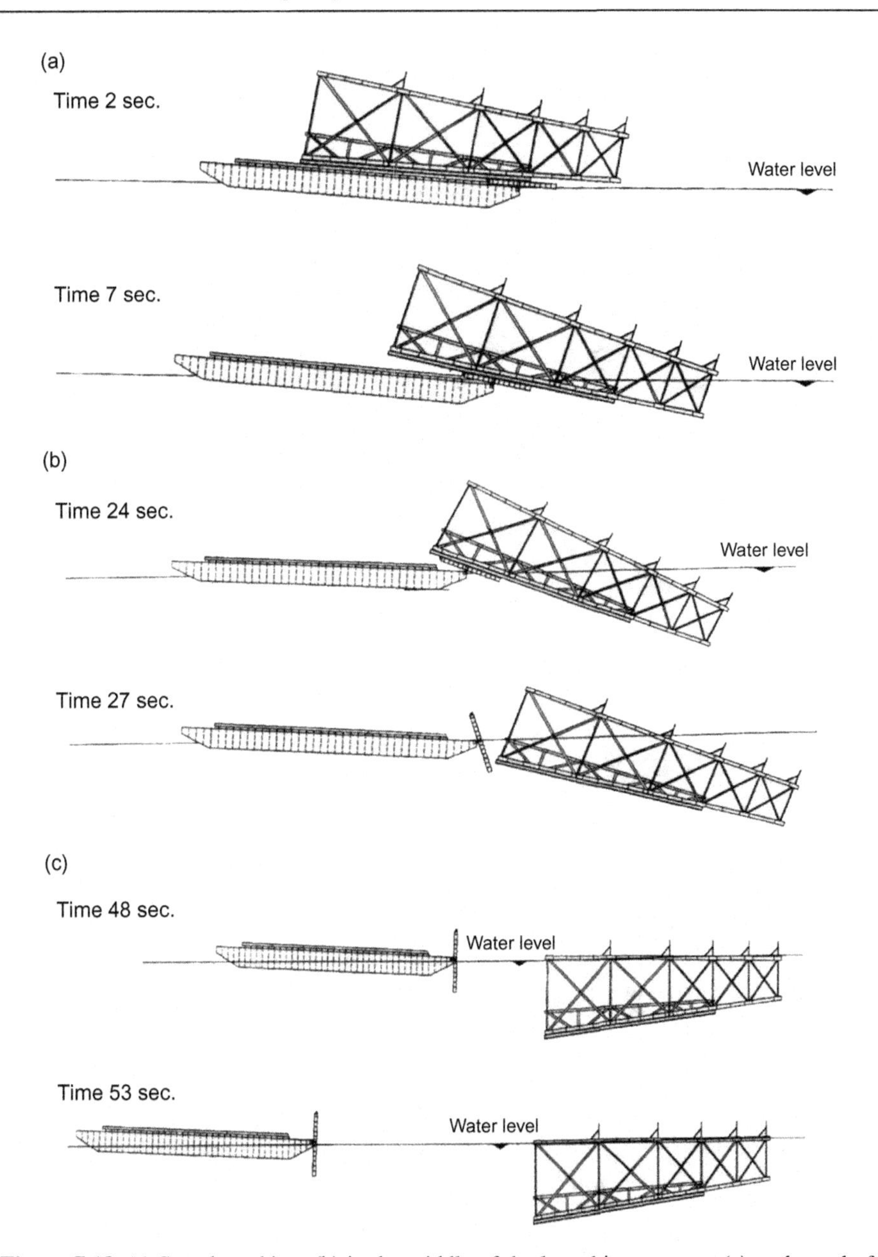

Figure 7.18 (a) Start launching; (b) in the middle of the launching process; (c) at the end of the launching process.

Figure 7.19 Lifting the jacket for installation.

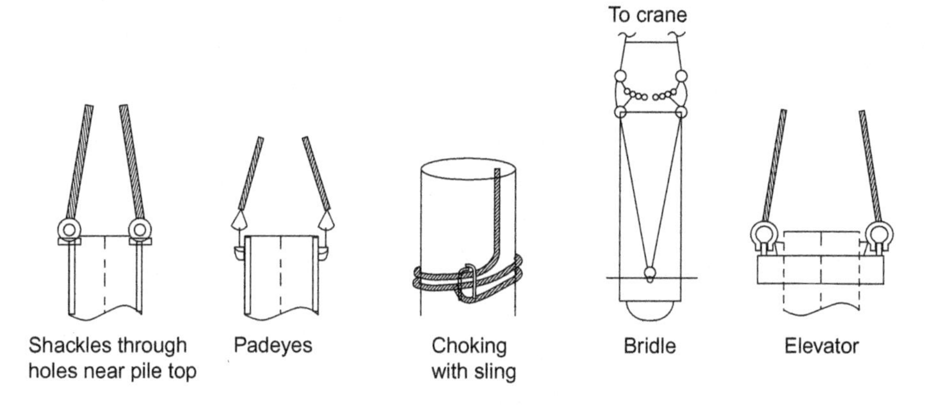

Figure 7.20 Pile lifting methods.

Figure 7.20 shows the different ways of providing lifting points for positioning pile sections. Padeyes are welded in the fabrication yard; their design should take into consideration the changes in load direction during lifting. Padeyes are then carefully cut before lowering the next pile section.

7.6 Lifting analysis

7.6.1 Weight control

The weight control report is about monitoring or removing weight and ensures that, at all stages throughout the life of a project, weight assessments are up to date and traceable through backup documentation.

From the beginning of front end engineering design (FEED) up to final platform construction, the weights and centers of gravity of topside and substructures are critical factors on which major decisions are made. Such decisions typically are as follows:

- The selection of construction pad at the fabrication yard.
- The loadout process.
- The selection of transportation barge,
- The definition of associated ballasting requirements,
- The selection of crane barge,
- The choice of the best lifting arrangement to facilitate installation.
- The direction of topside inventory changes during the platform operating phase to accommodate production changes and the selection of the optimum solution for platform abandonment, should this be necessary at the end of the platforms useful life.

7.6.2 Weight calculation

In addition to drilling loads and other temporary operating loads, the topside weight includes essentially permanent contributions from all associated engineering disciplines, including structural, mechanical, electrical, instrument, communication, HVAC, safety, and architectural. It is useful to categorize the weight elements associated with all disciplines other than structural and architectural as follows:

- Main equipment.
- Bulk materials.

Therefore, topside weight may be deemed to comprise the following six elements:

1. **1-Structural steel**. This typically includes modules; decks and all main framing; deck plating and stiffening; all equipment supports integrated in the deck structure; major facility supports such as pipe racks, walkways, service platforms, stairs, ladders, handrails; crane pedestal; and bridge supports. In general, structural steel consists of three main categories, which are
   - Primary steel.
   - Secondary steel.
   - Temporary steel. These items typically include sea fastenings bumpers, guides, lifting and rigging gear (i.e., slings, shackles, spreader beams), and secondary padeyes.
2. **Architectural**. This includes fire walls, ceilings. doors, windows, flooring and floor finishes, furniture, partitions, kitchen fixtures, fittings and utensils, appliances, toilet fixtures and accessories, acoustics, insulation, windshield cladding, weather louvers, heat shields and the like in the living quarters and emergency shelter.
3. **Main Equipment**. This includes all tagged mechanical, electrical, and instrument equipment as well as the skid weight, drip pa,n and all other items shown in the vendor package identification, on P&IDs, if applicable.
4. **Bulk Materials**. These typically include all mechanical, piping, and associated fittings and supports and electrical, communication, HVAC, and safety items that have not been listed under equipment. Piping bulk typically includes all process and utility piping, piping valves, pipe supports, trace heating, insulation, protective coatings, and all associated accessories that have not been included in the equipment skids. Electrical bulk typically includes all

electrical cables, cable trays, cable ladders, MCI, supports, light fixtures, junction boxes, and instrumentation. Bulk typically includes all instrument cable and all instruments shown on instrument diagram legends and all instrument piping (tubing) and associated supports including control valves; it also includes instrument workshop equipment, all untagged control room auxiliary equipment, and all instrument equipment not listed under equipment. HVAC bulk typically includes all ducts, duct supports, insulation, flow dampers, grills, and the like for ventilation and air conditioning. Safety bulk typically includes fire monitors, hose reels, deluge valves, fire and gas detectors, halon and CO_2 systems, potable fire extinguishers, life buoys, life jackets, life rafts, survival suits, and fireman equipment.

5. **Live load**. These are the temporary operating loads on the topside. They typically include bulk stores, laydown area loads, crane loads, helicopter, emergency shelter, luggage, consumable and personnel effects, and all these load defined in Chapter 2.
6. **Drilling loads**. Drilling loads, from the drilling required equipment and tools that are essentially temporary operating loads, are treated separately and are not covered by the live load allowance

7.6.3 *Classification of weight accuracy*

The three basic classifications of weight accuracy are as follows, and the allowance in weight for each project stage is proposed in Table 7.1.

- **Conceptual**. This is obtained from the initial estimates, possibly obtained from past projects. At the end of conceptual design, the structural weight estimates are based on the preliminary structure design by weight takeoff.
- **Detailed**. This is obtained from the weight takeoff and vendor information
- **Fabrication**. This is obtained from the approved drawings and final takeoff.

Weight information for all items on the platform should be recorded in a manner consistent with the following definition. Functional weight conditions are illustrated as follows:

- Dry is that condition in which a single weight item or a collection of items is characterized by its dead weight alone. This condition typically excludes any operating fluids or supplies, spares, maintenance, tools, and packing or temporary transportation materials. However, the recorded dry weight of that equipment delivered with lubricants or coolants or the like preinstalled are deemed to include such additional weights.
- Operating is that condition which exists when all equipment and bulks contain all relevant fluids and supply weights that occur under normal operating conditions.
- Hydrotest is that condition which generally occurs during topside hookup and commissioning and includes weight contributions from all permanent and temporary facilities together with all weight elements associated with hydrotesting.
- Loadout is that condition which exists during the activity of loading the platform (topside, substructure) out from the fabrication facility into the transportation barge.
- Lift weight is that condition which exists during lifting of the structure. The lift weight is deemed to include the weight of the structure together with the weight of all equipment and bulks actually being lifted. It includes all temporary lifting accessories, such as lifting slings, shackles, and support frames; tie-down and support beams; and any shipped-loose items temporarily placed on the structure during lifting, but excludes hookup spools, infill steel, or other items that are to be lifted separately.

Table 7.1 **The allowance in weight**

Project stage	Available data	Allowance percentage for topside	Allowance percentage for jacket
Conceptual design	Historic volumetric	15	10
	Vendor catalogue	15	10
	Vendor data/quotation	12	10
	Calculated from material takeoff (MTO)	10	10
	Historical weighed	5	10
	Design change allowance	5	5
	Fabrication change allowance	5	5
Detailed design	Vendor catalogue	15	5
	Vendor data or quotation	10	5
	Calculated MTO from (AFC drawings)	5	5
	Design change allowance	5	5
	Fabrication change allowance	5	5
Fabrication	Vendor data or quotation	5	3
	Design change allowance	0	0
	Fabrication change allowance	5	2

7.6.3.1 Allowances and contingencies

The allowance and contingencies of weight are usually based on the engineering firm experience. These contingency allowances are as follows, where the definition of allowance and contingency is obviously stated In Table 7.1.

The item accuracy allowance I_A appropriate to the beginning of each design phase is stipulated in Table 7.1. The average level of I_A should reduce with time through each design phase as a function of design maturity.

The value of the I_A depends on the degree of definition of each individual item or collection of items and the level of confidence in the weight estimate at any particular time. Three categories apply:

1. Conceptual design allowance (C_A).
2. Detailed design allowance (D_A).
3. Fabrication change allowance (F_A).

The detailed design allowance (D_A) provides for design changes during the detailed design phase. These design change are a normal part of the design activity and represent the optimization of the design in satisfying the preferred approach defined during the conceptual design of the project.

Fabrication allowance (F_A) provides for changes to design during the fabrication phase. These changes are a normal part of the fabrication activity and represent the optimization of the design in satisfying previously unidentified constraints arising during the fabrication phase. Examples are steel section substitutes and pipework, cable work, or ductwork rerouting as well as overrolling of plates, weld metal, and paint.

The fabrication change allowance is applied on a modular basis and has the same value for each functional weight condition.

The factored weight of an item or collection of items is obtained as

$$\text{Base weight of item} \times (1 + C_A) \times (1 + D_A) \times (1 + F_A)$$

7.6.3.2 Weight engineering procedures

Throughout the duration of the project, the lead structural engineer is responsible for gathering and recording the weight and center of gravity information from all disciplines and for generating the weight report. This person is additionally responsible for reporting the status of the weight and center of gravity of the platform components, such as topside modules, decks, bridge, and substructures to each department and to the project manager or engineer on a regular basis.

On the other hand, each department is responsible for furnishing all the necessary weight and center of gravity information of items within their discipline to the lead structural engineer. Each department is also responsible for updating such information at previously agreed-on regular intervals or when deemed necessary by the lead structural engineer

The limit of responsibility for supplying weight information is defined by department, as each department is responsible for furnishing information about the weight and center of gravity within its area.

Note that the realistic weight and weight allowances can be obtained from vendors by specifying in the request for quotations (RFQs) packages a requirement that vendors guarantee, within agreed limits, their quoted weights and center of gravity when submitting their quotations. The department should follow up the weight by recording its weight estimates, which should be kept in a tidy and proper manner. The records may be a quotation from vendors and marked-up drawings of weight takeoffs with piece mark numbers.

At the start of the project and as soon as the preliminary equipment layout drawings are established, each department uses the weight and center of gravity control form to furnish all the initial weight and center of gravity information to the structural engineer. The structural engineer is then responsible for generating the weight report. This is done using appropriate computer software. The engineer issues this report to each department and the project manager or engineering manager for their review and information. All assumptions, qualifications, and exclusions are included in the weight report.

At regular intervals for at least monthly during the project, each department furnishes updated weights and center of gravity information to the lead structural engineer, who is responsible to generate an updated weight report.

The responsible structural engineer is liable for gathering all information required for the weight control report and should draw management attention to any undesirable weight trends or problems and suggest corrective actions as appropriate.

7.6.4 *Loads from transportation, launch, and lifting operations*

The topside structure and jacket component are subjected to critical loadings during construction operations. Some jacket members and joints may be subjected to high bending and punching shear loads while braces and bents are assembled into a jacket in the fabrication yard. Analysis of such assembly loading conditions would require sequential simulation of jacket geometry and loads and the knowledge of the jacket assembly plan and procedures.

During the transportation of the jacket to the site on the barge, the jacket and tie-down braces, their connections, and the transportation barge are subjected to significant dynamic accelerations and inclined self-weight loads. These motions and resulting dynamic loads must be simulated in incremental loading sequences to determine and dimension the highest stressed components. The fact that some bracing may be needed only for the jacket transportation phase, some of these braces may have to be removed, before jacket is installed on site, to reduce in place wave loads.

During its launch to sea, the jacket is subjected to significant inertia and drag loadings. In general, most critical loading occurs as the jacket starts tilting around the launch beam and rapidly descends to sea. At this position, the titling beams exert high concentrated loads on the stiff bracing levels. These require a launch bracing system specially designed to distribute and reduce the launch forces. As the jacket hits the water plane and rapidly descends to the sea, leading jacket braces may experience high drag and inertia forces.

It is worth to mention that another critical factor in loading is the crane lift of the deck or jacket from the transportation barge.

In such lifting operations, deck and jacket members and connections may be loaded in directions different than their loading directions. Additionally, redundant or shorter (longer) lifting slings than planned may result in substantially different loads than those calculated for idealized conditions. Note that, in the case of a four-sling lift, if one sling is shorter than planned, three instead of four slings may carry the entire deck load. Such unplanned load distribution may also be caused by a center of gravity that may be at a location somewhat different than calculated. Lifting padeyes and lugs are components with high consequence of failure. A single padeye failure may result in the loss of the entire deck, jacket, and the crane. Such critical components and their connections to the structures lifted must be designed for higher safety factors. Safety factors for four or more against ultimate capacity are commonly used for padeyes, their connections to the structure, and the associated lifting gear.

7.6.5 *Lifting procedure and calculation*

Lifting forces are functions of the weight of the structural component being lifted, the number and location of lifting eyes used for the lift, the angle between each sling and the vertical axis, and the conditions under which the lift is performed.

All members and lifting point connections for the lifted component must be designed for the forces resulting from static equilibrium of the lifted weight and the sling tensions. Moreover, API RP2A recommends that, to compensate for any side movements, lifting

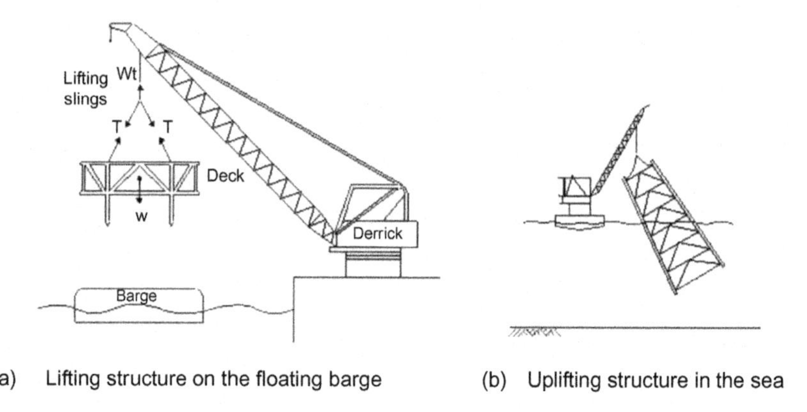

Figure 7.21 Installing the topside on the barge.

eyes and the connections to the supporting structural members should be designed for the combined action of the static sling load and a horizontal force equal to 5% this load, applied perpendicular to the padeye at the center of the pinhole. All these design forces are applied as static loads if the lifts are performed in the fabrication yard. On the other hand, if the lifting derrick or the structure to be lifted is on a floating vessel, then dynamic load factors should be applied to the static lifting forces.

In particular for lifts made offshore, API RP2A recommends two minimum values of dynamic load factors, which are 2.0 and 1.35. The factor 2.0 is to be considered in designing the padeyes as well as all members and their end connections framing the joint where the padeye is attached; while the factor 1.35 is used in designing all other members transmitting lifting forces. In loadout at sheltered locations, the corresponding minimum load factors for the padeye and the structural components, according to API RP2A, are 1.5 and 1.15, respectively.

Figure 7.21 presents the lifting of jacket and topside, and Figure 7.22 shows the lifting of the deck to put to the skid to start pullout into the barge.

Many terms are used in lifting, which are summarized in Table 7.2.

Jacket, topside, and living quarters lift analyses (onshore and offshore) should be performed based on the requirements of DNV rules. Stresses on all members and connections should be checked against API RP2A or AISC basic allowable stresses.

The weight contingency factor is a factor to allow for lift weight inaccuracies. For jacket structures, a minimum factor of 1.1 is used, unless the jacket is weighed at the end of the construction using load cells; in which case, this factor may be reduced to 1.03. Weight contingency factor is applied to the "net weight" and "rigging weight."

For any lifting requirement, the calculations carried out should include the following allowances, factors, and loads or equivalent weight contingency factors. Weight control should be performed by means of a well-defined, documented system, in accordance with current good practice, such as ISO Draft International Standard ISO/DIS 19901-5, "Petroleum and Natural Gas Industries—Specific

Figure 7.22 Transportation of the topside to the barge.

Requirements for Offshore Structures, Part 5: Weight Control during Engineering and Construction."

Where a limiting design sea state is derived by calculation or model tests, the limiting operational sea state should not exceed (0.7 × the limiting design sea state).

Unless operation-specific calculations show otherwise, for lifts by a single vessel, the dynamic amplification factors (DAFs) in Table 7.3 should be applied. Alternatively, the DAF may be derived from a suitable calculation or model test.

Where the lift is from a barge or vessel alongside the crane vessel, barge or vessel motions must be taken into account as well as the crane boom-tip motions:

$$\text{Lift weight} = \text{gross weight} \times \text{DAF} \tag{7.1}$$

For offshore lifts by two or more vessels, the lift weight computed in equation (7.1) is multiplied by a further DAF of 1.1.

Note that the dynamic amplification factors for the offshore lift presented in Table 7.3 is used when the operation is carried in a calm sea ($H_s < 2.5$ m). If, for any reason, the lifting is carried out in an adverse condition, this factor is recalculated based on the expected accelerations associated with the design sea state.

7.6.5.1 Calculated weight

- Class A weight control is needed if the project is weight or CoG sensitive for lifting and marine operations or has many contractors.
- Class B weight control definition applies to projects where the focus on weight and CoG is less critical for lifting and marine operations.
- Class C weight control definition applies to projects where the requirement for weight and CoG data are not critical.

Table 7.2 **Offshore lifting terminology**

Term	Definition
Allowance	An amount, expressed in terms of a percentage of the base weight, which experience on past projects has shown to be consumed during the various phases of project execution
Barge	The floating vessel, normally not propelled, on which the structure is transported; a ship or vessel
Base weight	The base weight of any individual item or collection of items is specified to a functional weight condition and is, at the time of estimating, the best available estimate of weight for that item or collection of items exclusive of all allowances and contingencies
Bending reduction factor	The factor by which the breaking load of a rope or cable is reduced to take account of the reduction in strength caused by bending round a shackle, trunnion, or crane hook
Breaking load	The load at which a rope or sling will break, the breaking load for a sling takes into account the termination efficiency factor
Cable-laid sling	A cable made up of six ropes laid up over a core rope, with suitable terminations each end
Certificate of approval	The formal document issued by an inspection company when, in its judgment and opinion, all reasonable checks, preparations, and precautions have been taken and an operation may proceed.
Consequence factor	A factor to ensure that main structural members have an increased factor of safety related to the consequence of their failure.
Contingency	An amount in tonnes to accommodate future changes in topside functional requirements instigated by management or by unexpected production changes
Crane vessel	The vessel, ship, or barge on which lifting equipment is mounted; it is considered to include crane barge, crane ship, derrick barge, floating shear-legs, heavy lift vessel, and semisubmersible crane vessel
Determinate lift	A lift where the slinging arrangement is such that the sling loads are statically determinate, and are not significantly affected by minor differences in sling length or elasticity
Dynamic amplification factor	The factor by which the gross lift weight is multiplied, to account for dynamic loads and impacts during the lifting operation
Grommet	A single length of unit rope laid up six times over a core to form an endless loop

(*Continued*)

Table 7.2 **(Continued)**

Term	Definition
Factored weight	The factored weight of any individual item is characterized by its base weight multiplied by the product of all relevant allowances
Gross weight	The calculated weight of the structure to be lifted including contingencies or the weighed weight including weighing allowance
Hook load	The summation of the lift weight and the rigging weight
Indeterminate lift	Any lift where the sling loads are not statically determinate
Lift point	The connection between the rigging and the structure to be lifted; it may include padear,'padeye, or trunnion
Lift weight	The gross weight in addition to the allowance for dynamic effects
Loadout	The transfer of topside or jacket from land onto a barge by horizontal movement or lifting
Loadout, lifted	A loadout performed by crane
Minimum required breaking load	The minimum allowable value of breaking load for a particular lifting operation
Net weight	The calculated or weighed weight of a structure, with no contingency or weighing allowance
Padear	A lift point consisting of a central member, which may be of tubular or flat plate form, with horizontal trunnions around which a sling or grommet may be passed
Padeye	A lift point consisting essentially of a plate, reinforced by cheek plates if necessary, with a hole through which a shackle may be connected
Rigging	The slings, shackles, and other devices, including spreaders, used to connect the structure to be lifted to the crane
Rigging weight	The total weight of rigging, including slings, shackles, and spreaders
Rope	The unit rope from which a cable laid sling or grommet may be constructed, made from either six or eight strands around a steel core.
Safe working load	The safe working load of a sling, shackle or lift point is the maximum load that the sling may raise, lower, or suspend under specific service conditions
Sea fastenings	The system used to attach a structure to a barge or vessel for transportation
Skew load factor	The factor by which the load on any lift point or pair of lift points is multiplied to account for sling mismatch in a statically indeterminate lift

(Continued)

Table 7.2 (Continued)

Term	Definition
Sling breaking load	The breaking load of a sling is the calculated breaking load reduced by the termination efficiency factor or bending reduction factor, as appropriate
Sling eye	A loop at each end of a sling, usually formed by an eye splice or mechanical termination
Splice	That length of sling where the rope is connected back into itself by tucking the tails of the unit ropes back through the main body of the rope after forming the sling eye
Spreader bar (frame)	A structure designed to resist the compression forces induced by angled slings, by altering the line of action of the force on a lift point into a vertical plane
Termination efficiency factor	The factor by which the breaking load of a wire or cable is multiplied to take account of the reduction of breaking load caused by a splice or other end termination
Trunnion	A lift point consisting of a horizontal tubular cantilever, around which a sling or grommet may be passed; an upending trunnion is used to rotate a structure from horizontal to vertical or vice versa, and the trunnion forms a bearing around which the sling, grommet, or another structure will rotate

Table 7.3 DAF in different locations

Gross weight, *W*, in tons	Offshore	Onshore	Onshore	
			Moving	Static
$W \leq 100$	1.30	1.15	1.15	1.0
$100 < W \leq 1000$	1.20	1.10	1.10	1.0
$1000 < W \leq 2500$	1.15	1.05	1.05	1.0
$W > 2500$	1.10	1.05	1.05	1.0

Unless it can be shown that a particular structure and specific lift operation are not weight or CoG sensitive, then Class A weight control is needed. If the 50/50 weight estimate, then a reserve of no less than 5% should be applied. The extremes of the CoG envelope should be used. A reserve of no less than 3% should be applied to the final weighed weight.

$$\text{Gross weight} = \text{calculated or weighed weight} \times \text{reserve} \tag{7.2}$$

7.6.5.2 Hook load

The hook load is calculated in the case of the loading on a padeye or the structure. The lift weight as defined before should be used. Loads in slings and the total loading on the crane should be based on hook load, where

$$\text{Hook load} = \text{lift weight} + \text{rigging weight} \tag{7.3}$$

Rigging weight includes all items between the padeyes and the crane hook, including slings, shackles, and spreaders as appropriate. Note that the definition of *padeye*, here, is taken, as generally, to include any type of lift point, including padear, trunnion, or other type.

The shackle type is provided by the marine company responsible for the installation. The configuration of the shackle is in Figure 7.23, with its important

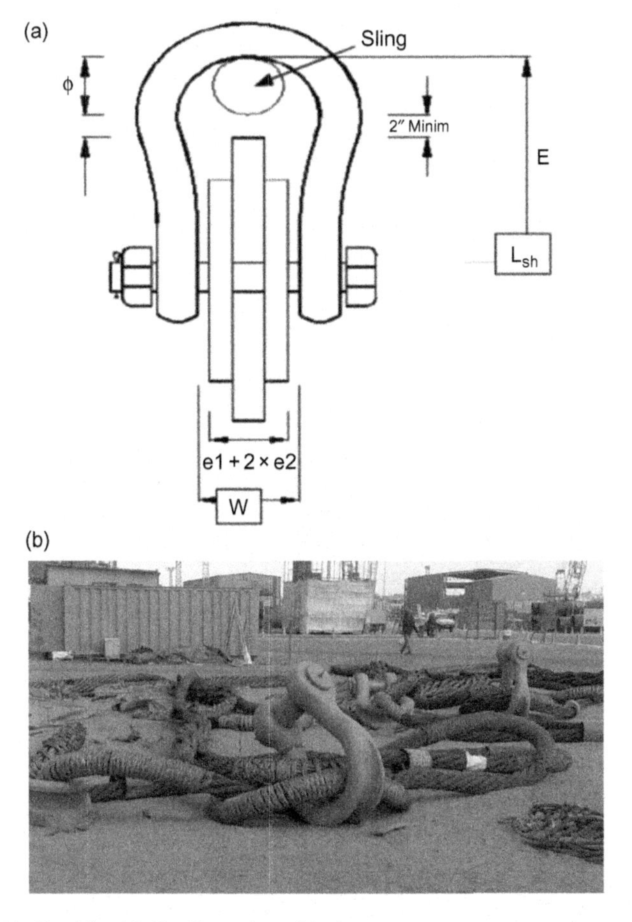

Figure 7.23 (a) Shackle: (a) Configuration; (b) shape.

dimensions, pin diameter and shackle length and width, as the design of the padeye depends on these dimensions. This will illustrated in examples.

7.6.5.3 Skew load factor

For indeterminate four-sling lifts using matched slings, a skew load factor (SKL) of 1.25 is applied to each diagonally opposite pair of lift points in tum. For determinate lifts, the SKL may be taken to be 1.0.

$$\text{Vertical pad eye load} = \text{padeye resolved lift weight} \times \text{SKL} \tag{7.4}$$

where the pad eye resolved lift weight is the vertical load at each padeye, taking into account lift weight and center of gravity only. When the allowable center of gravity position is specified as a cruciform or other geometric shape, the most conservative center of gravity position within the allowable area should be taken, until the position can be determined with confidence.

7.6.5.4 Resolved padeye load

The responsibility of the engineering firm is to design the padeye and check the attached structure member. In the case of a heavier load, the trunnion is used, as shown in Figure 7.24.

The resolved padeye load is the vertical padeye load divided by the sine of the sling angle;

$$\text{Resolved padeye load} = (\text{vertical padeye load})/\sin\Theta \tag{7.5}$$

where Θ is the sling angle, the angle between the sling and the horizontal plane.

Figure 7.24 Trunnion.

Provided the lift point is correctly orientated to the sling direction, a horizontal force equal to 5% of the resolved padeye load is applied, acting through the centerline and along the axis of the pinhole or trunnion.

If the lift point is not correctly orientated to the sling direction, then the computed force acting along the axis of the pinhole or trunnion plus 5% of the resolved padeye load is applied.

Example 7.2

A new helideck with dimensions of $12.2 \times 12.5 \times 7.0$ m and a gross weight of 44.20 tons is being lifted. Weight calculations have been provided:

Gross weight = 44.20 tons
Weight contingency factor = 5%
DAF = 1.3
Lift weight = $44.20 \times 1.05 \times 1.3 = 60.4$ tons

According to Figure 7.25, we know that the length between the pick points, L = 11.630, and the width, W = 5.280.

By knowing the angle of inclination 60° so the minimum sling length SL = 12.8 m and sling height HR = 11.1 m. The rigging weight is 2.0 tons.

Hook load = $60.4 + 2 \times 1.3 = 63$ tons

Figure 7.26 presents a sketch of the padeye, to present the main terminology, where

R = the main plate radius
Tp = main plate thickness
r = cheek plate radius
t = check plate thickness
dh = hole diameter
H = lever arm from section 1 and section2

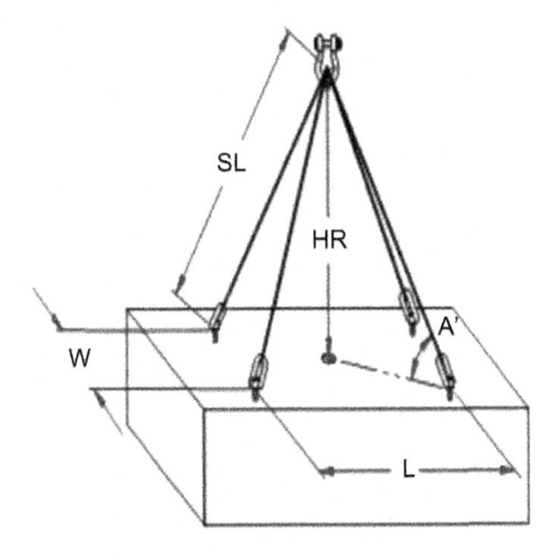

Figure 7.25 Four-sling configuration.

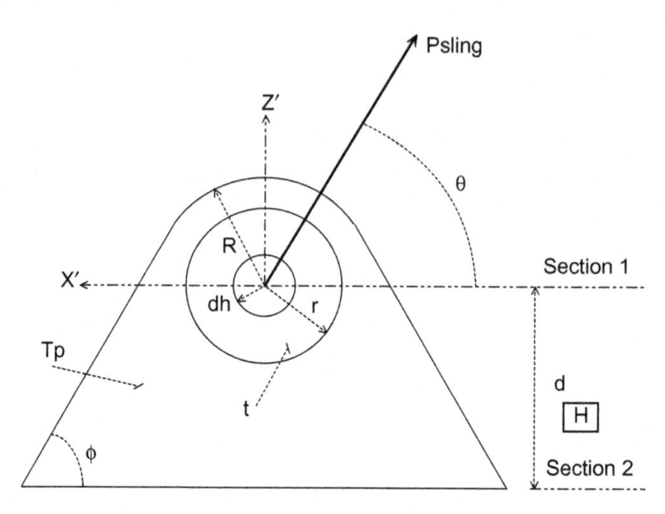

Figure 7.26 Sketch of padeye for calculation terminology.

ϕ = slope of padeye slides to the horizontal
θ = sling angle
X′ = local padeye *x*-axis
Y′ = local padeye *y*-axis (out of plane)
Z′ = local padeye *z*-axia
Section 1 = through padeye hold
Section 2 = at base of padeye

Example 7.3

Consider the design of a padeye. The input data follow. For the padeye forces,

DAF = 2.0
Static sling force, $P_s = 49.1$ kN = 11.04 kips
Sling angle degree in horizontal, $\theta = 17° = 0.30$ rad
Dynamic sling force, $P = P_s \times \text{DAF} = 49.1 \times 2 = 98.2$ kN = 22.08 kips
Component of dynamic force in horizontal direction, $P'_x = P\cos\theta = 93.9$ kN = 21.11 kips
Out of plane component of force (=25% of total dynamic force), $= 0.25P = 24.6$ kN = 5.53 kips
Vertical component of total dynamic force, $P'_z = P\sin\theta = 28.7$ kN = 6.45 kips

For the padeye dimensions,

Shackle capacity, SC = 5.9 tons
Bow diameter, $d_b = 57.9$ mm = 2.3 in.
Pin diameter, $d_p = 25.4$ mm = 1.0 in.°
Inside width of shackle, $w = 36.5$ mm = 1.4 in.
Inside length of shackle, $L_{sh} = 84.1$ mm = 3.3 in.

For the padeye properties and their dimensions,

Hole diameter, $d_h = 28$ mm = 1.1 in.
Main plate thickness, $T_p = 30$ mm = 1.18 in.

Main plate radius, $R = 65$ mm $= 25.6$ in.
Cheek plate thickness, $t = 0$ mm $= 0$ in.
Cheek plate radius, $r = 55$ mm $= 2.17$ in.
Cheek plate weld size, $t_w = 0.1$ mm $= 0.0025$ in.
Distance between section 1 and section 2, $H = 100$ mm $= 3.94$ in.
$\phi = 80° = 1.40$ rad
Allowable shear stress (AISC), $F_y = 345$ MPa $= 50$ ksi
Weld tension stress, $F_{wt} = 410$ MPa $= 59.46$ ksi
$\nu = 0.3$
Young's modulus, 200000 MPa = 29,005.53 ksi

On the initial geometry check, the shackle clearance is as follows:

Minimum recommended inner width clearance, $C_w(\min) = 0.2w = 7.3$ mm $= 0.29$ in.
Maximum recommended inner width clearance, $C_w(\max) = 0.4w = 14.6$ mm $= 0.58$ in.
Shackle inner width clearance, $C(w) = w - (T_p + 2t) = 7$ mm $= 0.26$ in.
Shackle inner length clearance, $C(L) = L_{sh} - (R - (d_h/2) = 33$ mm $= 1.30$ in.

The weld clearance is as follows:

Minimum cheek plate weld clearance, $C(\text{weld}) = 0.0$
Weld inspection clearance, $C(\text{inspection}) = 45$ mm $= 1.77$ in.

The design for bearing requires several stress checks. A *standard bearing check*:

Allowable bearing stress (AISC), $F_b = 0.9x\ F_y = 310.3$ MPa $= 45$ ksi
Bearing area, $A_b = d_p(T_p + 2x\ t) = 762\ mm^2 = 1.18\ in.^2$
Actual bearing stress, $f_b = 1000x\ P/A_b = 128.9$ MPa $= 18.69$ ksi
Utilization factor, UF = 0.42

A Hertz stress check finds

Allowable Hertz stress ($2F_y$), $F_{Hz} = 2Fy = 689.5$ MPa $= 100.0$ ksi
Actual Hertz stress, $f_{Hz} = (PE\ (d_h - d_p)/\pi\ (1 - \upsilon^2)((T_p + 2t)d_h d_p 1000))^{0.5} = 915$ MPa = 132.70 ksi
UF = 1.33

A check of contact stress (DNV) finds

$f_{cs} = 23.7 \times (P/d_h \times T_p)^{0.5} = 23.7(49.1 \times 1000/(30 \times 28))^{0.5} = 181.2$ MPa
UF (DNV) $= f_{cs}/F_y = 181/345 = 0.53$

In checking for *shear stress*, the cheek plate shear stress is found to be

Allowable shear stress, $F_v = 0.4F_y = 137.9$ MPa
$P_{cheek} = [t/(T_p + 2t)]P$
Shear area of padeye, $A_v = 0.5\ (2\pi r t_w)$
Shear stress, $f_v = (P_{cheek}/A_v)1000 = 0.0$ MPa
UF $= f_v/F_v = 0.0$ MPa

A check of *weld shear* stress on the cheek plate finds the following:

Weld allowable shear stress, $F_w = 0.3\ F_{wt} = 123.0$ MPa $= 17.84$ ksi
$P_{cheek} = 0.0$
Shear area of weld, $A_w = 0.5 \times [(2\pi r)(t_w/(2)^{0.5}] = 12\ mm^2 = 0.02\ in.^2$

Shear stress, $f_w = P_{\text{check}}/A_w = 0.00$ MPa
UF $= f_w/F_w = 0.00$ MPa

A pin *"pullout" shear stress* check finds

$F_v = 0.4F_y = 137.9$ MPa $= 20$ ksi
Pullout shear force (double failure plane), $V = P/2 = 49.10$ kN $= 11.04$ kips
Shear area of padeye, $A_v = T_p(R - d_h/2) + 2t[r - (d_h/2)] = 1530\ \text{mm}^2 = 2.37\ \text{in.}^2$
$f_v = 1000V/A_v = 32.1$ MPa $= 4.65$ ksi
UF $= f_v/F_v = 0.23$
As per DNV (2006), $F_v = 0.33F_y = 115$ MPa
UF (DNV) $= 32.1/115 = 0.28$

The checks of padeye stress are done by section. For section 1, through the padeye hole, the tension check finds

$F_t = 0.45F_y = 155.1$ MPa $= 22.5$ ksi
$P'_z = 28.7$ kN $= 6.45$ kips
$A_t = 2T_p[(t/\sin\phi) - d_h] + 2t(2r - d_h) = 3120\ \text{mm}^2 = 4.84\ \text{in.}^2$
$f_t = P'_z/A_t\ 1000 = 9.2$ MPa
UF $= f_t/F_t = 0.06$

and the shear check, which conservatively assumes that only the padeye main plate takes shear, finds

$F_v = 0.4F_y = 137.9$ MPa $= 20.0$ ksi
$V = P'_x = 93.90$ kN $= 21.11$ kips
Cross-sectional area (main plate only), $A_p = T_p[(2R/\sin\phi) - d_h] = 3120\ \text{mm}^2 = 4.84\ \text{in.}^2$
Shear stress at section 1, $f_s = V/A_p = 30.1$ MPa $= 4.36$ ksi
UF $= f_s/F_V = 0.22$

The properties at section 2, the section at the padeye connection to the existing steelwork, are

Plate length, $L = 2(R/\sin\phi + d/\tan\phi) = 167$ mm $= 6.59$ in.
$A_2 = L \times T_p = 5018\ \text{mm}^2 = 7.78\ \text{in.}^2$
Moment of inertia about weak axis, $I_{yy} = L(T_p)^3/12 = 3.76 \times 10^5\ \text{mm}^4 = 0.91\ \text{in.}^4$
Moment of inertia about strong axis, $I_{xx} = T_p(L)^3/12 = 1.17 \times 10^7 = 28.11\ \text{in.}^4$
Section modulus about weak axis, $Z_{yy} = L(T_p)^2/6 = 2.51 \times 10^4\ \text{mm}^3 = 1.53\ \text{in.}^3$
Section modulus about strong axis, $Z_{xx} = T_p(L)^2/6 = 1.4 \times 10^5\ \text{mm}^3 = 8.54\ \text{in.}^3$

A combined stress check finds

$f_a = P'_z/A_2 \times 1000 = 5.7$ MPa $= 0.83$ ksi
$f_{bx} = P'_x(d) \times 1000/Z_{xx} = 67.1$ MPa $= 9.73$ ksi
$f_{by} = P'_y(d) \times 1000/Z_{yy} = 98.0$ MPa $= 14.22$ ksi
Allowable AISC tensile stress, $F_a = 0.6f_y = 206.9$ MPa $= 30$ ksi
Allowable AISC stress for in-plane bending, $F_{bx} = 0.66\ F_y = 227.5$ MPa $= 33.0$ ksi
Allowable AISC stress for out of plane bending, $F_{by} = 0.75F_y = 258.6$ MPa $= 37.50$ ksi
UF $= f_a/F_a + f_{bx}/F_{bx} + f_{by}/F_{by} = 0.70$

A Von Mises's stress check finds

$f_{bx} = 67.1$ MPa
$f_{by} = 98.0$ MPa
Average shear stress, f_{ve}(average) $= P'_x/A_2 \times 1000 = 18$ MPa
Equivalent Von Mises, $f_{eq} = [(f_a + f_{bx} + f_{by})^2 + 3(f_{ve})^2]^{0.5} = 173.93$ MPa
Allowable Von Mises, $F_e = 310.3$ MPa
UF = 0.56

Table 7.4 presents a summary of the utilization factors:

Table 7.4 Summary of utilization factors in Example 7.3

Category	UF
Standard bearing	= 0.42
Hertz bearing stress	= 1.33 (worst case)
Cheek plate shear	= 0.00
Cheek plate weld shear	= 0.00
Pin "pull-out" shear	= 0.23
Tension section1	= 0.06
Shear-section 1	= 0.22
Combined bending and axial, section 2	= 0.70

Example 7.4

Consider the design of a Padeye with a cheek plate. The input data follow. For the padeye forces,

DAF = 2.0
Static sling force, $P_s = 107.2$ kN = 24.09 kips
Sling angle degree in horizontal, $\theta = 45° = 0.79$ rad
Dynamic sling force, $P = P_s \times \text{DAF} = 107.2 \times 2 = 214.3$ kN = 48.18 kips
Component of dynamic force in horizontal direction, $P'_x = P \cos\theta = 112.6$ kN = 25.31 kips
Out-of-plane component of force (=25% of total dynamic force), $P'_y = 53.6$ kN = 12.04 kips
Vertical component of total dynamic force, $P'_z = P \sin\theta = 182.3$ kN = 40.99 kips

The shackle properties are:

Shackle capacity, SC = 5.9 tons
Bow diameter, $d_b = 57.9$ mm = 2.3 in.
Pin diameter, $d_p = 32$ mm = 1.3 in.
Inside width of shackle, $w = 47$ mm = 1.9 in.
Inside length of shackle, $L_{sh} = 84.1$ mm = 3.3 in.

Input data : the padeye properties and their dimensions are:

Hole diameter, $d_h = 33$ mm = 1.30 in.
Main plate thickness , $T_p = 25$ mm = 0.98 in.
Main plate radius, $R = 60$ mm = 2.36 in.
Cheek plate thickness, $t = 8$ mm = 0.31 in.
Cheek plate radius, $r = 40$ mm = 1.57 in.

Cheek plate weld size, $t_w = 8$ mm = 0.31 in.
Refer to Figure 7.26, $H = 100$ mm = 3.94 in.
$\phi = 60° = 1.05$ rad
Allowable shear stress (AISC), $F_y = 345$ MPa = 50 ksi
Weld tension stress, $F_{wt} = 410$ MPa = 59.46 ksi
$\nu = 0.3$
Young's modulus, $E = 200000$MPa = 29,005.53 ksi

On the geometry check, the shackle clearance is as follows:

Minimum recommended inner width clearance, C_w (min) = 9.4 mm = 0.37 in.
Maximum recommended inner width clearance, C_w (max) = 18.8 mm = 0.74 in.
Shackle inner width clearance, $C(w) = 6$ mm = 0.24 in.
Shackle inner length clearance, $C(L) = 41$ mm = 1.60 in.

The weld clearance is as follows:

Minimum cheek plate weld clearance; C(weld) = 0.0
Weld inspection clearance; C(inspection) = 27 mm = 1.06 in.

Standard bearing check finds the following:

Allowable bearing stress (AISC), $F_b = 0.9xF_y = 310.3$ MPa = 45 ksi
Bearing area, $A_b = d_p(T_p + 2xt) = 1312$ mm^2 = 2.03 in.^{2}
Actual bearing stress, $f_b = 1000xP/A_b = 163.6$ MPa = 23.69 ksi
Utilization factor, k UF = 0.53

The Hertz stress check finds

Allowable Hertz stress ($2F_y$), $F_{Hz} = 2F_y = 689.5$ MPa = 100.0 ksi
Actual Hertz stress, $f_{Hz} = [PE(d_h - d_p)/\pi\ (1 - \upsilon^2)((T_p + 2t)d_h d_p 1000)]^{0.5} = 587.2$ MPa = 85.17 ksi
UF = 0.85

A check of contact stress (DNV) finds

$f_{cs} = 23.7x(P/d_h x T_p)^{0.5} = 23.7[214.3 \times 1000/(33 \times 41)]^{0.5} = 298.27$ MPa
UF (DNV) $= f_{cs}/F_y = 298.27/345 = 0.87$

In checking for shear stress, the cheek plate shear stress is found to be

Allowable shear stress, $F_v = 0.4F_y = 137.9$ MPa
$P_{cheek} = [t/(T_p + 2t)]P = 41.81$ kN
Shear area of padeye, $A_v = 0.5\ (2\pi r t_w) = 1005$ mm^2 = 1.56 in.^{2}
Shear stress, $f_v = (P_{cheek}/A_v)1000 = 41.6$ MPa = 6.03 ksi
UF $= f_v/_{FV} = 0.3$ MPa

A check for weld shear stress on the cheek plate finds

Weld allowable shear stress, $F_w = 0.3F_{wt} = 123.0$ MPa = 17.84 ksi
$P_{cheek} = 41.81$ kN = 9.40 kips
Shear area of weld, $A_w = 0.5 \times [(2\pi r)(t_w/(2)^{0.5}] = 711$ mm^2 = 1.1 in.^{2}
Shear stress, $f_w = P_{cheek}/A_w = 58.8$ MPa = 8.53 ksi
UF $= f_w/F_w = 0.48$ MPa

A check of pin "pullout" shear stress finds

$F_v = 0.4F_y = 137.9$ MPa = 20 ksi
Pullout shear force (double failure plane), $V = P/2 = 107.15$ kN = 24.09 kips
Shear area of padeye, $A_v = T_p(R - d_h/2) + 2t[r - (d_h/2)] = 1464$ mm^2 = 2.27 in.^{2}
$f_v = 1000xV/A_v = 73.2$ MPa = 10.62 ksi
UF $= f_v/F_v = 0.53$
As per DNV (2006), $F_v = 0.33F_y = 113.8$ MPa
UF(DNV) = 73.2/113.8 = 0.64

The checks of padeye stress are done by section. For, section 1, through the padeye hole, the tension check finds

$F_t = 0.45F_y = 155.1$ MPa = 22.5 ksi
$P'_z = 182.35$ kN = 40.99 kips
$A_t = 2T_p[(t/\sin\phi) - d_h] + 2t(2r - d_h) = 3391$ mm^2 = 5.26 in.^{2}
$f_t = P'_z/A_t x1000 = 53.8$ MPa = 7.80 ksi
UF $= f_t/F_t = 0.35$

and shear check, which conservatively assumes that only the padeye main plate takes shear, finds

$F_v = 0.4F_y = 137.9$ MPa = 20.0 ksi
$V = P'_x = 112.58$ kN = 25.31 kips
Cross-sectional area (main plate only), $A_p = T_p[(2R/\sin\phi) - d_h] = 2639$ mm^2 = 4.09 in.^{2}
Shear stress at section 1, $f_s = V/A_p = 42.7$ MPa = 6.19 ksi
UF $= f_s/F_V = 0.31$

The properties at section 2, the section at the padeye connection to the existing steelwork), are

Plate length, $L = 2(R/\sin\phi + d/\tan\phi) = 225$ mm = 8.86 in.
$A_2 = Lx\ T_p = 5629$ mm^2 = 8.73 in.^{2}
Moment of inertia about weak axis, $I_{yy} = L(T_p)^3/12 = 2.93 \times 10^5$ mm^4 = 0.70 in.^{4}
Moment of inertia about strong axis, $I_{xx} = T_p(L)^3/12 = 2.38 \times 10^7 = 57.14$ in.^{4}
Section modulus about weak axis, $Z_{yy} = L(T_p)2/6 = 2.35 \times 10^4$ mm^3 = 1.43 in.^{3}
Section modulus about strong axis, $Z_{xx} = T_p(L)3/6 = 2.11 \times 10^5$ mm^3 = 12.89 in.^{3}

A combined stress check finds

$f_a = P'_z/A_2 x1000 = 32.4$ MPa = 4.7 ksi
$f_{bx} = P'_x(d)x1000/Z_{xx} = 40.0$ MPa = 5.80 ksi
$f_{by} = P'_y(d)x1000/Z_{yy} = 171.3$ MPa = 24.85 ksi
Allowable AISC tensile stress, $F_a = 0.6f_y = 206.9$ MPa = 30 ksi
Allowable AISC stress for in-plane bending, $F_{bx} = 0.66F_y = 227.5$ MPa = 33.0 ksi
Allowable AISC stress for out of plane bending, $F_{by} = 0.75F_y = 258.6$ MPa = 37.50 ksi
UF $= f_a/F_a + f_{bx}/F_{bx} + f_{by}/F_{by} = 0.99$

A Von Mises's stress check finds

$f_{bx} = 40.0$ MPa = 5.8 ksi
$f_{by} = 171.3$ MPa = 24.85 ksi
Average shear stress, f_{ve} (average) $= P'_x/A_2 x1000 = 20$ MPa = 2.90 ksi

Equivalent Von Mises, $f_{eq} = [(f_a + f_{bx} + f_{by})^2 + 3(f_{ve})^2]^{0.5} = 246.12$ MPa $= 35.70$ ksi
Allowable Von Mises, $F_e = 310.3$ MPa $= 45$ ksi
UF = 0.79

Table 7.5 presents a summary of the utilization factors.

Table 7.5 Summary of utilization factors for Example 7.4

Category	UF
Standard bearing	= 0.53
Hertz bearing stress	= 0.85 (worst case)
Cheek plate shear	= 0.30
Cheek plate weld shear	= 0.48
Pin "pullout" shear	= 0.53
Tension section 1	= 0.35
Shear-section 1	= 0.31
Combined bending and axial UF, section 2	= 0.99

The shackle data from the Crosby catalogue from Crosby Group (1987) is a sample summary, shown in Tables 7.6–7.10.

Table 7.6 Lifting gear shackle gear type S

Shackle capacity, WLL, tons	Bow diameter, D_b, in.	Pin diameter, d_p, in.	Inside width, w, in.	Inside length, L, in.
2	33	16	21	48
3.2	43	19	27	61
4.7	51	22	32	72
6.5	58	25	37	84
8.5	68	29	43	95
9.5	74	32	46	108
12	83	35	52	119
13	92	38	57	133
17	98	41	60	146

Table 7.7 Small bow shackles, quality grade M

Shackle capacity, WLL, tons	Bow diameter, D_b, in.	Pin diameter, d_p, in.	Inside width, w, in.	Inside length, L, in.
5	67	32	51	114
6.2	76	35	57	127
7.5	83	38	60	140
9.2	89	44	67	152
10	98	48	73	165
12	105	51	79	178

Table 7.8 **Bullivants shackle: Small bow shackles, quality grade L2**

Shackle capacity, WLL, tons	Bow diameter, D_b, in.	Pin diameter, d_p, in.	Inside width, w, in.	Inside length, L, in.
6	83	38	60	140
7	89	44	67	152
8	98	48	73	165
9.5	105	51	79	178
11	114	54	86	191
12	121	57	92	203

Table 7.9 **Bullivants shackle: Large bow shackles, quality grade M**

Shackle capacity, WLL, tons	Bow diameter, D_b, in.	Pin diameter, d_p, in.	Inside width, w, in.	Inside length, L, in.
5.8	83	35	57	137
11	121	51	86	206

Table 7.10 **Crosby shackle: Forged anchor shackles**

Shackle capacity, WLL, tons	Bow diameter, D_b, in.	Pin diameter, d_p, in.	Inside width, w, in.	Inside length, L, in.
6.5	2.28125	1	1.4375	3.3125
12	3.25	1.375	2.03125	4.6875

7.6.5.5 Sling force

The sling force is the vertical padeye load plus the sling weight (per sling) divided by the sine of the sling angle:

$$\text{Sling force} = (\text{vertical padeye load} + \text{sling weight})/\sin\theta$$

The minimum safety factor on a sling or grommet breaking load after resolution of the load based on center of gravity position and sling angle and consideration of the factors should be no less than 2.25.

7.6.5.6 *Crane lift factors*

For a two-crane lift, the resolved load at each crane is multiplied by the following factors, based on a DNV report:

Center of gravity factor = 1.03 tilt factor = 1.03

$$\text{Crane resolved lift weight} = (\text{statically resolved lift weight into each crane}) \times (\text{center of gravity factor}) \times (\text{tilt factor}) \tag{7.6}$$

For a two-crane lift, with two slings to each hook, the load resolved to each pad eye should be multiplied by a yaw factor:

Yaw factor = 1.05

$$\text{Padeye resolved lift weight} = (\text{crane resolved lift weight, resolved to each padeye}) \times (\text{yaw factor}) \tag{7.7}$$

Two-crane lifts with other rigging arrangements require special consideration.

7.6.5.7 *Part sling factor*

Where a two-part sling passes over, around, or through a shackle, trunnion, padear, or crane hook, other than at a termination, the total sling force should be distributed into each part in the ratio 45:55:

Sling load = sling force × 0.55 (for two-part slings)

7.6.5.8 *Termination efficiency factor*

The breaking load of a sling ending in an eye splice is assumed to be the calculated rope-breaking load multiplied by a factor as follows:

For hand splices = 0.75
For resin sockets = 1. 00
Swage fittings, ergo "superloop" = 1.00

Other methods of termination require special consideration:

$$\text{Sling-breaking load} = \text{rope-breaking load} \times \text{termination efficiency factor} \tag{7.8}$$

7.6.5.9 *Bending efficiency factor*

When any rope is bent around a shackle, trunnion, padear, or crane hook, the breaking load is assumed to be the calculated rope-breaking load multiplied by a bending efficiency factor:

$$\text{Bending efficiency factor} = 1 - 0.5/(P_d/r_d)0.5. \tag{7.9}$$

where, P_d is the pin diameter and r_d is the rope diameter.

This results in the Table 7.11 of bending efficiency factors.

Table 7.11 Bending efficiency factor

P_d/r_d	<0.8	0.8	0.9	1.0	1.5	2.0	3.0	4.0	5.0
Factor	Not Advised	0.44	0.47	0.50	0.59	0.65	0.71	0.75	0.78

Figure 7.27 Grommet shape.

7.6.5.10 Grommets

Grommets require special consideration to ensure that the rope-breaking load and bending efficiency have been correctly taken into account.

The core of a grommet should be discounted when computing breaking load. The breaking load of each part of a grommet is therefore usually taken as six times the unit rope-breaking load, with a factor to account for the spinning losses in cabling. This factor is normally taken as 0.85:

$$\text{Grommet BL (each part)} = 0.85 \times 6 \times \text{breaking load of unit rope} \tag{7.10}$$

Typically, a grommet, as shown in Figure 7.27, is used with one end over the crane hook and the other end connected to a padeye by a shackle. The bending efficiency factors at each end may differ, and the more severe value should be taken.

Bending efficiency is derived as before, where rope diameter is the single part grommet diameter. The total breaking load of the grommet used in this manner is

2 × (single part grommet BL) × (more severe bending efficiency factor)

7.6.5.11 Shackle safety factors

The minimum shackle-breaking load, where this can be reliable determined, should be no less than the minimum required sling-breaking load:

The minimum shackle Safe Working Load (SWL) > sling force/DAF

whichever results in the larger required shackle size.

Where the shackle is at the lower end of the rigging, the weight of the rigging components above the shackle (including DAF and taking account of the sling angle) may be deducted from the sling force.

7.6.5.12 Consequence factors

The consequence factors in Table 7.12 are further applied to the structure, including the lift points and their attachments into the structure.

These consequence factors are applied based on the calculated lift point loads after consideration of all the factors. If a limit state analysis is used, then the additional factors sare applied. Lifting calculations flow chart with the various factors and their application are illustrated in Figure 7.28.

7.6.6 Structural calculations

Structural calculations, based on the load factors discussed previously, include adequate load cases to justify the structure. For example, for an indeterminate, four-point lift, the following load cases should normally be considered;

1. Base case, using lift weight, resolved to the lift points but with no skew load factor.
2. Lift weight, with skew load factor applied to one diagonal.
3. Lift weight, with skew load factor applied to the other diagonal.

In all cases, the correct sling angle and point of action and any offset or torsional loading imposed by the slings should be considered.

Table 7.12 Consequence factors

Lift points including spreaders	1.35
Attachments of lift points to structure	1.35
Members directly supporting or framing into the lift points	1.15
Other structural members	1.00

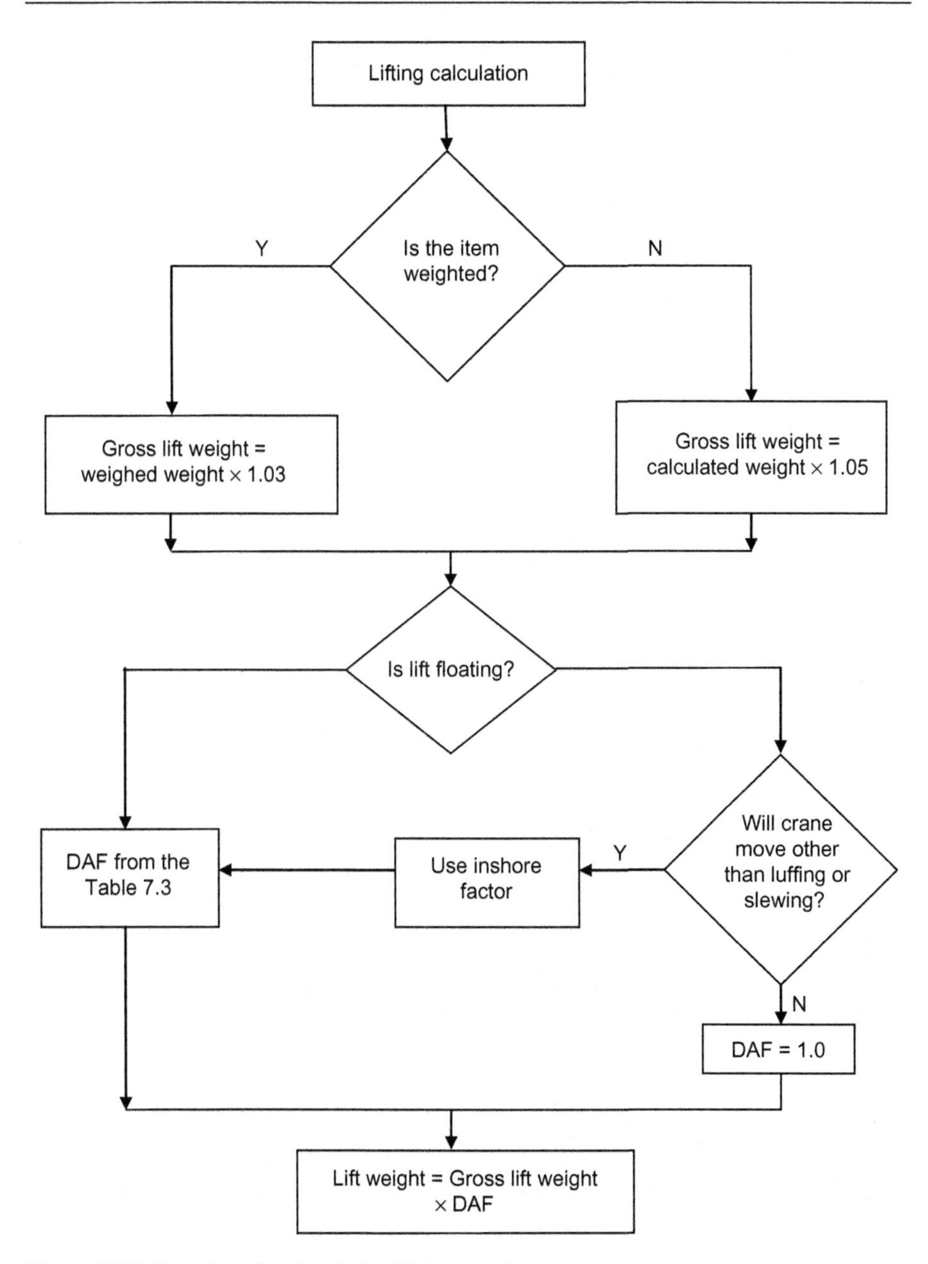

Figure 7.28 Procedure for calculating lifting weight.

The overall structure should be analyzed for the loadings. The primary supporting members are analyzed using the most severe loading, with a consequence factor of 1.15 applied, as in Table 7.12.

An analysis of the lift points and attachments to the structure should be performed, using most severe load and a consequence factor of 1.35. The 5% side load should also be applied, as should any torsional load resulting from the 45:55 two-part sling loading, if applicable.

Where the lift point forms a structural node, the calculations should also include the loads imposed by the members framing into it.

Spreader bars or frames, if used with load cases, should be similarly treated. A consequence factor of 1.35 is applied to lift points, and 1.15 is applied to members directly supporting the lift points, as presented in Table 7.12.

Stress levels should be within those permitted by the latest edition of a recognized and applicable offshore structures code. The loading should be treated as a normal serviceability level functional load with associated load and resistance or safety factors as in a working stress code. The one third increase for environmental loadings is not allowed. Similarly, for an Load Resistance Factor Design (LRFD) or partial factor code, the load factor would be greater than that used for ultimate conditions.

A limit state analysis or load resistance factor design approach may be applied according to a recognized code, provided that the total load factor is no less than the product of all the required factors, multiplied by a further factor of 1.30.

The material reduction factor should be no less than

Elastic design of steel structures, 1.15.
Plastic design of steel structures, 1.30.

7.6.7 Lift point design

In addition to the structural requirements described in the previous section, the following should be taken in to account in the lift point design.

Adequate clearance is required between cheek plates or inside trunnion keeper plates, to allow for ovalization under load. In general, the width available for the sling should be no less than ($1.25D + 25$ mm), where D is the nominal sling diameter. However, the practical aspects of the rigging and derigging operations may demand a greater clearance than this.

In general, for fabricated lift points, the direction of loading should be in line with the plate rolling direction. Lift point drawings should show the rolling direction.

Thickness loading of lift points and their attachments to the structure should be avoided if possible. If such loading cannot be avoided, the material used should be documented to be free of laminations, with a recognized through-thickness designation.

Pinholes should be bored or reamed and designed to suit the shackle proposed. Adequate spacer plates should be provided to centralize the shackles.

Cast padears should be designed taking into account the geometrical considerations, the stress analysis process, the manufacturing process, and quality control.

The extent of nondestructive testing should be submitted for review. Where repeated use is to be made of a lift point, a procedure should be presented for reinspection after each lift.

7.6.8 Clearances

The clearance around the lifting object and crane vessel should be studied in the lifting procedure. The required clearance depends on the nature of the lift, the proposed limiting weather conditions, the arrangement of bumpers and guides, and the size and motion characteristics of the crane vessel and the transport barge. Subject to these factors, for offshore lifts, the clearances discussed in the following subsections should normally be maintained at each stage of the operation.

7.6.8.1 Clearances around the lifted object

The clearance between any part of the lifted object (including spreaders and lift points) and the crane boom is no less than 3 m.

The vertical clearance between the underside of the lifted object and any other previously installed structure, except in the immediate vicinity of the proposed landing area, is no less than 3 m.

The distance between the lifted object and other structures on the same transport barge should be no less than 3 m.

The horizontal clearance between the lifted object and any other previously installed structure, unless purpose-built guides or bumpers are fitted, is more than or equal to 3 m. The 3-m clearance also is reasonable between traveling block and fixed block at maximum load elevation.

7.6.8.2 Clearances around the crane vessel

Nobel Denton (2009) recommends, in mooring the crane vessel adjacent to an existing platform, allowing a clearance of 3 m between any part of the crane vessel and the platform and 10 m between any anchor line and the platform.

When the crane vessel is dynamically positioned, a 5-m nominal clearance between any part of the crane vessel and the platform is needed.

The clearance between the crane vessel and seabed, after taking account of tidal conditions, vessel motions, increased draft, and changed heel or trim during the lift, is 3 m.

The clearances around the mooring lines and anchors stated later are given as guidelines to good practice. The specific requirements and clearances should be defined for each project and operation, taking into account particular circumstances, such as

- Water depth.
- Proximity of subsea assets.
- Survey accuracy.

- The control ability of the anchor handling vessel.
- Seabed conditions.
- Estimated anchor drag during embedment.
- The probable weather conditions during anchor installation.

Operators and contractors may have their own requirements, which may differ from those stated here, and should govern if more conservative.

Clearances should take into account the possible working and standoff positions of the crane vessel, and the moorings should never be laid in such a way that they could be in contact with any subsea asset. This may be relaxed when the subsea asset is a trenched pipeline, provided it can be demonstrated that the mooring will not cause frictional damage or abrasion.

In any case, moorings should not be run over the top of a subsea completion or wellhead. Whenever an anchor is run out over a pipeline, flow line, or umbilical, the anchor should be securely stowed on the deck of the anchor handling vessel. In circumstances where either gravity anchors or closed stem tugs are used and anchors cannot be stowed on deck, the anchors should be doubly secured through the additional use of a safety strap or a similar method.

The vertical clearance between any anchor line and any subsea asset should be no less than 20 m in water depths exceeding 40 m, and 50% of water depth in depths of less than 40 m.

Clearance between any mooring line and any structure other than a subsea asset should be more than 10 m.

When an anchor is placed on the same side of a subsea asset as the crane vessel, it should not be placed closer to the subsea asset than 100 m.

When the subsea asset lies between the anchor and the crane vessel, the final anchor position should be more than 200 m from the subsea asset.

During lifting operations, crossed mooring situations should be avoided wherever practical. Where crossed moorings cannot be avoided, the separation between active catenaries should be no less than 30 m in water depths exceeding 100 m, and 30% of water depth in water depths less than 100 m.

If any of the clearances are impractical because of the mooring configuration or seabed layout, a risk assessment should be carried out and special precautions taken as necessary.

Figure 7.28 presents the procedure for calculating the lifting weight; the procedure for calculating the loads on the padeye is illustrated in Figure 7.29. After obtaining the padeye loads, the structural members and padeye are required to be checked through a procedure presented in the flowchart in Figure 7.30. Figure 7.31 presents the procedure for checking the rigging facilities. The barge operator should have a chart that defines the crane's capability to lift the structure to or from the barge; a sample chart is shown in Figure 7.32.

Figure 7.32 presents the lifting capacity for the crane on a barge, which is the relation between boom radius, hook height, and the lifting capacity. This chart should be included in the lifting procedure, which is usually delivered from the construction company and reviewed by the engineering firm and the company representative.

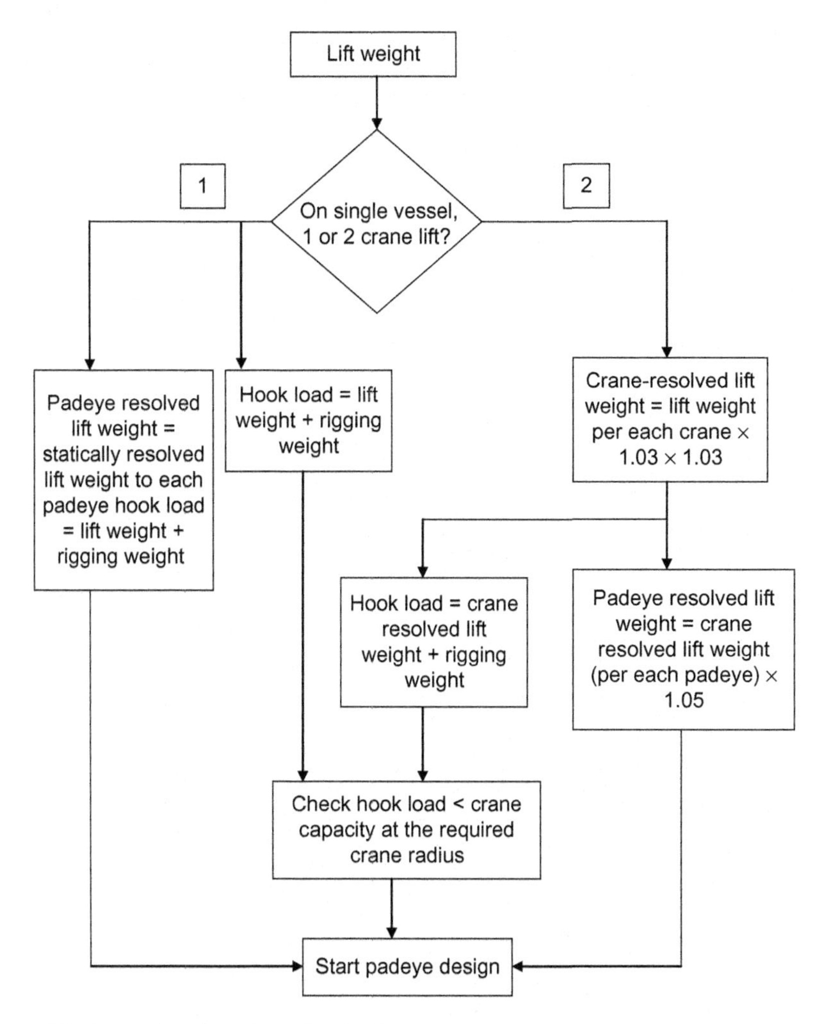

Figure 7.29 Procedure for calculating loads on the padeye.

Example 7.5

Consider a lifting load and spreader bar with gross weight 20 tons. Calculate the load in the padeye, the main structure carrying the lifting, and another structure member.

Dynamic amplification factor (DAF) = 1.2
Skew load factor (SKL) = 1.1
Sling angle = 60°
Center of gravity shift factor = 1.03
Tilt factor = 1.03
Yaw factor = 1.05
Weight contingency = 1.03
Consequence factor padeye = 1.30

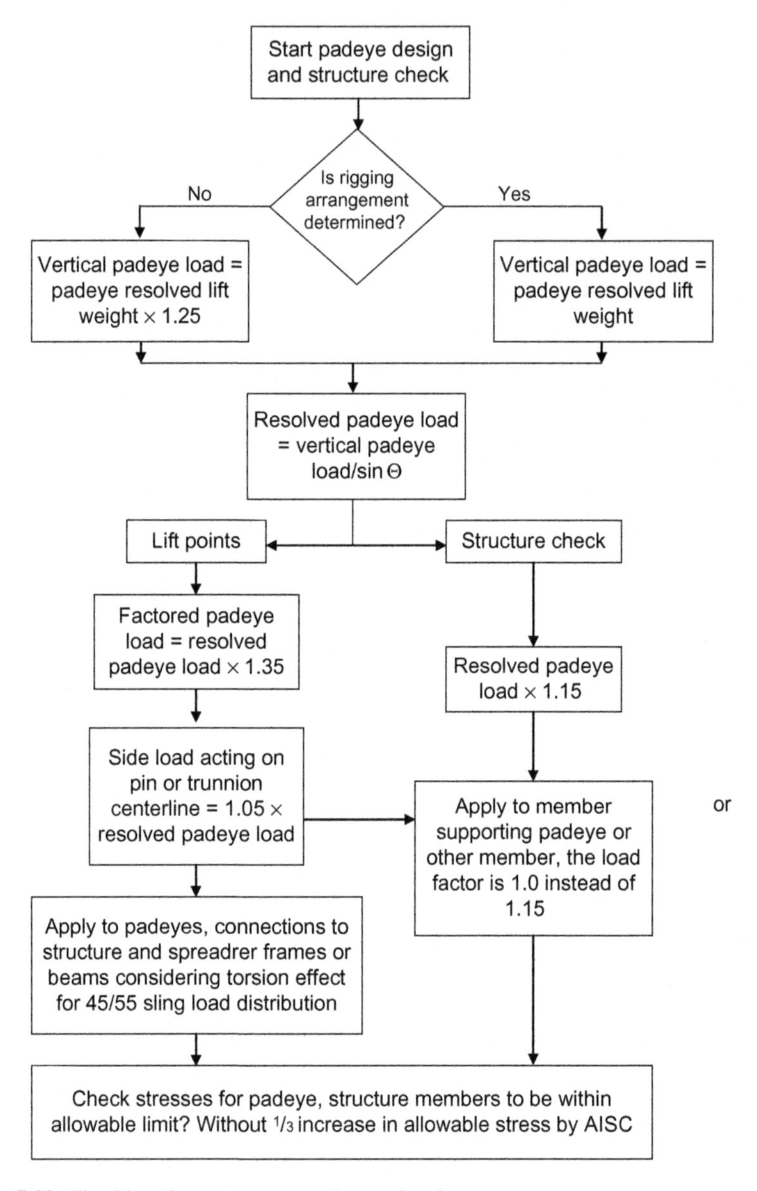

Figure 7.30 Checking the structure member and padeye.

Consequence factor for main structure = 1.15
Consequence factor for the other structure member = 1.0
Total lift factor for the padeye design = 1.2 × 1.1 × 1.03 × 1.03 × 1.05 × 1.03 × 1.30 = 1.97
Applied load for the padeye design = 1.97 × 20 = 39.4 tons
Total lift factor for the main structure = 1.2 × 1.1 × 1.03 × 1.03 × 1.05 × 1.03 × 1.15 = 1.74
Applied load for the main structure design check = 1.74 × 20 = 34.8 tons

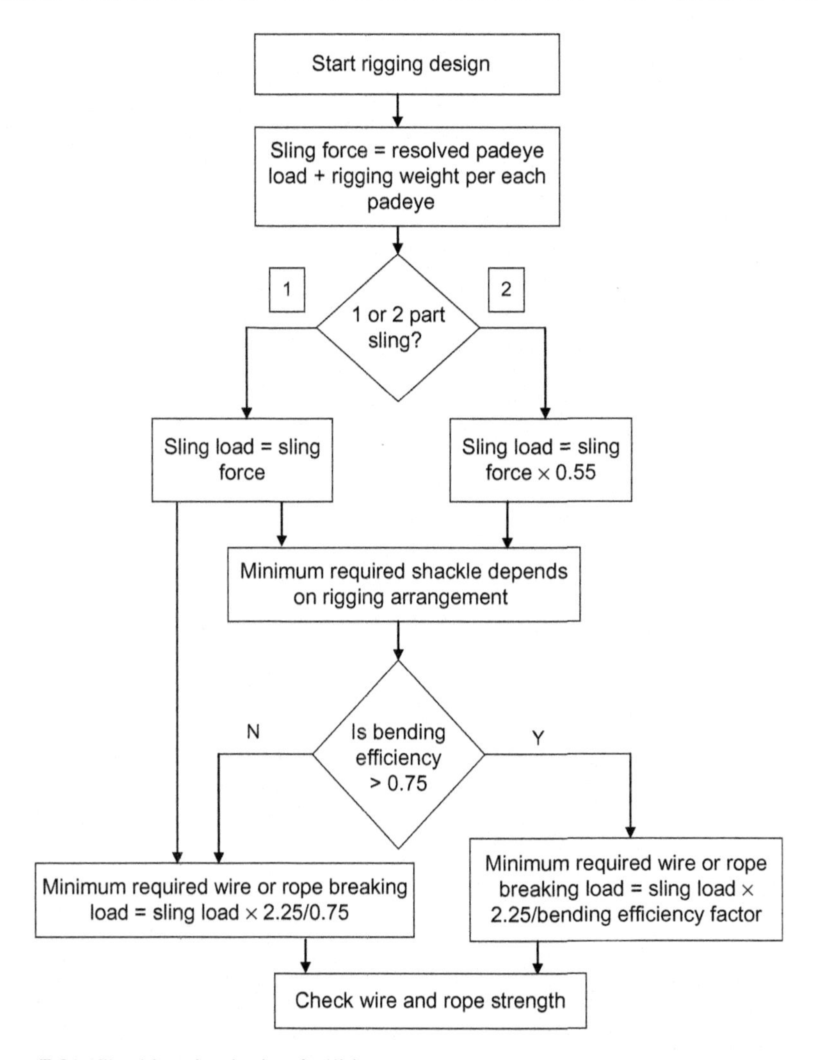

Figure 7.31 Checking the rigging facilities.

Total lift factor for the otherstructure member $= 1.2 \times 1.1 \times 1.03 \times 1.03 \times 1.05 \times 1.03 \times 1.0 = 1.51$

Applied load for the other structure member check $= 1.52 \times 20 = 30.3$ tons

7.6.9 *Lifting calculation report*

Calculations should be presented for the structure to be lifted, demonstrating its capacity to withstand, without overstress, the loads imposed by the lift operation, with the load and safety factors and the load cases.

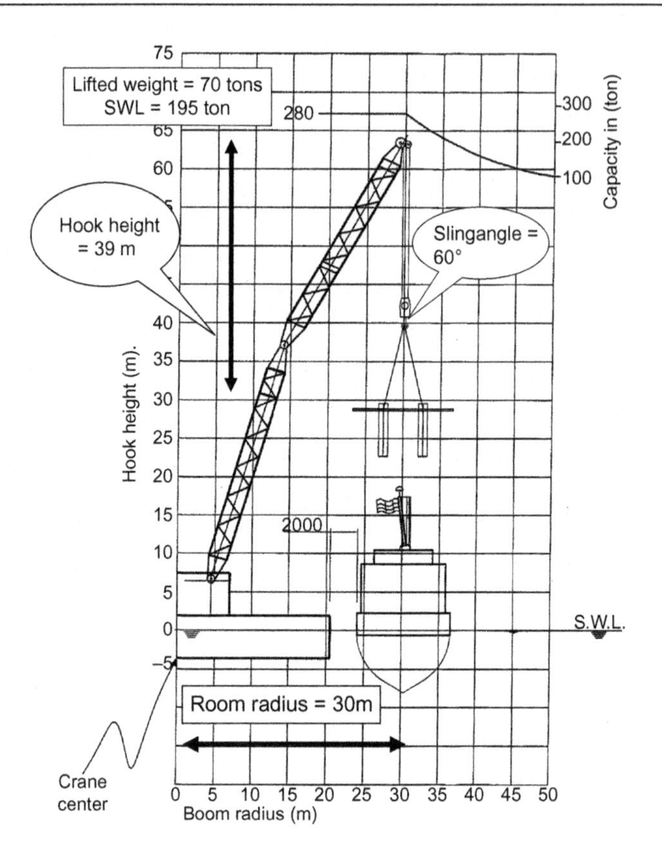

Figure 7.32 Lifting capacity chart of the crane.

The calculation package for lifting should contain the following as a minimum requirements:

1. Plans, elevations, and sections showing main structural members.
2. The structural model. This should account for the proposed lifting geometry, including any offset of the lift points.
3. The weight and center of gravity.
4. The steel grades and properties.
5. The load cases imposed.
6. The codes used.
7. A tabulation of member unity checks or a statement that unity checks are less than 0.8.
8. Justification or proposal for redesign for any members with a unity check in excess of 1.0.

A similar analysis is presented for spreader bars, beams, and frames. An analysis or equivalent justification are presented for all lift points, including padeyes, padears, and trunnions, to demonstrate that each lift point and its attachment into the structure are adequate for the loads and factors set out.

A proposal should be presented, showing

1. The proposed rigging geometry, showing dimensions of the structure, center of gravity position, lift points, crane hook, sling lengths, and angles, including shackle dimensions and "lost" length around hook and trunnions.
2. A computation of the sling and shackle loads and required breaking loads, taking into account the safety factors as described.
3. A list of actual slings and shackles proposed, tabulating
   - Position on structure.
   - Sling or shackle identification number.
   - Sling length and construction diameter, minimum breaking load for slings, SWL, and minimum breaking load (MBL) for shackles.
   - Direction of lay.
4. Copies of inspection or test certificates for all rigging components.

Manufactured shackles should deliver a test certificate, which should not be more than 5 years old, and if not new, a report of an inspection by a competent person since the last lift.

For each sling and shackle, a detailed record of all previous lifts should be presented, including the date and calculated load for each lift.

7.6.9.1 The crane vessel

Information should be submitted on the crane vessel and the crane. This includes, as appropriate,

- Vessel general arrangement drawings and specifications.
- Details of registry and class.
- Mooring system and anchors.
- Operating and survival drafts.
- Crane specifications and operating curves.
- Details of any ballasting operations required during the lift.

The mooring arrangement for the operation and standoff position should be submitted. This should include the lengths and specifications of all mooring wires and anchors, and a mooring plan showing adequate horizontal clearances on all platforms, pipelines, and any other seabed obstructions. An elevation of the catenary for each mooring line, for upper and lower tension limits, should demonstrate adequate vertical clearance over pipelines.

Further reading

American Petroleum Institute, 2007. Fixed Offshore Structure Design, WSD. API, API RP2A.

Gerwick, Ben C., 2007. Construction of Marine Offshore Structure. CRC Press, San Francisco, California, USA.

The Crosby Group Inc., 1987. General Catalogue. The Crosby Group Inc., p. 16.

Delton, N., 2009. Guidelines for Lifting Operations by Floating Crane Vessels.

Det Norske Veritas, 2008. Standard Specifications for Offshore Containers. No. 2.7-1. DNV.

International Organization for Standardization, 2004. Petroleum and Natural Gas Industries – Offshore Structures. Part 2. Fixed Steel Structures. ISO, ISO/DIS 19902, Amsterdam.

Elreedy, M.A., 2012. Offshore Structures, Design, Construction and Maintenance. Elsevier.

SACS Software

8.1 Introduction

SACS software is the most famous offshore structure analysis software on the market with SESAM and USFOS. Staad and SAP2000 are used to implement the offshore module. Note that USFOS is used mainly in pushover analysis.

As SACS is the most traditional software, this chapter illustrates using this software step by step. This is not an easy software to start using, so you should have strong theoretical background from the previous chapters.

The software depends in a collection of modules that should be used in each analysis. The main program carries the nodes, members, and loads on it, and other modules do the subroutine used for every analysis you need to perform. Figure 8.1 presents a summary for in-place analysis and the check of the joint can and the dynamic analysis, which is illustrated clearly in the following sections.

8.2 In-place analysis

The first step is to do in-place analysis. Then, you can cover dynamic analysis, seismic, and pushover; the last campaign of analysis is to check the structure during the installation phase.

The first step in SACS is to develop the name of the project as in Figure 8.2 and define the location of the folder for this new project. Note that organizing the folder is very essential and important, as you will run a lot of input and output files during the analysis.

Figure 8.3 shows that you have a three options, which modify an existing model that you performed before, create a new one, or just open the last one.

After you select to create a new model, a menu appears, as in Figure 8.4, to ask about start from blank or use the existing library and choose the units. A wizard is available for fixed offshore platforms, so it is easy to use structure definition wizard.

Start building the structure model through the Structure Definition dialog box, define the jack/pile using the following settings in the Elevations tab, as shown in Figure 8.5. The following input data as an example:

- Working Point Elevation: 6.0 m
- Pile Connecting Elevation: 4.6 m
- Water Depth: 75 m

Marine Structural Design Calculations. DOI: http://dx.doi.org/10.1016/B978-0-08-099987-6.00008-8

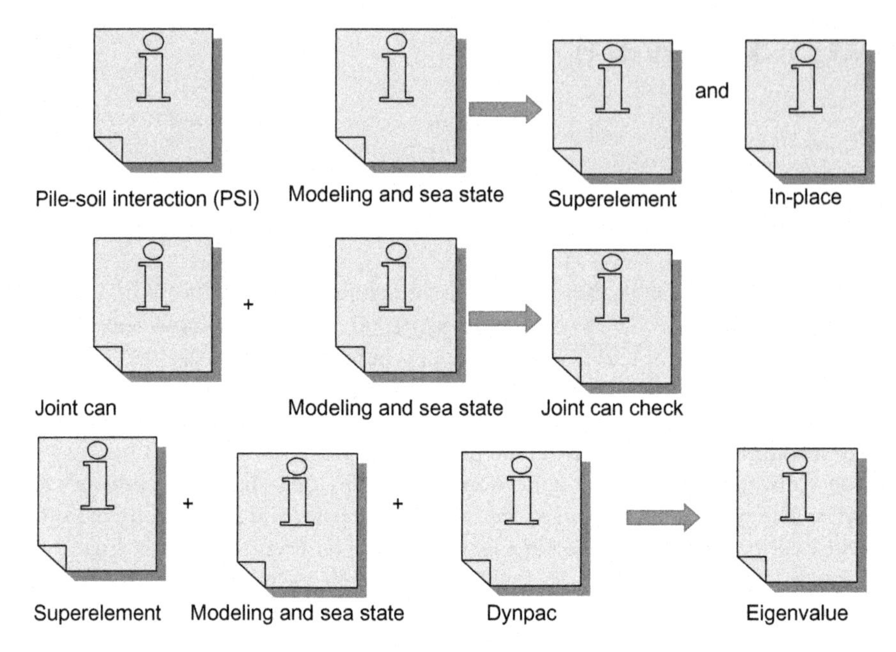

Figure 8.1 Software overview.

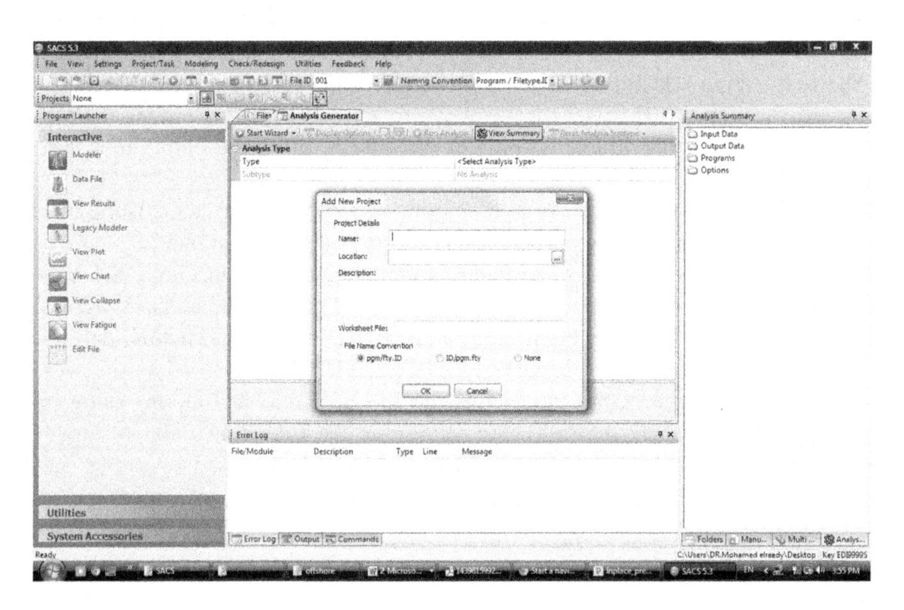

Figure 8.2 Project start menu.

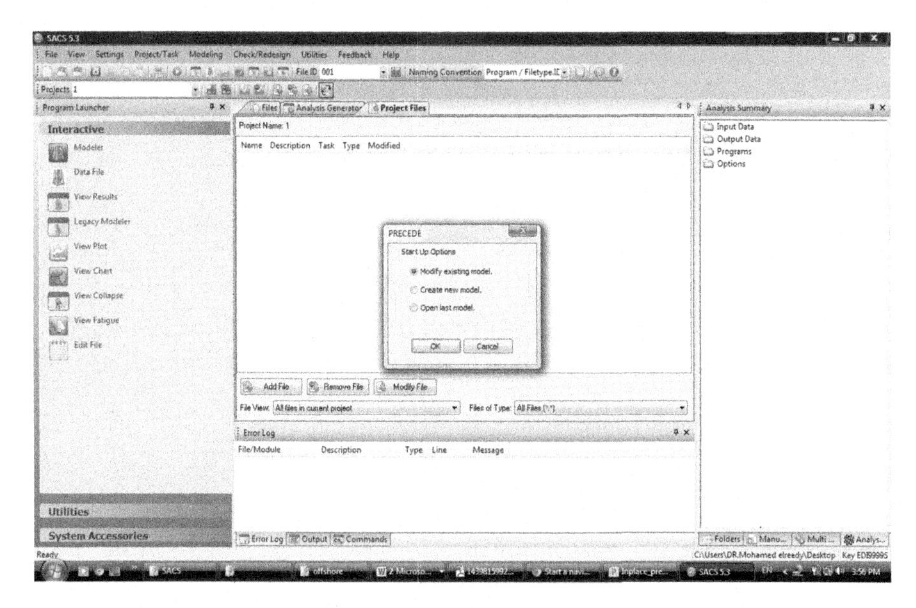

Figure 8.3 Options to start.

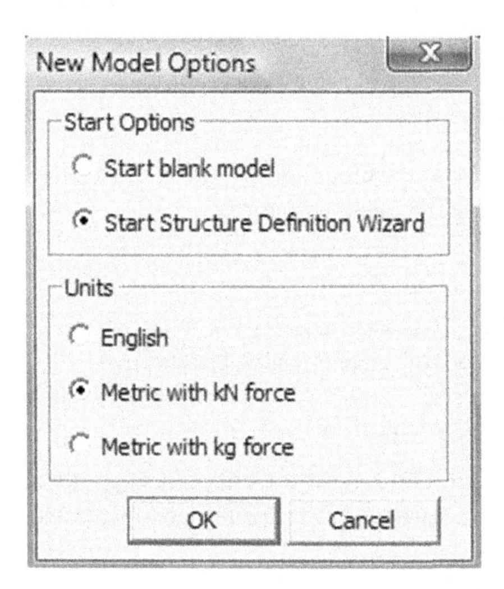

Figure 8.4 Units menu.

- Mudline Elevation: −75 m
- Pile Stub Elevation: − **75**
- Leg Extension Elevation
- Generate Seastate Hydrodynamic Data: (checked)
- Other elevations: −57, −38, −21, −3, 4

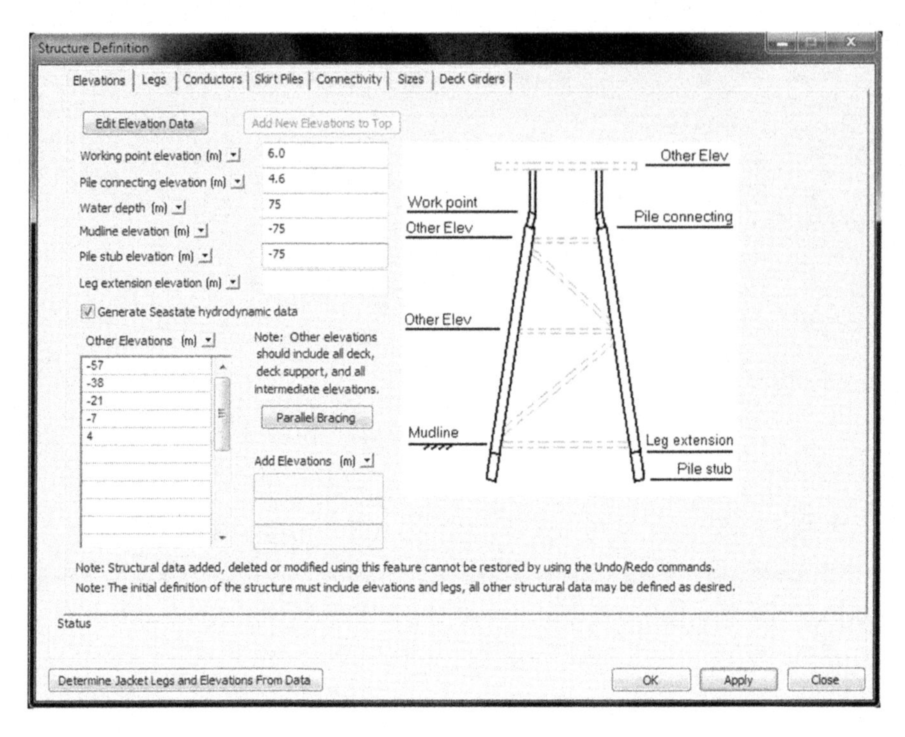

Figure 8.5 Start structure wizard.

After that, click on the Legs tab, as in Figure 8.6, to enter the data for the jacket legs and set the following data as an example:

- Number of legs: 4
- Leg type: Ungrouted
- Leg spacing at working point: X1 = 16 m, Y1 = 10 m.
- Row Labeling: Define the row label to match the drawing.
- Pile/Leg Batter: Row 1 (leg 1 and leg 3, 1st Y Row) is single batter in Y; Row 2 (leg 2 and leg 4, 2nd Y Row) is double batter

For the conductor data, Click on the Conductors tab, then click on the Add/Edit Conductor Data button to enter the data for the conductors, as in Figure 8.7:

- Number of Conductor Well Bays: 1
- Top Conductor Elevation: 15.3 m
- First conductor number: 1
- Number of conductors in X direction: 3
- Number of conductors in Y direction: 2
- Coordinate of LL Corner: X = −4.5 m, Y = −1.0 m. This is the location of the first conductor.
- Distance Between Conductors: 2.0 m in both X and Y directions
- Disconnected elevations: −75 and 4.0 m

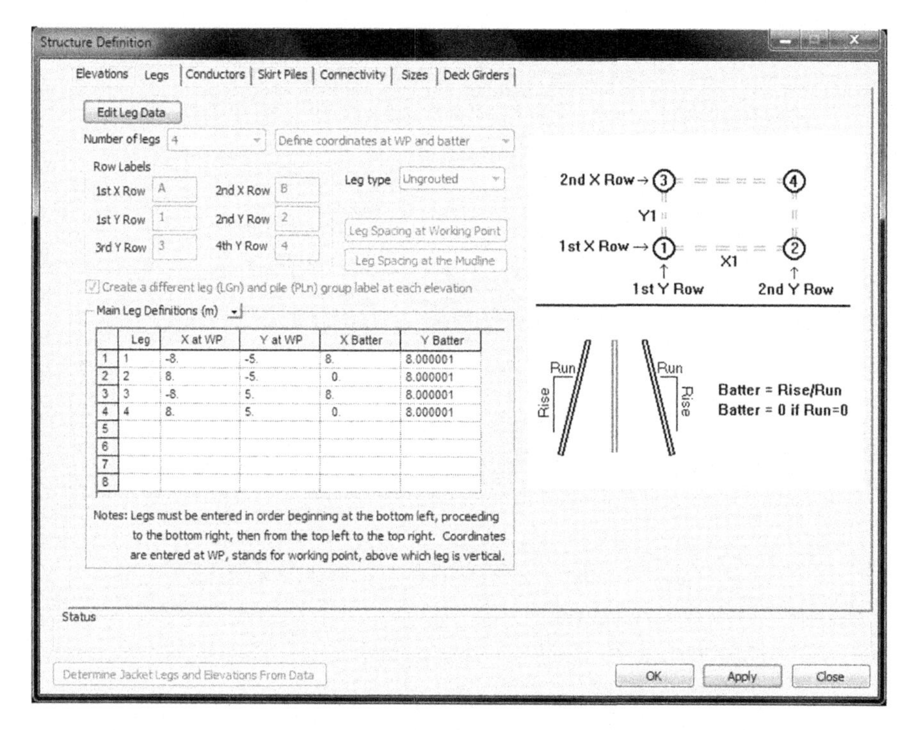

	Leg	X at WP	Y at WP	X Batter	Y Batter
1	1	-8.	-5.	8.	8.000001
2	2	8.	-5.	0.	8.000001
3	3	-8.	5.	8.	8.000001
4	4	8.	5.	0.	8.000001
5					
6					
7					
8					

Figure 8.6 Platform configuration input data.

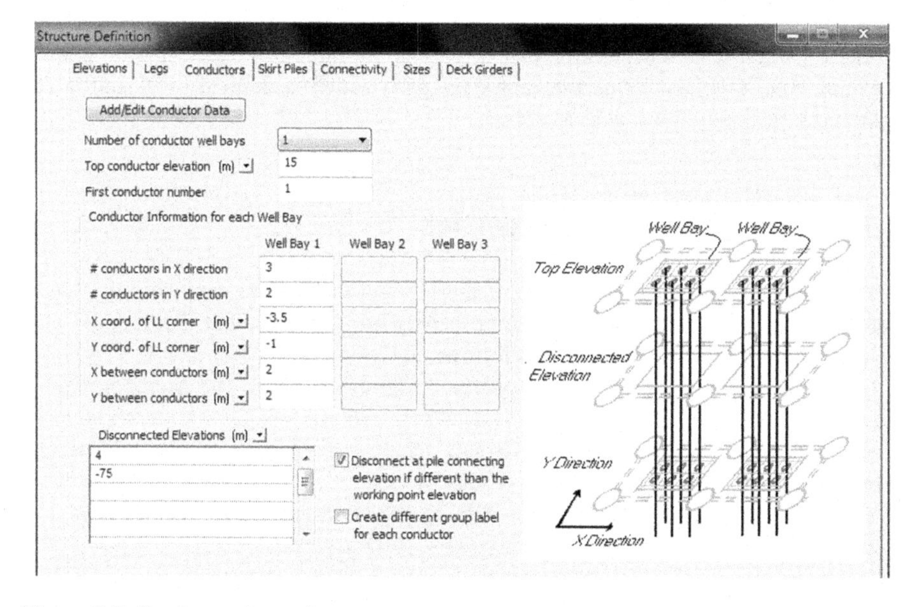

Figure 8.7 Conductors input data.

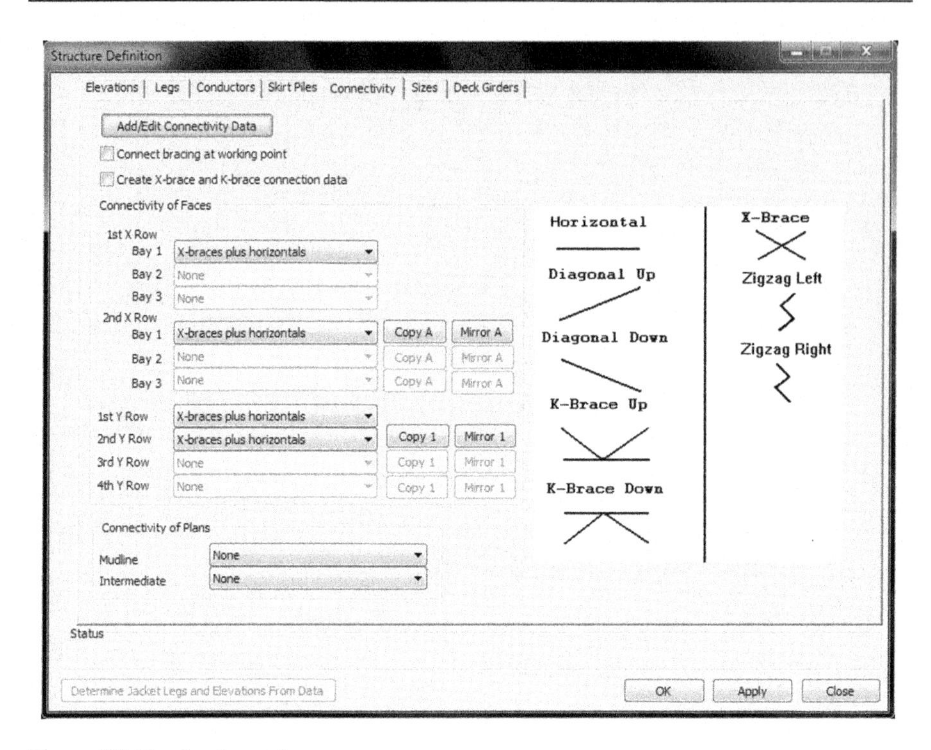

Figure 8.8 Bracing input data.

Then, press the Connectivity tab, as in Figure 8.8, to choose the bracing system for the jacket.

The topside deck configuration can be choose by the Deck Girder tab, as shown in Figure 8.9. You can define the cantilever part, which is defined by X and Y for right and left, upper and lower, as shown.

8.3 Defining member properties

It is the time to define the properties of the leg and the bracing members that were created using the jacket wizard. These properties are shown on the Figure 8.10, and the user entered them in the way that follows

On the Precede toolbar, select Property > Member Group. The Member Group Manage dialog box appears.

Highlight LG1 in the Undefined Groups window, then click on the Add button to define the section and material properties of LG1.

To enter multiple segments in the Add Member Group dialog box, follow this procedure:

- Enter the Segment 1 Parameters: D = 48.5 in., T = 1.75 in., you can choose between the units through the cell itself.
- Fy = 34.5 kN/cm^2, Segment Length = 1.0 m, Flooded Member

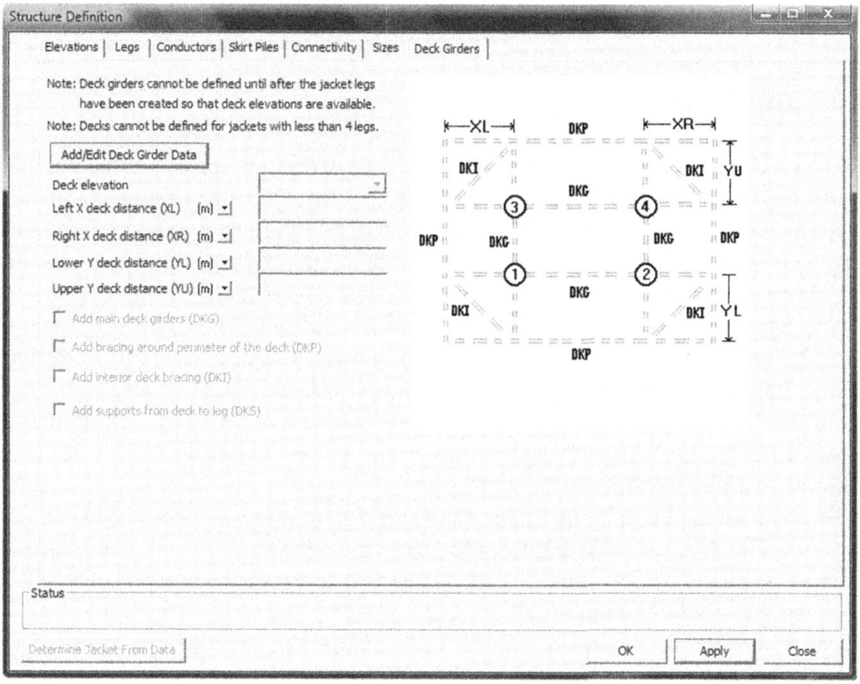

Structure Definition

Elevations | Legs | Conductors | Skirt Piles | Connectivity | Sizes | Deck Girders

Note: Deck girders cannot be defined until after the jacket legs have been created so that deck elevations are available.

Note: Decks cannot be defined for jackets with less than 4 legs.

Add/Edit Deck Girder Data

Deck elevation

Left X deck distance (XL) (m)

Right X deck distance (XR) (m)

Lower Y deck distance (YL) (m)

Upper Y deck distance (YU) (m)

Add main deck girders (DKG)

Add bracing around perimeter of the deck (DKP)

Add interior deck bracing (DKI)

Add supports from deck to leg (DKS)

XL DKP XR DKI DKI YU DKG 3 4 DKP DKG DKG DKP 1 2 DKG DKI DKI YL DKP

Status

Determine Jacket From Data OK Apply Close

Figure 8.9 Deck wizard.

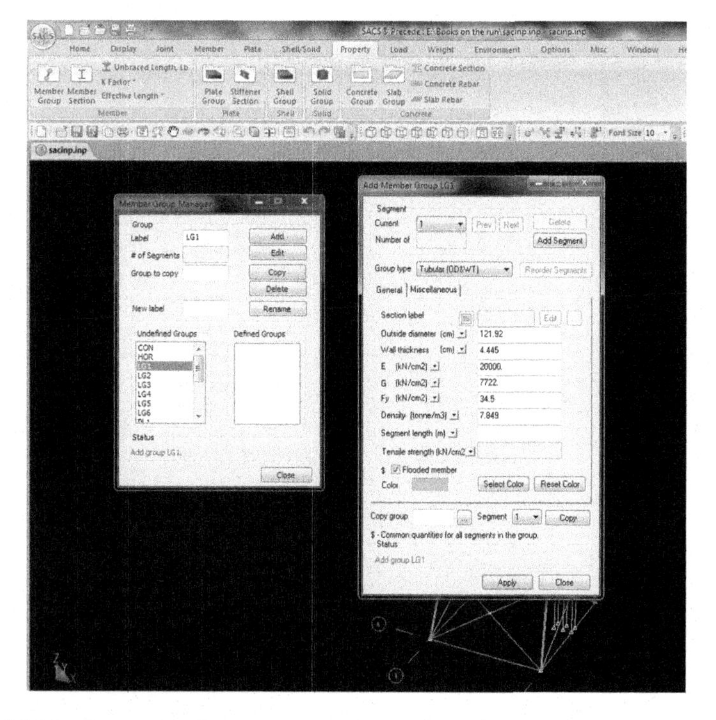

Figure 8.10 Members properties.

8.4 Input the load data

After you define the geometry, the next step is to define the loads and the environmental load as wave, current, and wind. Illustrated here is the environmental load.

Go to Environment > Loading > Seastate to define the wave, current, wind, and dead/buoyancy load parameters. The data can be found in the design specifications, and the images that follow show the details of load case 6, as in Figure 8.11.

You have five tabs, then two for each Wave I and Wave II, Wind I and Wind II, Current I and Current II, Dead, and Drag. As shown in Figure 8.11, this is for load case 6 the wave height = 14.1 m and time period = 12.2 s, the direction = 0.0. These data are in an extreme wave for 100 years. For Wave II, as shown in Figure 8.12, the initial crest wave = −15 m and the wave position step size is 1.5 with total number for crest = 20.

So, the load case 6 is assigned the following for an extreme 100-year load at 0.0 direction. For Wind I, as in Figure 8.13, the wind velocity = 24 m/s, the reference height = 10 m; this is by default. According to the code, the reference height is 10 m and the wind direction = 0.0

For Wind II, as shown in Figure 8.14, define the area that is affected by the wind.

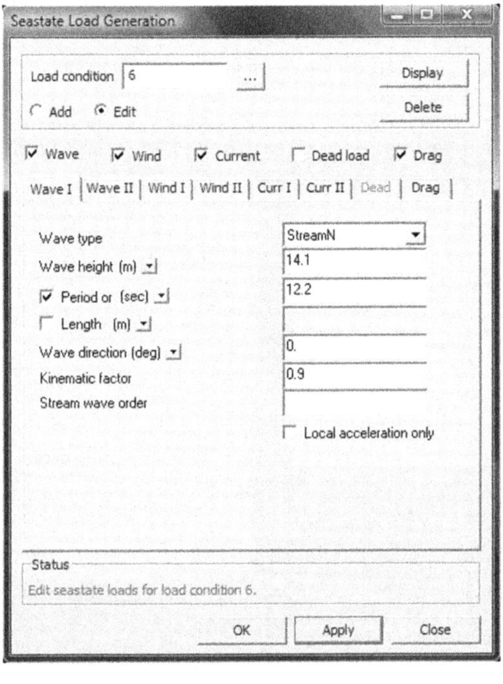

Figure 8.11 Wave input data.

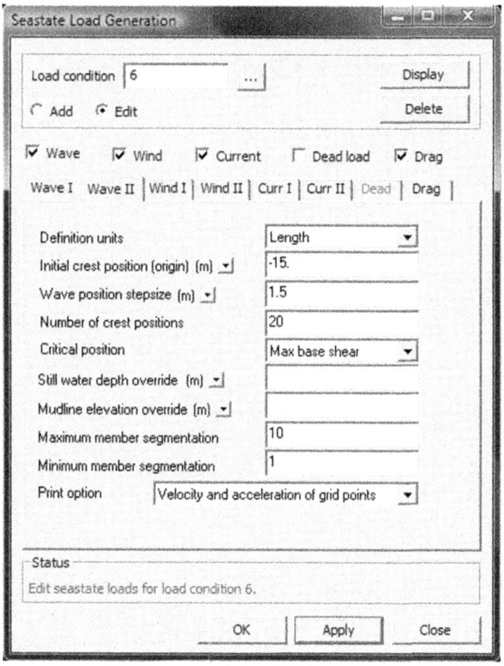

Figure 8.12 Continuation of wave input data.

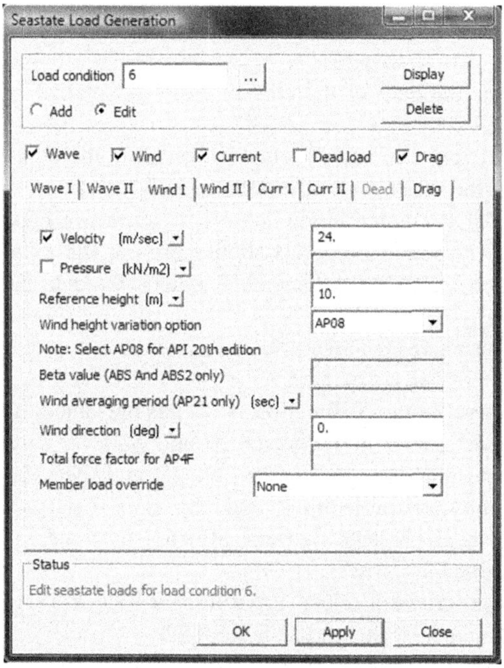

Figure 8.13 Wind input data.

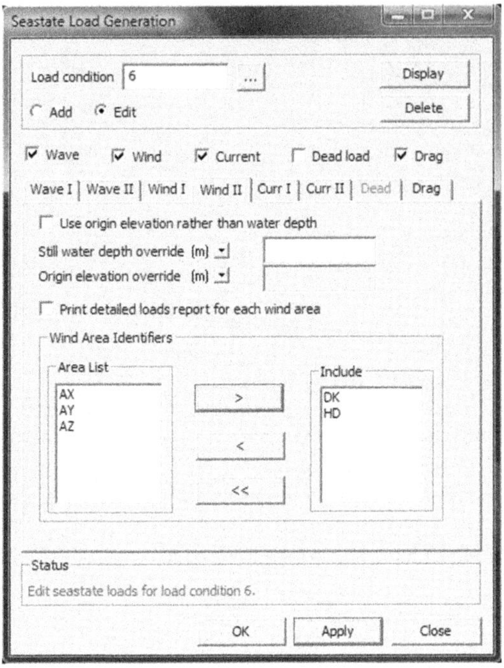

Figure 8.14 Wind area.

For Current, enter the data of the current from the seabed to the sea level. As shown in Figure 8.15.

In the drag coefficient tab, as in Figure 8.16, define the area that is affected by the drag on it in 0.0 direction, so it affects the area AX.

A wave load of 45° is entered in load case 7, as shown in Figure 8.17.

So, eight cases with eight directions should present the extreme condition and another eight cases in eight directions should include the data under operating conditions over one year.

As shown in Figure 8.18, the area affected by wind and drag at 45° includes AX and AY.

In area of drag effect in the Y direction is AY, as presented in Figure 8.19.

Define the area as shown in Figures 8.20 and 8.21 by Environment > Global Parameter > Area and Volume. Define the area ID and the data type for the wind area. Define the area of projections and the area centroid of effect, as in Figure 8.21. Figure 8.21 presents the area affected by wind in the helideck. The projected area is defined as 11 m^2.

To define marine growth, select Environment > Global Parameters > Marine growth and enter the data, as shown in Figure 8.22.

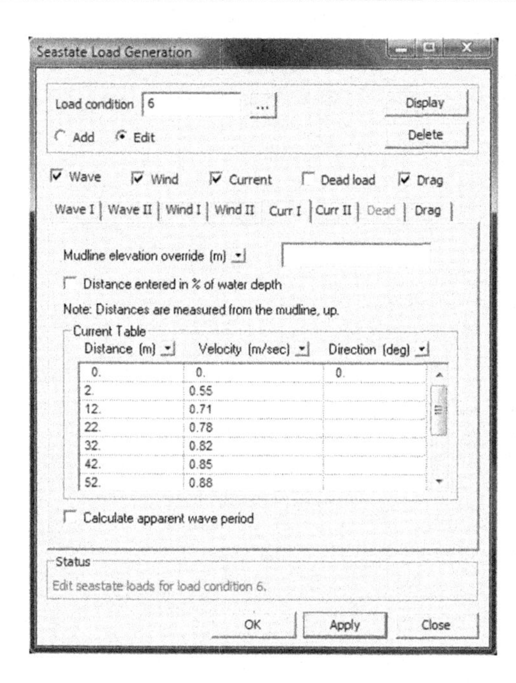

Figure 8.15 Current data.

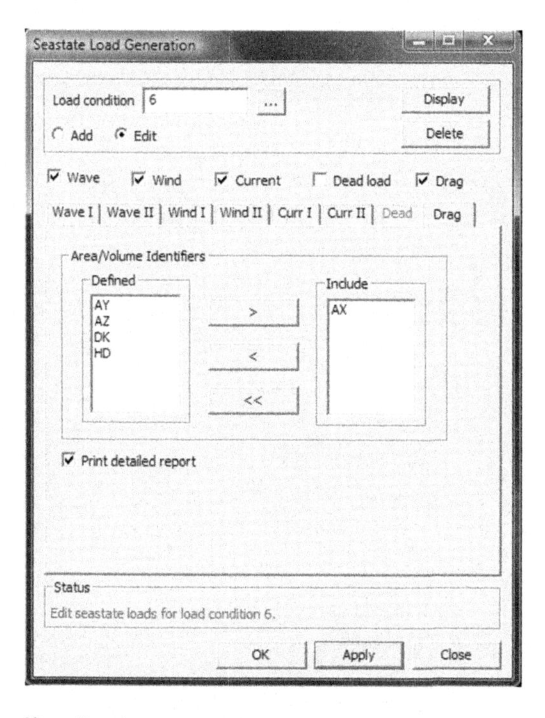

Figure 8.16 Drag effect direction.

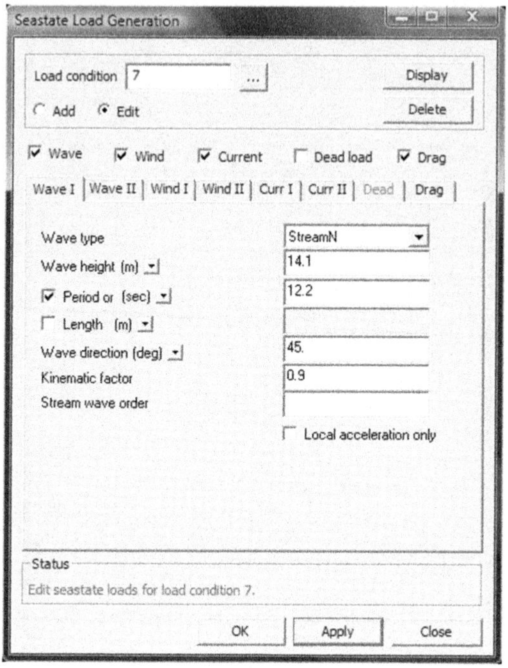

Seastate Load Generation
Load condition 7
Display
Add Edit
Delete
Wave Wind Current Dead load Drag
Wave I | Wave II | Wind I | Wind II | Curr I | Curr II | Dead | Drag
Wave type StreamN
Wave height (m) 14.1
Period or (sec) 12.2
Length (m)
Wave direction (deg) 45.
Kinematic factor 0.9
Stream wave order
Local acceleration only
Status
Edit seastate loads for load condition 7.
OK Apply Close

Figure 8.17 Wave load in 45°.

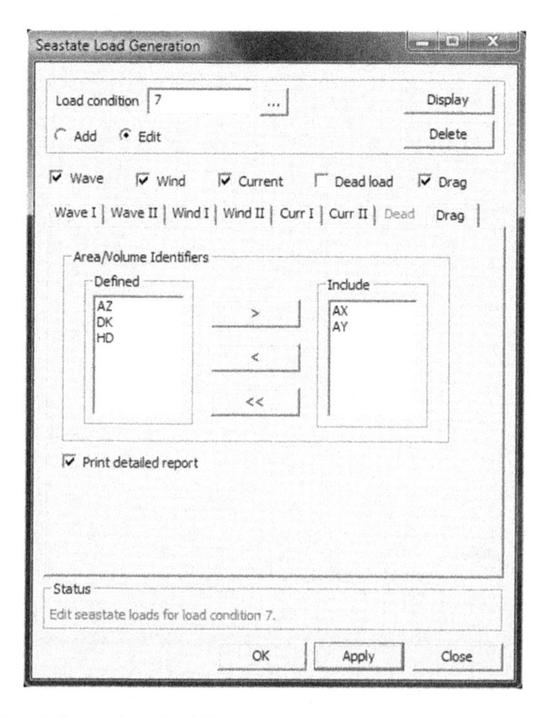

Figure 8.18 Wind and drag effect in 45°.

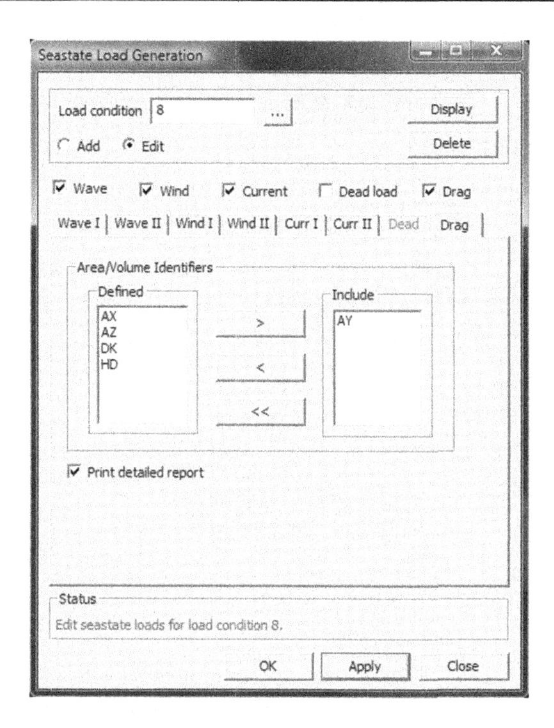

Figure 8.19 Drag effect in Y direction.

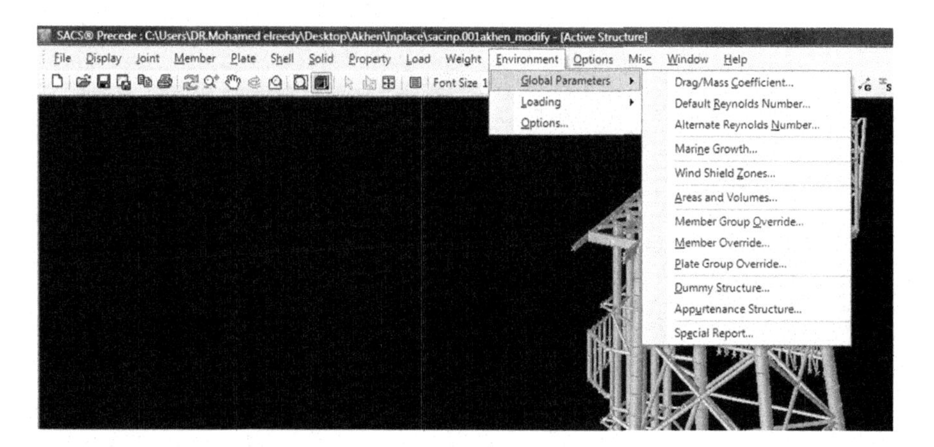

Figure 8.20 Define the area.

The hydrostatic collapse data are inserted in the menu presented in Figure 8.23. The data are the code check, the water depth, and the ring design option; they are selected to be external; and define geometric of the ring.

Wind Area, Submerged Area, and Submerged Volume Data
Area group ID HD # 2
Enter AL if ID is to be used in all LCs.
Display 1
Delete Active
Add new area group Add new component Edit component
Data type Wind area
Area type Flat surface, rectangular coordinates
Select Joints
Select Areas
Note: Areas selection is performed by placing the cursor next to the perimeter of the area.
Area Specification
X projection (m2) 11.
Y projection (m2)
Z projection (m2)
Total area (m2)
Horizontal angle (deg)
Vertical angle (deg)
(m)
X 5.75
Y 0.
Z 23.22
Select Centroid
Calculate Centroid
Shape factor 2.
Distribution joints 3425
Joint List
Save Load Screen
Status
Edit Area group HD, component 1.
Reset OK Apply Close

Figure 8.21 Helideck wind area.

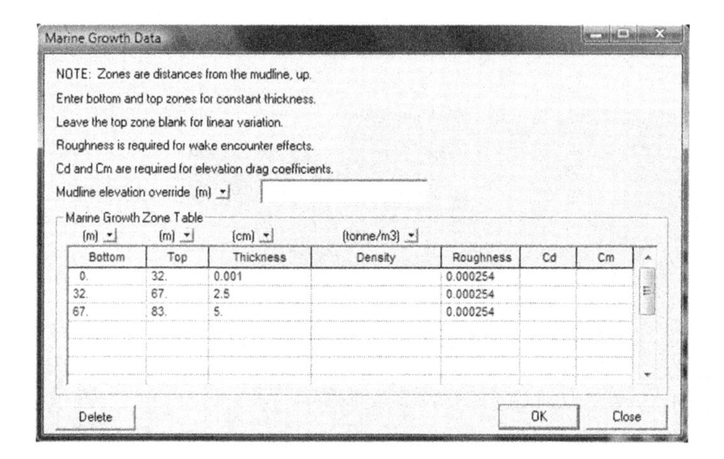

Figure 8.22 Marine growth data.

After you build the model, save the file, noting that this software is different from other software in that it put the extension first then the file name; so, after you save the file, it will be sacinp.file name, for example.

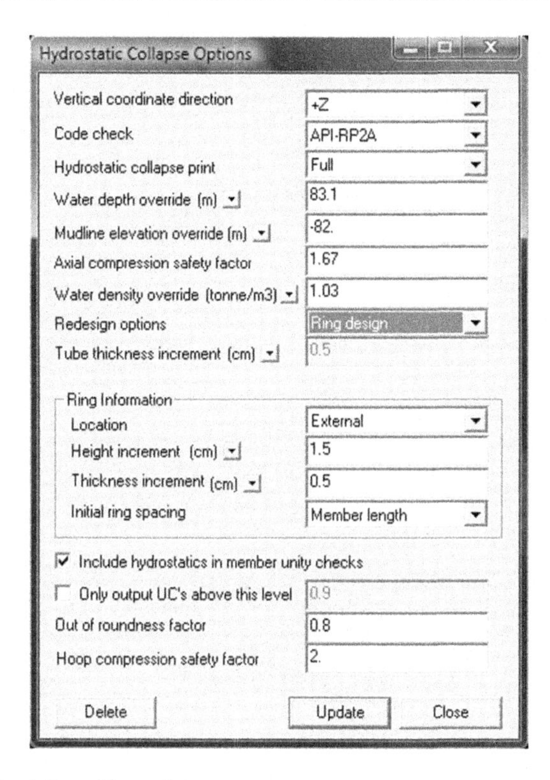

Figure 8.23 Hydrostatic collapse data.

8.4.1 Joint can

To check the joint can, as in Figure 8.1, build a subroutine to define the joint can options you need to check. The joint can editor is shown in Figure 8.24.

To model the joint can, first, follow these steps:

1. Click on Data File from SACS Executive screen, as presented in Figure 8.25.
2. Select Create new data file. And click on OK
3. Select Units as Metric, forces, click on the Post icon, select Joint can, and then click on Select.
4. Make selections as shown in Figure 8.26:

   Joint Check Option: API
   - Units: MN
   - Allowable Limit (Thicker Can Option): R
   - Leave others no change

   Reports Tab, as in Figure 8.27:
   - Strength Analysis Report Option: Summ
   - Print Load Path Report: Checked
   - Suppress Warnings: Checked
   - UC Order: Full

   Joint Can Output Report Options: Full

   In Figure 8.28, it will choose the default, which is the most suitable selection.

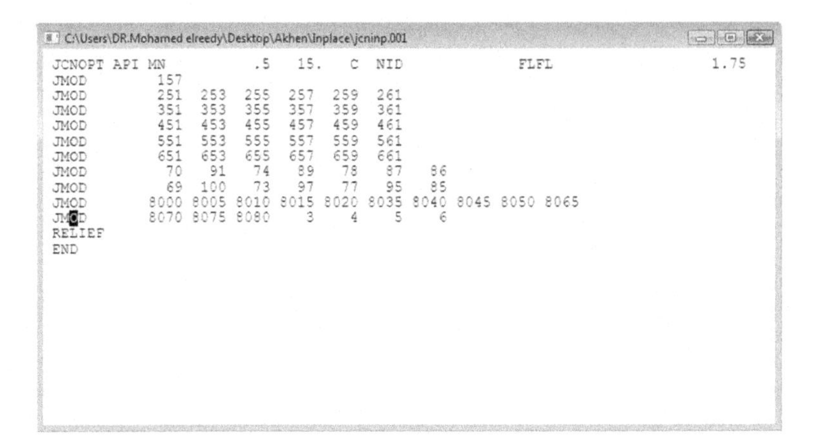

Figure 8.24 Joint can editor.

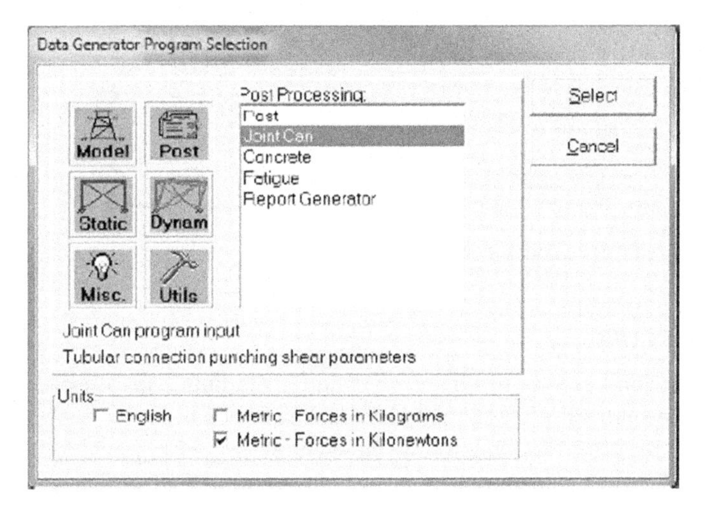

Figure 8.25 Joint can data generator.

5. In Figure 8.29, click on No for Select output load cases?
6. In Figure 8.30, click on No for Change any of SACS allowable stress modifier?
7. As shown in Figure 8.31, click on No for Change SACS yield stresses?
8. In Figure 8.32, click on No for Change GRUP yield stresses?
9. In Figure 8.33, click on No for Change joint yield stresses?
10. In Figure 8.34, click on No for Specify the allowable weld stress for overlapped braces?
11. The Brace Stresses Calculation at Surface of Chord dialog box opens, select Yes, then click on Next, as in Figure 8.35.
12. In Figure 8.36, click on No for Choose specific joints for analysis?
13. In Figure 8.37, click on No for Override the chord thickness for punching shear analysis?

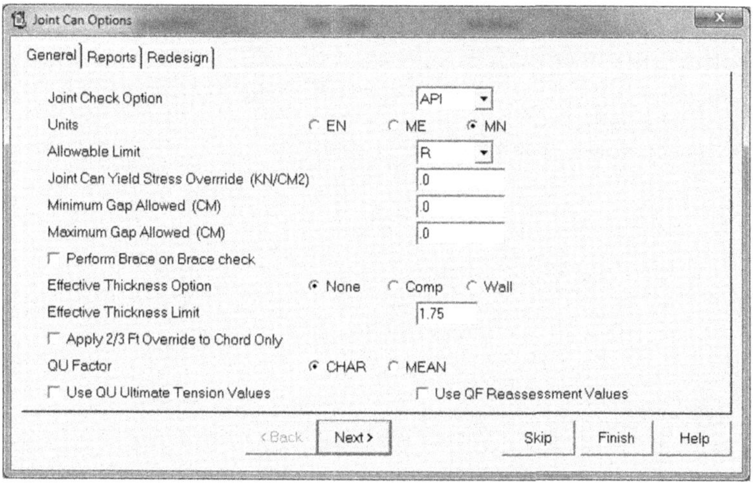

Figure 8.26 Joint can option.

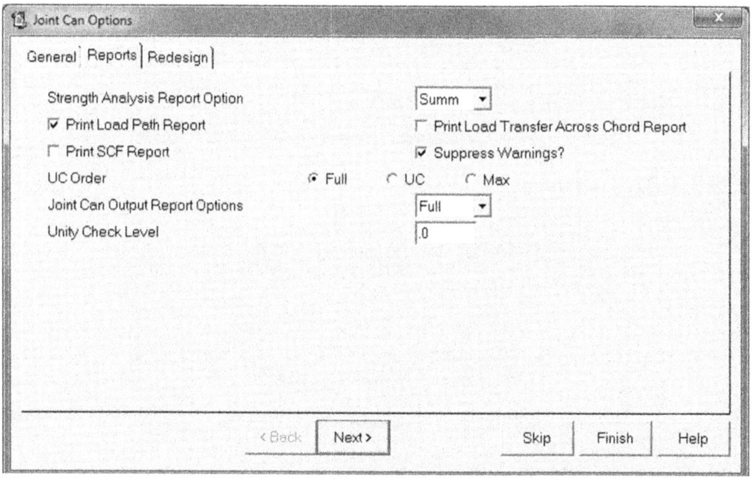

Figure 8.27 Joint can reports.

14. In the dialog box in Figure 8.38, click on No for Set the Brace to Chord angle limit?
15. In Figure 8.39, click on No for Change SACS load combinations?
16. Save the file as jcninp.filename.

The another method, valid for revisions 5.3 and above, in the analysis option, is to choose Tubular joint check. A menu will open so you can put the data on it directly, as shown in Figures 8.40 and 8.41.

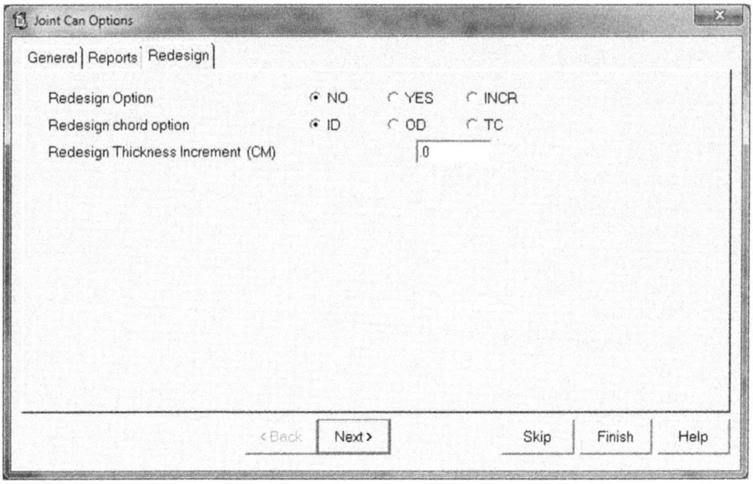

Figure 8.28 Joint can redesign option.

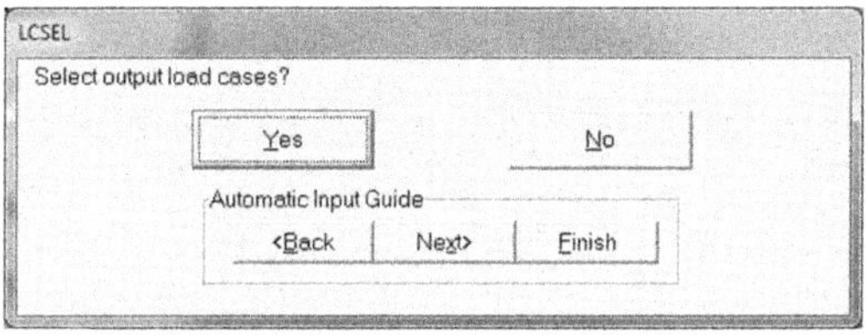

Figure 8.29 Load cases options.

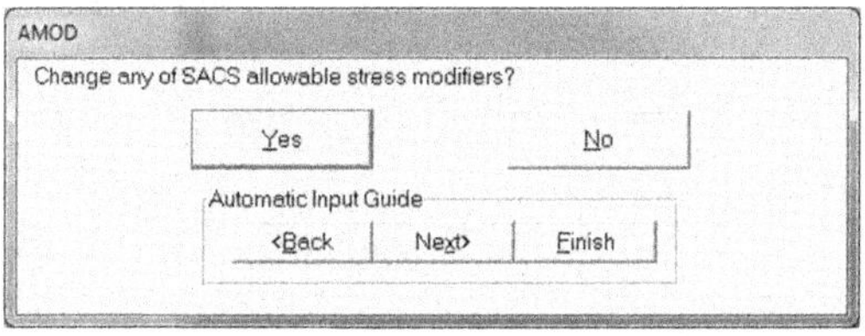

Figure 8.30 Allowable stress modifies.

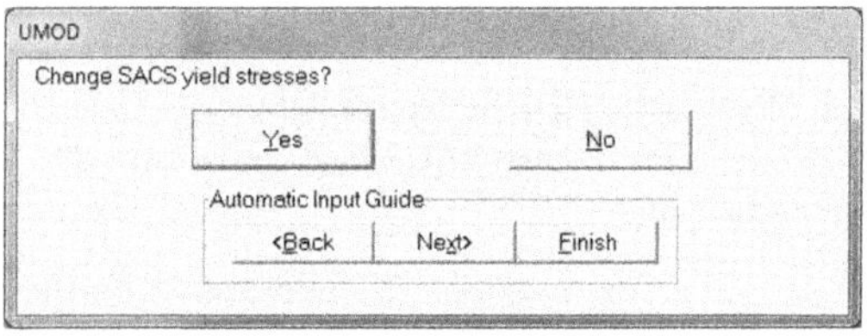

Figure 8.31 SACS yield stresses values.

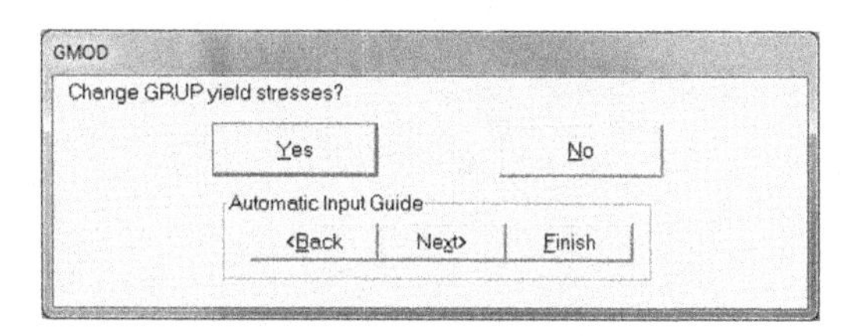

Figure 8.32 Change group yield stress.

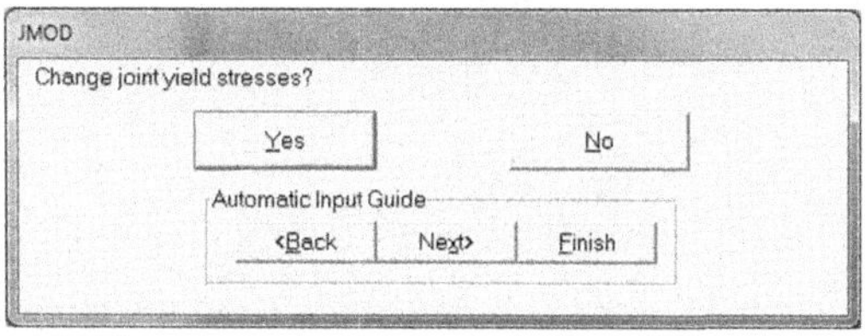

JMOD

Change joint yield stresses?

Yes No

Automatic Input Guide

<Back Next> Finish

Figure 8.33 Change joint yield stress.

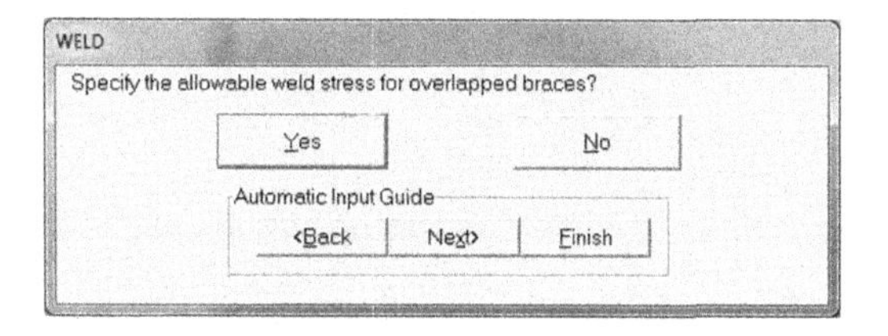

Figure 8.34 Change allowable weld joint yield stress.

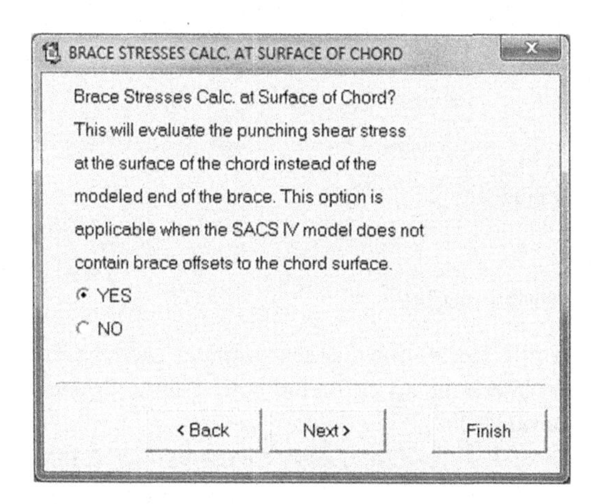

Figure 8.35 Brace stresses.

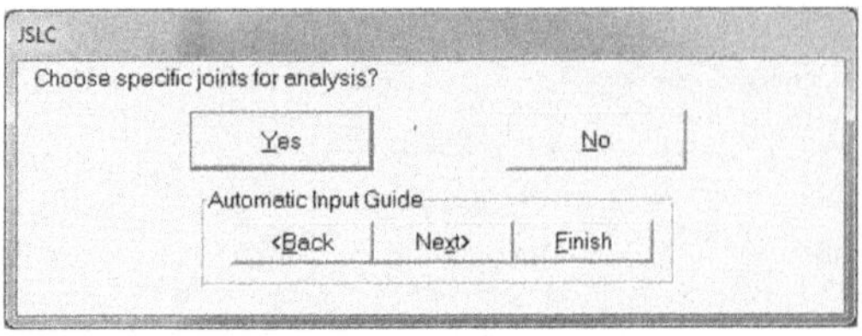

Figure 8.36 Joint analysis option.

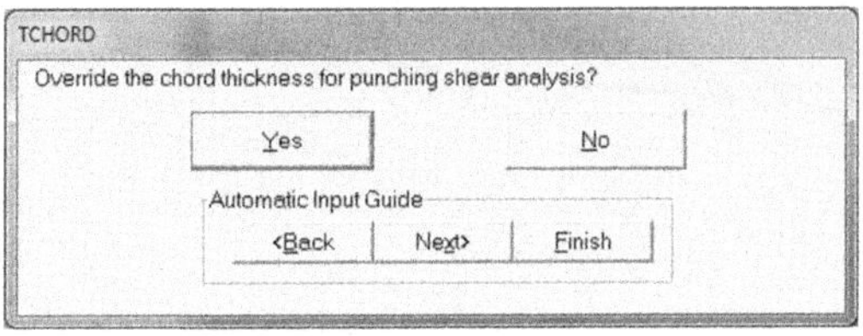

Figure 8.37 Chord thickness option.

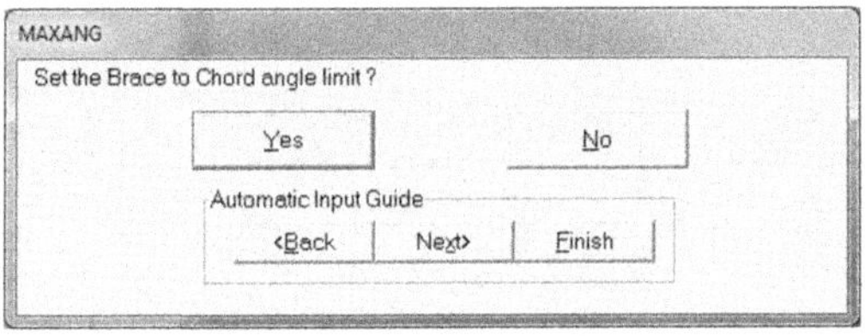

Figure 8.38 Brace to chord angle limit.

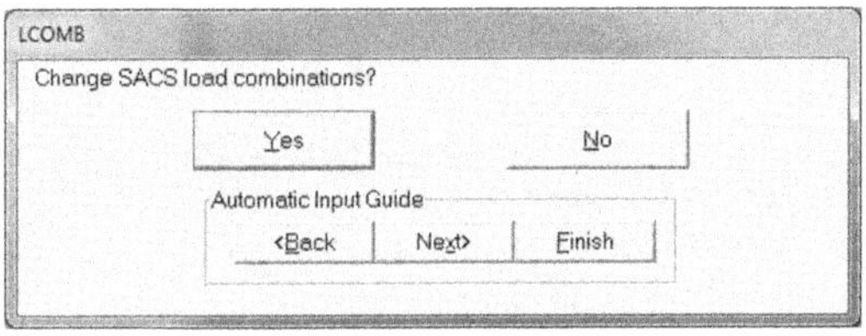

Figure 8.39 Load combination.

8.4.2 The foundation model

The simulation of a pile foundation in the computer model takes into account the pile stiffness and the lateral behavior of the soil. The nonlinear approach for pile-soil interaction is considered.

Geotechnical data of P-Y, T-Z, and Q-Z curves extracted from the geotechnical report, as described in the pile foundation design report, are used in the analysis to obtain a rigorous solution of the soil-pile-structure interaction.

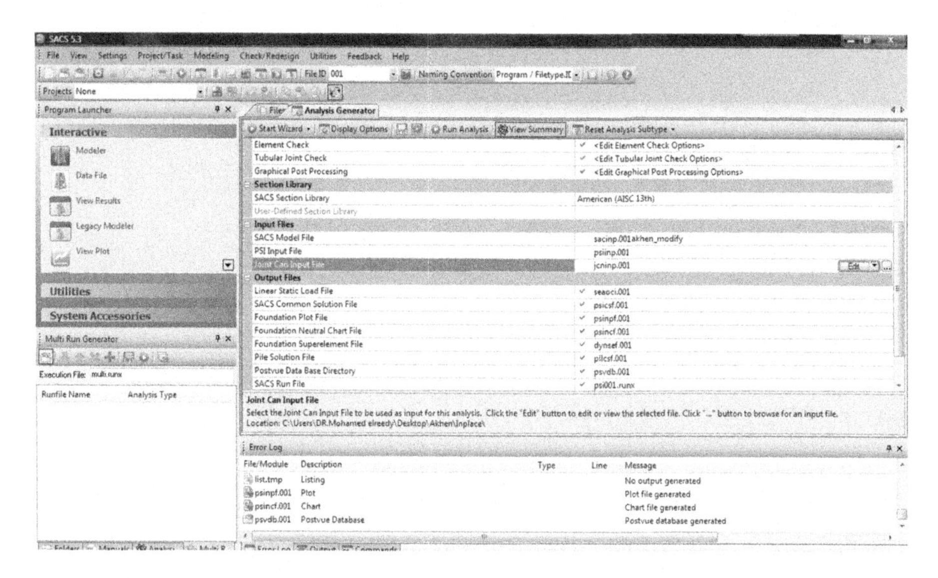

Figure 8.40 Joint can input.

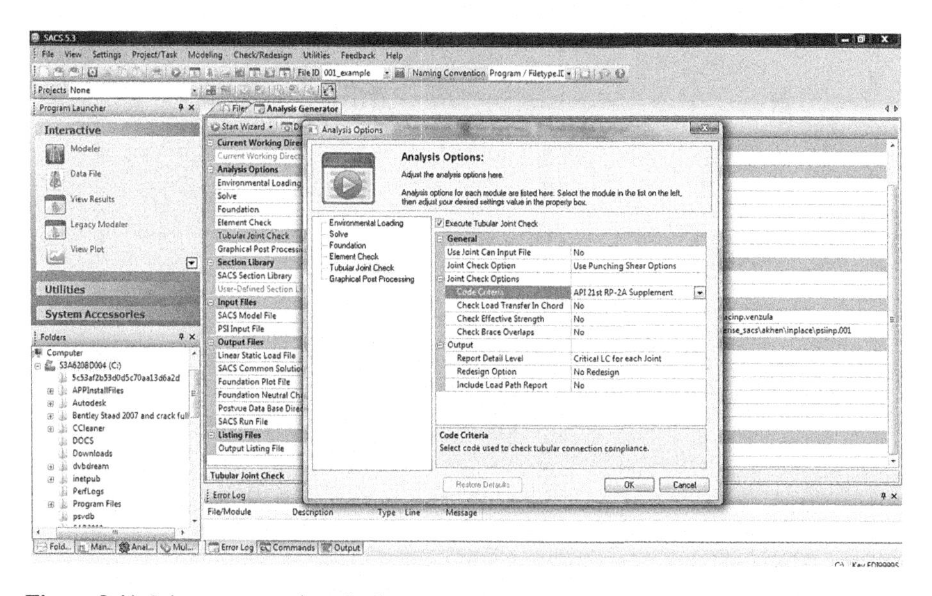

Figure 8.41 Joint can second method.

Iterative analysis is carried out by the PSI program until reaching the pile head displacement and rotation convergence. Thereafter, PSI extracts the final pile head loads and analyzes the pile. Being nonlinear, the analysis is carried out for the combination as in basic load cases. So, you need to enter the soil data in the subroutine in the data executive file, as shown in Figure 8.42.

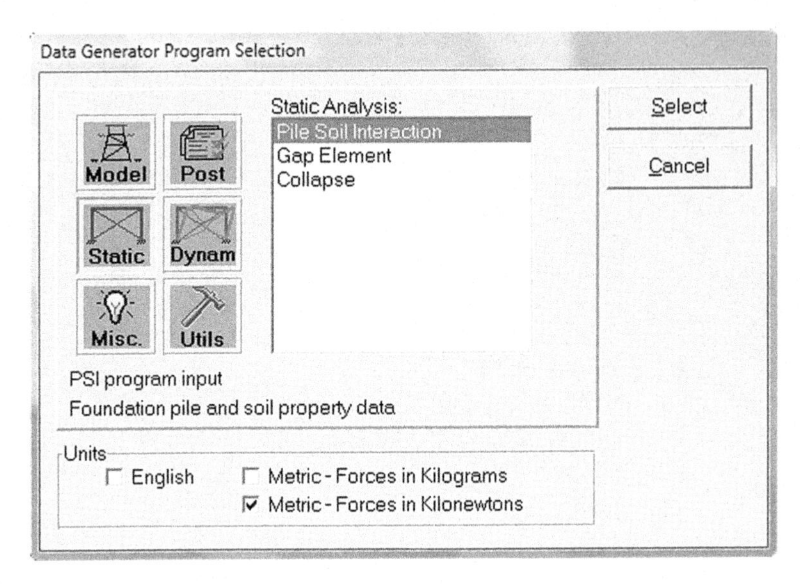

Figure 8.42 Pile-soil interaction menu.

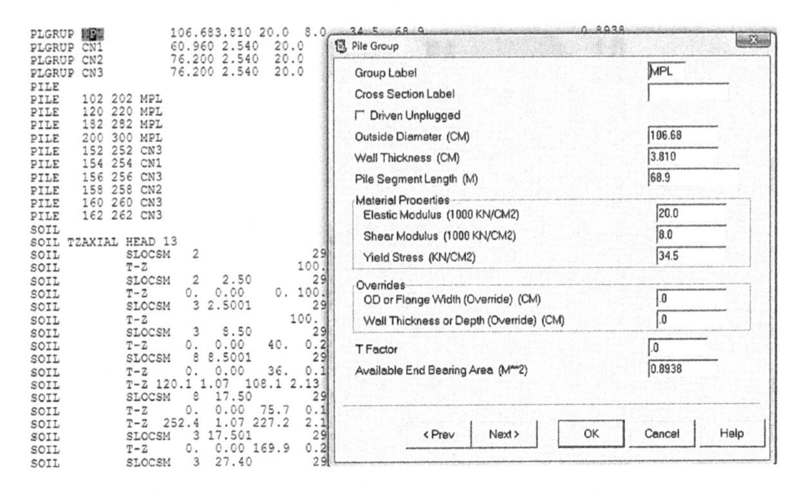

Figure 8.43 Pile properties input data.

A series of menus ask for your input. Then, the main menu about the pile group appears, as shown in Figure 8.43. The pile data about its diameter, length, and thickness are defined. The last line is about the end bearing area, which you should have calculated. based on the end bearing area under consideration.

The soil data are presented in the following figures. Figure 8.44 provides the number of soil strata available in soil geotechnical report.

Figure 8.44 T-Z soil axial data input.

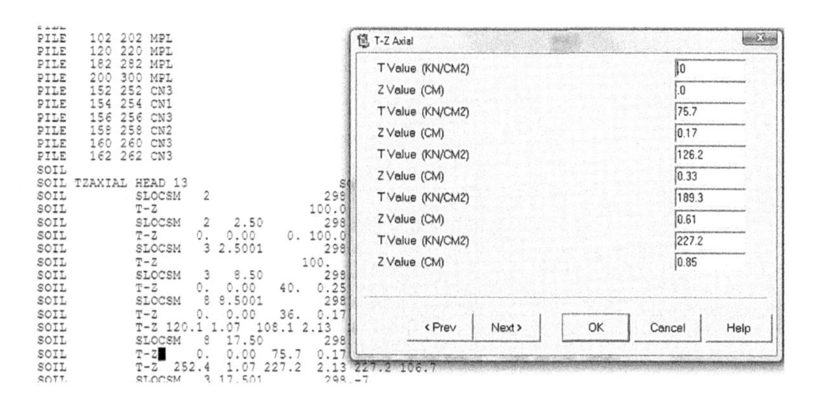

Figure 8.45 T-Z axil stratum soil data.

Figure 8.46 T-Z axial input data.

As previously, you define the number of strata, and Figure 8.45 presents the menu that provides the number of points for each strata. For this example, you should have eight points of data between T and Z.

The input of T in kN/cm^2 with the Z in cm is presented in Figure 8.46. Figure 8.47 shows the T-Z axial bearing, and Figure 8.48 shows the T-Z axial bearing strata. For the bearing strata, the value of T and Z for different point will be input in the menu as shown in Figure 8.49.

Figure 8.50 presents the menu of the input data and its limitations. In the data generator file in this figure define two strata and the ID of the soil, Sol1.

T-Z Axial Bearing
Number of Soil Strata 2
Z-Factor .0
Soil Table ID SOL1
Max. No. of Points in Any T-Z Curve 0
< Prev Next > OK Cancel Help

Figure 8.47 T-Z axial bearing.

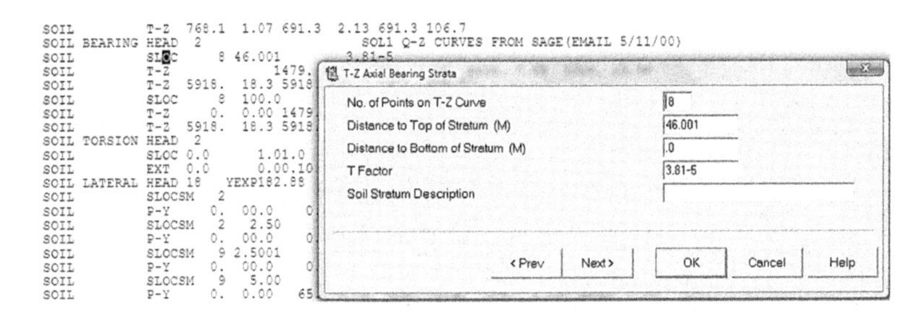

Figure 8.48 T-Z axial bearing strata.

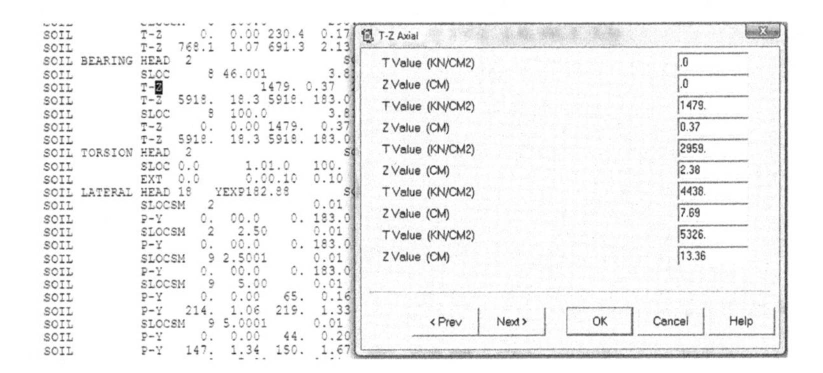

Figure 8.49 T-Z axial bearing soil data.

The top and bottom elevation of strata is shown in Figure 8.51.

The torsion adhesion data in the second strata top and bottom are the same and = 0.1 kN/cm^2, as shown in Figure 8.52.

To start input the data for P-Y curve, define the number of strata, which in this example = 18, as in Figure 8.53.

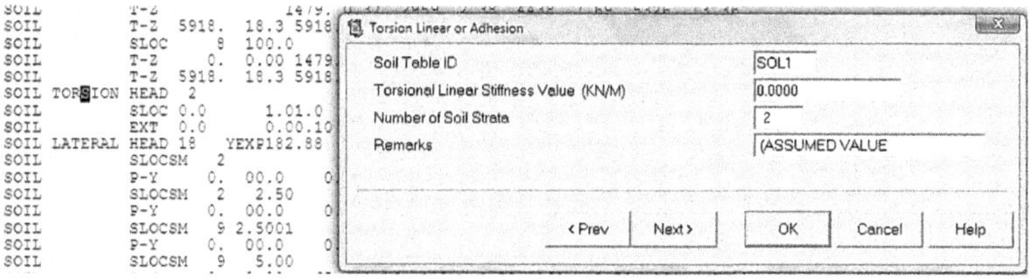

Figure 8.50 Torsion head data.

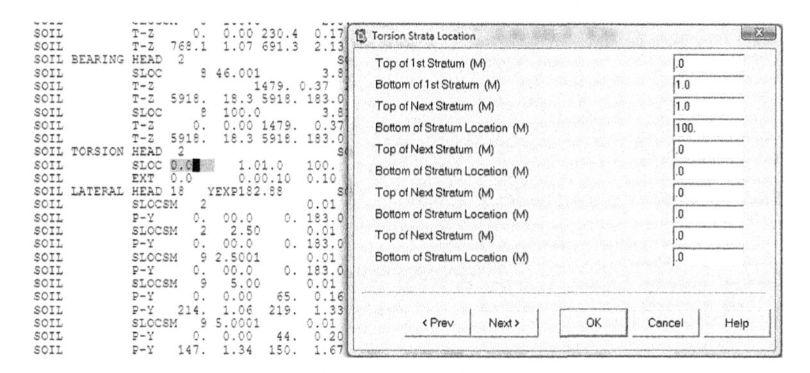

Figure 8.51 Locations for the two strata.

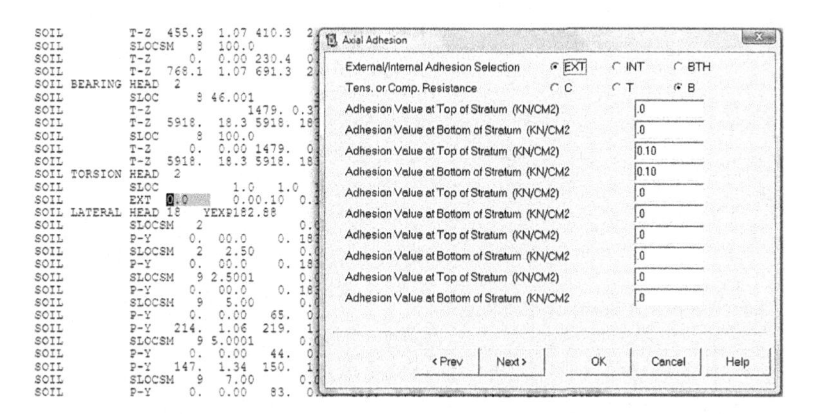

Figure 8.52 Torsion adhesion data.

For each strata, define the number of points for which you have data from the geotechnical report, as in Figure 8.54. Then, enter the data of P-Y for the nine points you defined. The data are as shown in Figure 8.55.

After that, save the file as psiinp.filename. The data you entered can be shown from the left menu in revision 5.3 Utilities > Plot soil data. Figures 8.56, 8.57 and 8.58 present the input data.

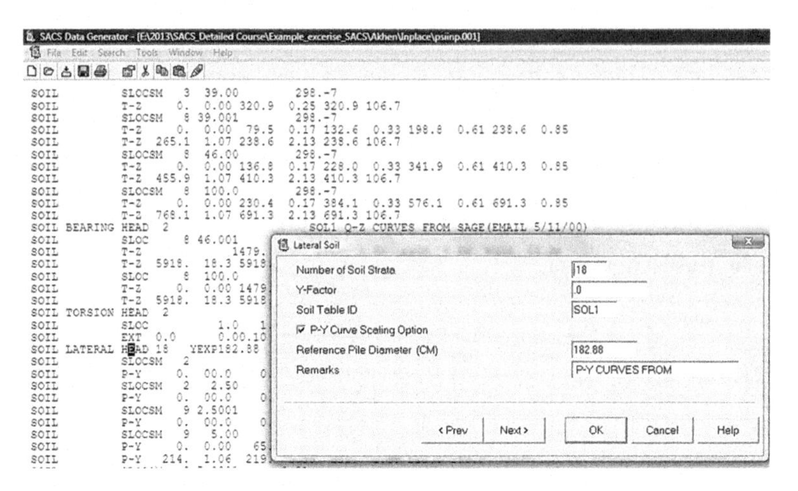

Figure 8.53 Number of soil strata for P-Y data.

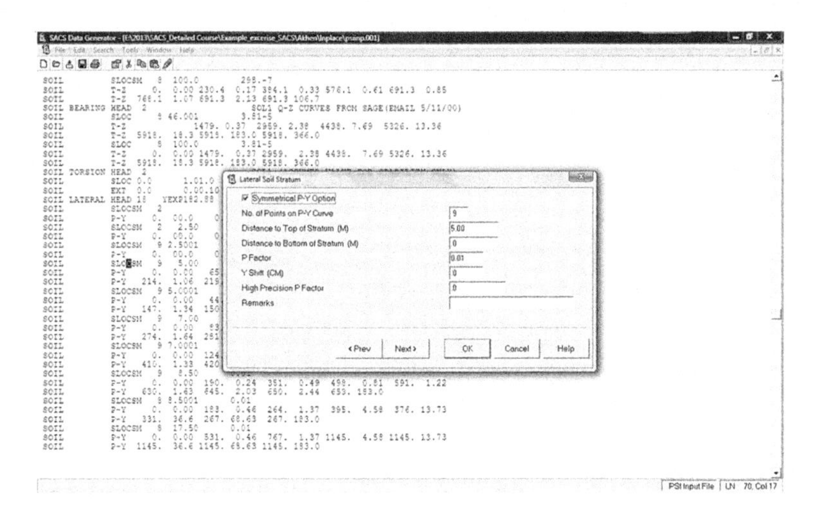

Figure 8.54 Number of point on P-Y curve.

8.5 Output data

The output data is presented by the postvue icon. The following Figure 8.59 presents the output drawing. When you select a member, it is identified in the menu on the right by its nodes, from Member > Review Member. Select the member, as shown in Figure 8.59.

The member design calculation is presented by going through the menu of the member and choosing Review Member. If you need comprehensive data, click on the Detailed Report button as in Figure 8.60.

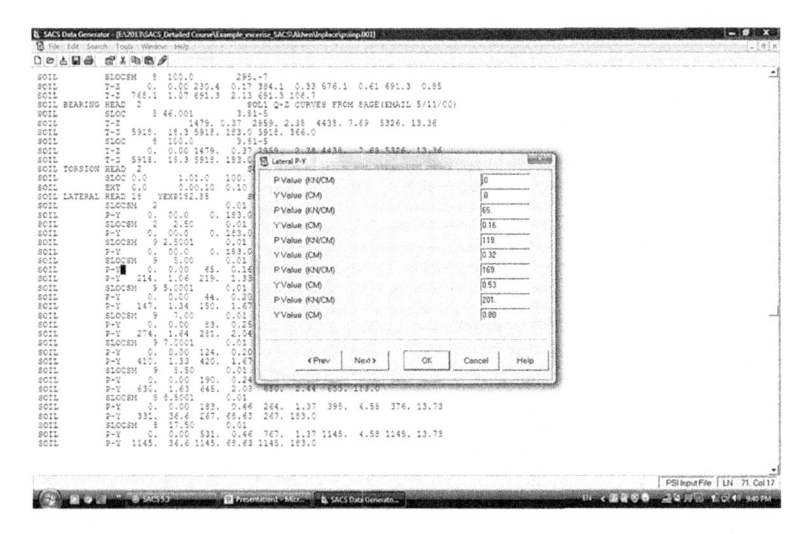

Figure 8.55 P-Y curve input data.

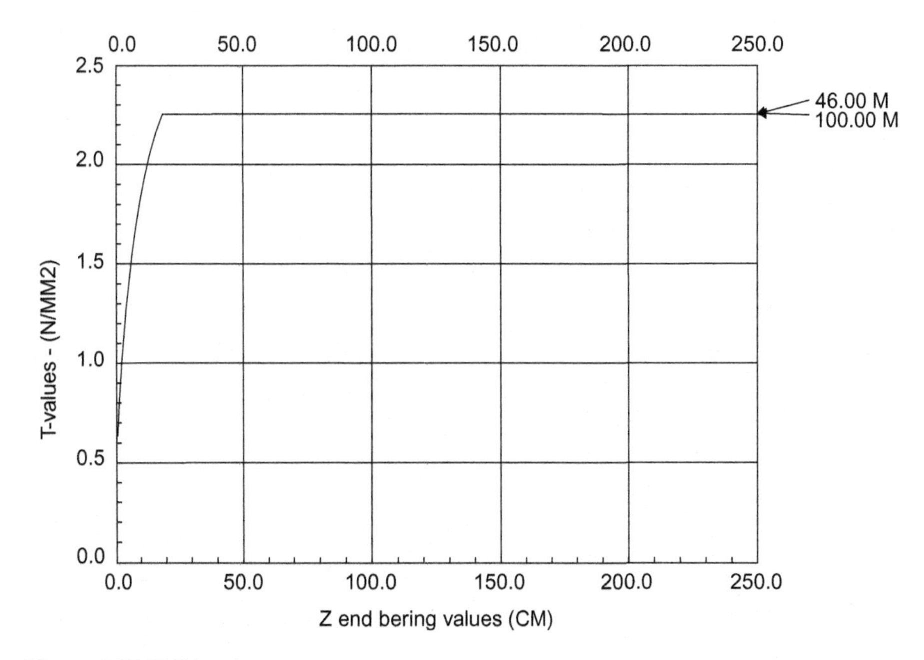

Figure 8.56 T-Z bearing curve.

To study the normal force, shearing, and bending moment for each member, select Member > Member Diagram, then select the member to study, as shown in Figure 8.61. The normal forces, shearing force, and bending moment are presented for each member, as shown in this Figure 8.62.

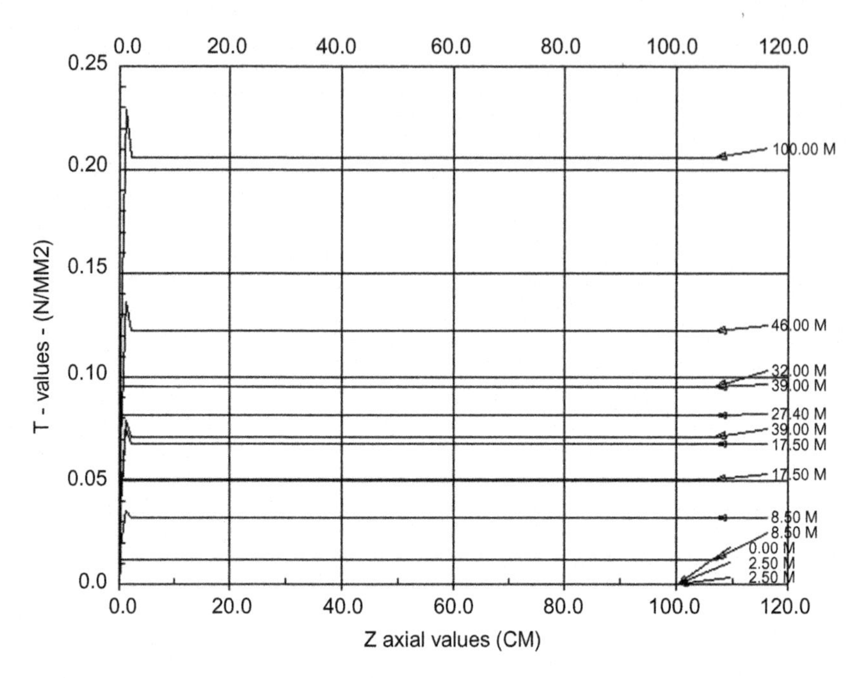

Figure 8.57 T-Z axila curve.

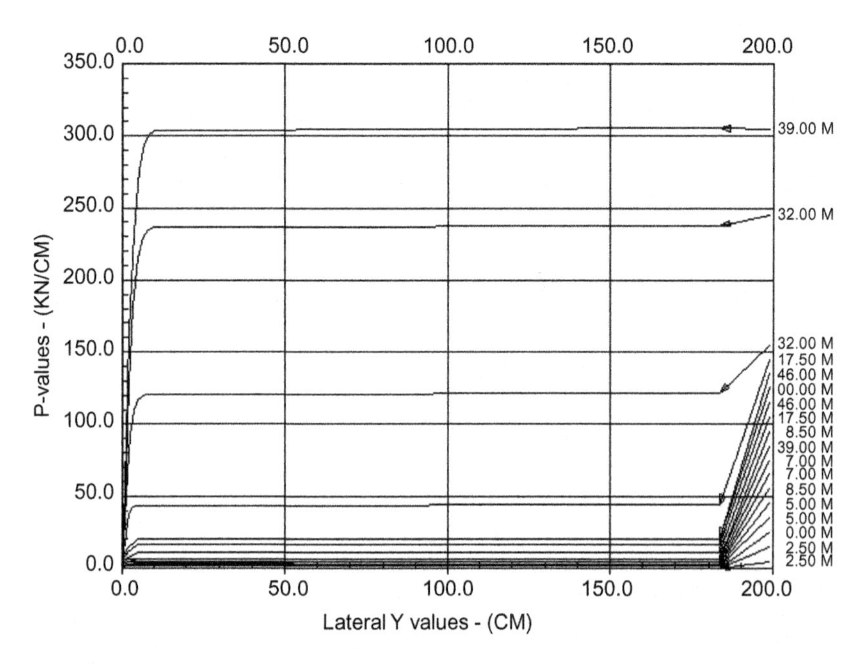

Figure 8.58 P-Y curve for different strata.

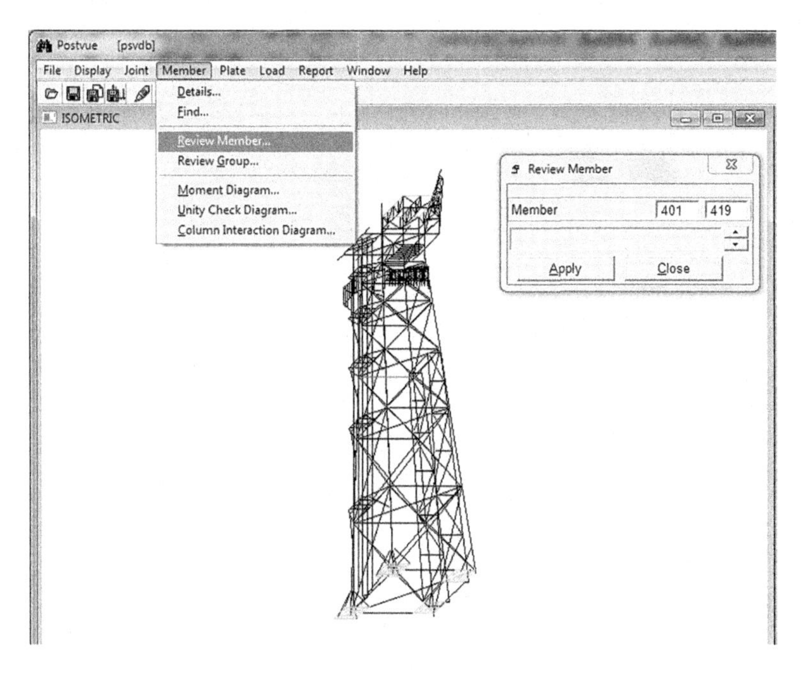

Figure 8.59 Review Member data.

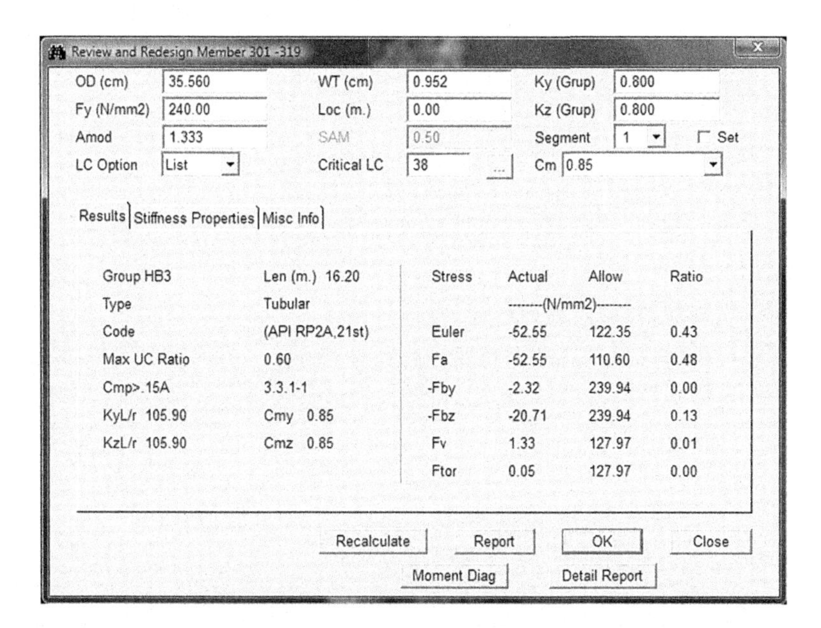

Figure 8.60 Example of member design output data.

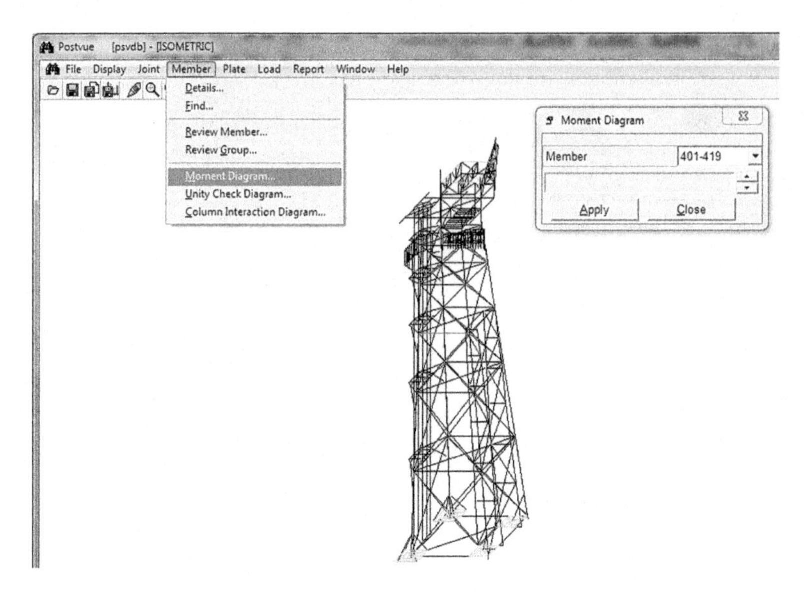

Figure 8.61 Member Diagram.

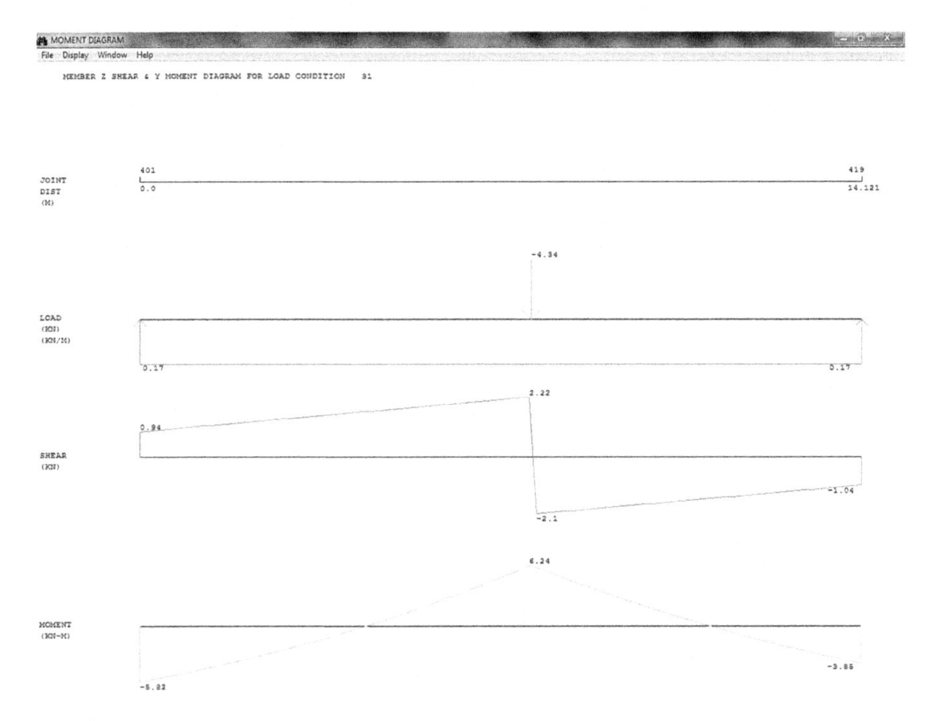

Figure 8.62 Straining action.

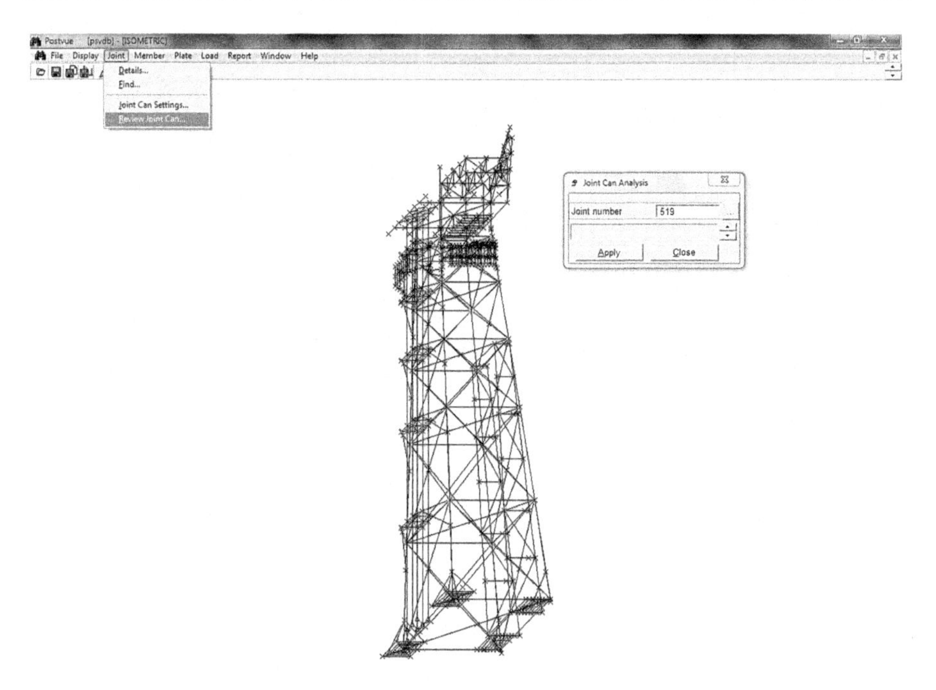

Figure 8.63 Joint can output data.

To review the joint can, select Review Joint Can from the Joint List menu and select the joint via Joint > Review Joint Can, as shown in Figure 8.63. The joint can calculation is presented in the following figure, and through the Member Review, you can obtain the detailed report.

The joint can complete data analysis is shown in Figure 8.64.

The deformation of the structure or the unit check of the member can be obtained by going through Display > Shape. From the menu in Figure 8.65, you can obtain the deformed shape

8.6 Dynamic analysis

The Dynpac module is used for dynamic analysis. In most cases, you must extract the first 10 mode shapes to properly simulate the dynamic response of the platform. The Degree of freedom (DOF) is retained for some joints, where the dynamic equations are generated. These joints are selected from the mass for each level and under the decks. Superelements are used to represent the soil-pile stiffness commensurate with the expected level of loading.

The DAF is calculated from the following API equation:

$$\mathrm{DAF} = \frac{1}{\sqrt{\left[\left(1-\frac{T_p^2}{T_n^2}\right)^2 + 4\left(\in \frac{T_p}{T_n}\right)^2\right]}} \tag{8.1}$$

Review and Redesign Joint 438

Chord Properties		Brace Properties			
Chord Jnt	419	Brace Jnts	499	LC Option	List
Diam (cm)	50.800	Diam (cm)	50.800	Amod	1.333
Thick (cm)	2.062	Thick (cm)	0.952	Load cond.	39
Fy (N/mm2)	344.800	Fy (N/mm2)	240.000		

Review Overlaps

Punching Results | Classification

Code	API 21st Suppl 3	Brace Stresses (N/mm2)			
Max UC Ratio	0.428	Type	Actual	Allow	Ratio
Chord SRSS (N/mm2)	-15.808	Fa	60.574	162.338	0.373
Joint Type	X	OPB	17.620	320.802	0.055
Gap (cm)		IPB	2.159	461.788	0.005
Chord angle	92.240	Strength	120.000	194.854	0.616

Note: Changes will not be saved, use Report.

Recalculate | Report | Close

Detail Report

Figure 8.64 Joint can design.

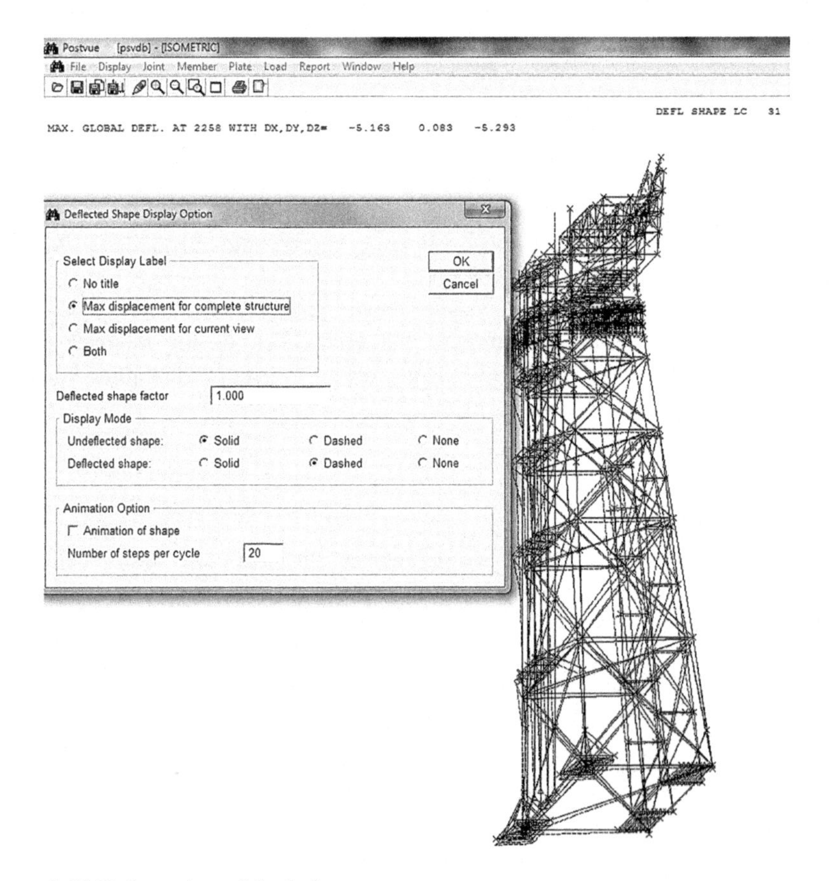

Figure 8.65 Deformation of the jacket.

where

T_p = platform natural period
T_n = wave period
ε = damping factor (2%)

Note that, if the values of T_p/T_n are less than 0.2, the DAF is taken as 1.0.

The DAF calculations consider the two longest natural periods for the first two cantilever modes of the platform under the respective loading conditions.

8.6.1 Eigenvalue analysis

To acquire the dynamic characteristics of the platform, a modal analysis is performed using the Dynpac module of the SACS. Normally, you extract the first 10 mode shapes to properly simulate the dynamic responses of the platform. Start the analysis start via Dynamic > Extract Mode Shapes, as shown in Figure 8.66.

The model described in the previous step is used in conjunction with the created superelement after applying the modifications or modeling aspects found hereafter. Masses of all modeled members were generated as consistent masses by Dynpac.

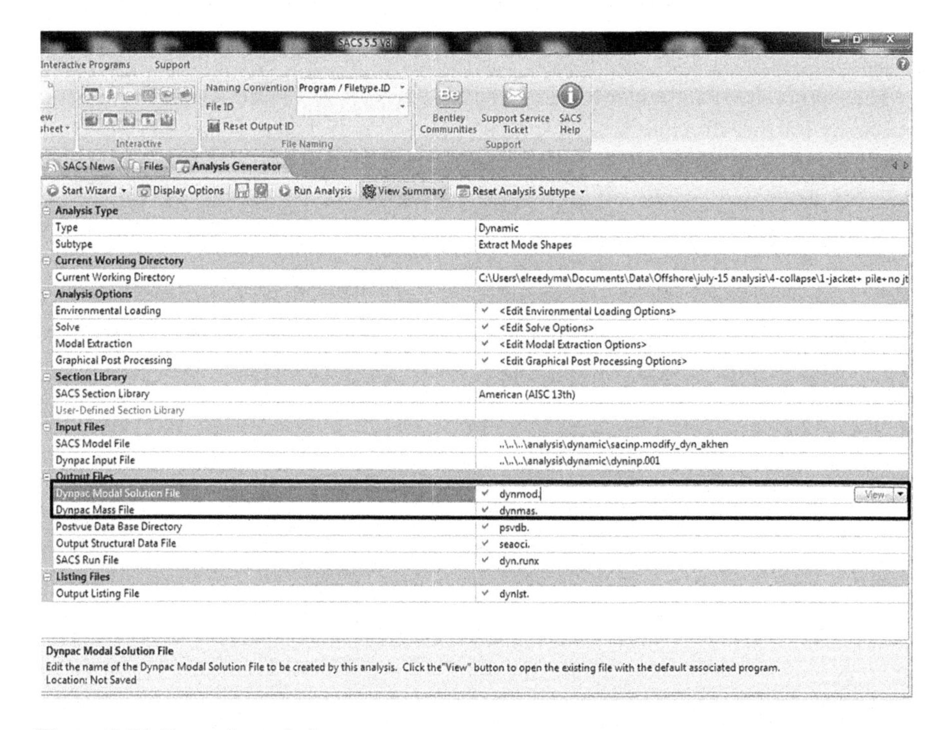

Figure 8.66 Dynamic analysis.

In most cases, a 5–7% increase is applied to the members mass density for contingency.

Masses of unmodeled loads are converted by Dynpac to lumped or distributed masses.

The mass of the marine growth is computed automatically by Dynpac. The full marine growth thickness is considered. The marine growth usually has density values from 1308 kg/m^3 to 1400 kg/m^3.

Water added a mass to members below the water surface. The mass of water is assumed to move in association with the platform as it vibrates. For tubular members, Dynpac automatically generates consistent added masses based on the mass of water displaced by the submerged member and the marine growth covering it, if any.

Proper overrides were introduced in the Dynpac input file to neglect any added mass to the wishbone and pile members. The Dynpac input file can be accessed through menus to put in the general data by data gen editing, as shown in Figure 8.67, or through the Edit Model Extraction Option in revision 5.3 and higher.

Entrapped mass is the mass of water enclosed by the flooded members. Such mass is automatically generated by Dynpac. It should be noted that the MSL water depth is used in the added mass and entrapped mass calculations.

The joint fixity of some joint that is connected to the big mass has 222,000, based on the software requirement. The location of the joint should be retained for dynamic analysis, as in Figure 8.68.

The output from the eigenvalue analysis is shown in Figure 8.69. The natural frequency for this structure is 2.78 s. You obtain it from dynlst. filename. There are other output files, which are dynmod.filename and dynmas.filename, as shown in Figure 8.66.

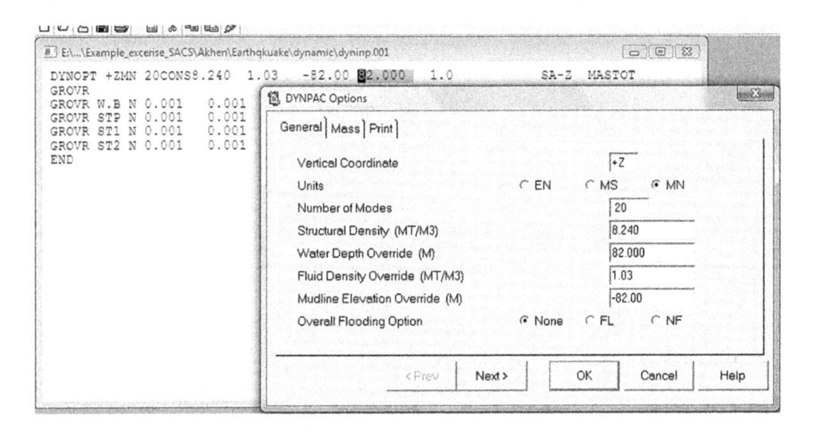

Figure 8.67 Dynpac input file.

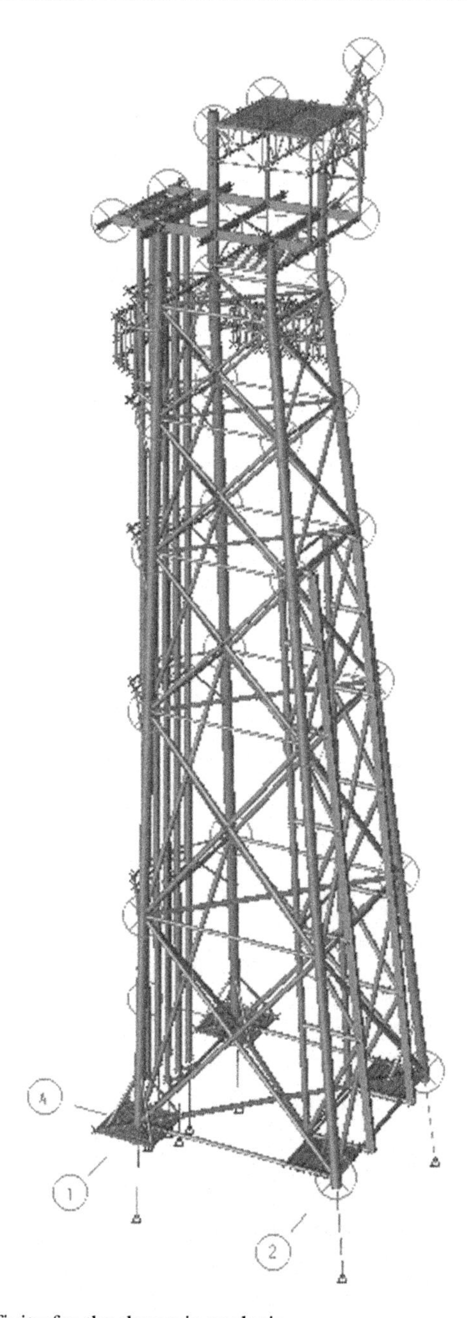

Figure 8.68 Special fixity for the dynamic analysis.

MODE	FREQ.(CPS)	GEN. MASS	EIGENVALUE	PERIOD(SECS)
1	0.359532	1.4398206E+03	1.9595940E-01	2.7813967
2	0.365019	1.4329760E+03	1.9011232E-01	2.7395865
3	0.947352	1.7228982E+02	2.8223946E-02	1.0555741
4	1.256868	7.9027864E+01	1.6034694E-02	0.7956283
5	1.299310	1.2357525E+02	1.5004265E-02	0.7696393
6	1.510276	5.9006293E+01	1.1105236E-02	0.6621308
7	1.639177	6.9291067E+01	9.4273318E-03	0.6100624
8	1.840840	2.2106880E+01	7.4749478E-03	0.5432303
9	2.359410	4.2520544E+01	4.5502290E-03	0.4238347
10	2.455534	4.9211527E+01	4.2009595E-03	0.4072435
11	2.593701	2.5382744E+02	3.7653068E-03	0.3855494
12	2.732897	7.8084096E+02	3.3915147E-03	0.3659121
13	2.786183	7.3910558E+01	3.2630296E-03	0.3589140
14	3.473479	1.4400715E+02	2.0994756E-03	0.2878958
15	3.921317	2.5125141E+01	1.6473137E-03	0.2550164

Figure 8.69 Output of eigenvalue analysis.

8.7 Seismic analysis

The seismic analysis is performed using a response spectrum method through the SACS computer package and comprises the following sub analyses:

- Generation of foundation superelement.
- Analysis under the gravity loads.
- Performance of an eigenvalue analysis to obtain the dynamic characteristics of the platform as described.
- Earthquake spectral response analysis.
- Combined analysis of gravity and earthquake loadings.

The treatment of the water-added mass of members below the water surface, overrides, and entrapped mass are as just discussed.

Dynamic spectral response analysis, in accordance with API RP2A, was adopted to obtain the internal forces in the structural elements due to the structure's response to the effective ground acceleration.

Seismic loading was generated using the API RP2A normalized response spectra for soil type C factored by the ALE effective ground acceleration of 0. g from seismic studies, considering a damping factor of 5%.

The dynamic response file, dyrinp.filename, is presented in Figure 8.70.

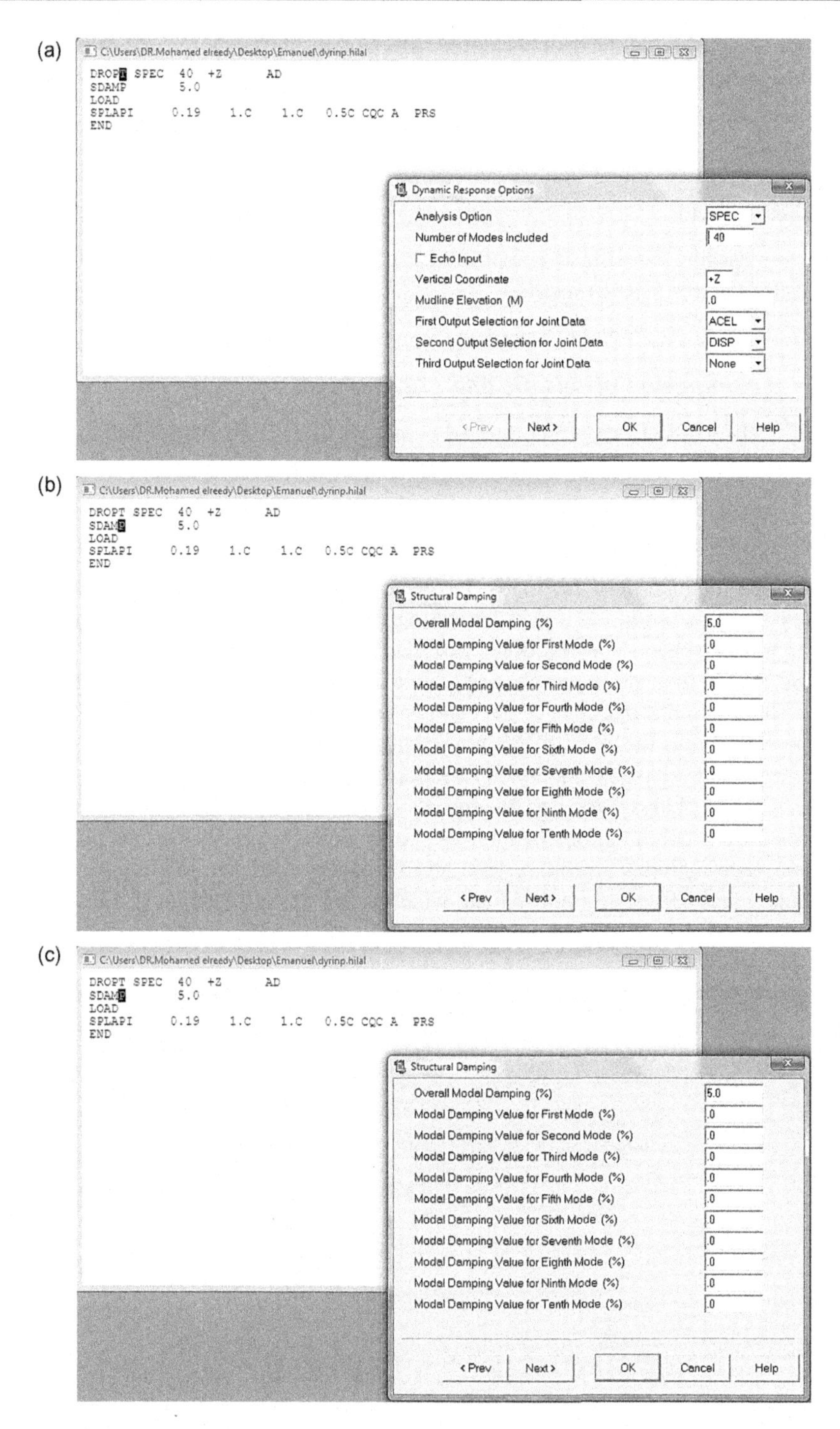

Figure 8.70 Seismic analysis input data: (a) first line; (b) second line, damping ratio; (c) spectral analysis input data.

Modal responses in each of the three orthogonal directions (X, Y, Z) are combined using the complete quadratic combination (CQC) method, using the following factors on the orthogonal components of the spectra:

- X-direction factor.
- Y-direction factor.
- Z-direction factor.

These resulting directional responses are combined using the square root of the sum of the squares (SRSS) method.

8.7.1 *Combination of seismic and gravity loads*

Gravity and seismic loads are combined using the ISO 19902 partial action factors, thus yielding the following two combinations:

- 1.1 Permanent gravity loads + 1.1 variable gravity loads + 0.9 seismic loads.
- 0.9 Permanent gravity loads + 0.8 variable gravity loads + 0.9 seismic loads.

Since earthquake-induced loads have no sign, each of these combinations requires the generation of two subcombinations, one assumes that seismic axial stresses are in tension and the other assumes that those axial stresses are in compression. Those load combinations are generated using SACS Combine module.

The file dynip has the lines of data shown in Figure 8.70. Figure 8.70(b) presents the line containing the overall damping ratio, which is 5%. The spectral analysis input and output data ares illustrated in Figure 8.71.

The seismic analysis by SACS requires a combined gravity and seismic load, performed by a combination module, as shown in Figure 8.72. The file, combinp. filename, data includes a primary load and secondary load; the load factors are 1.1 for primary and 0.9 for secondary load, which is the seismic load, as in Figure 8.73.

The combination of load by SRSS is as in Figure 8.74.

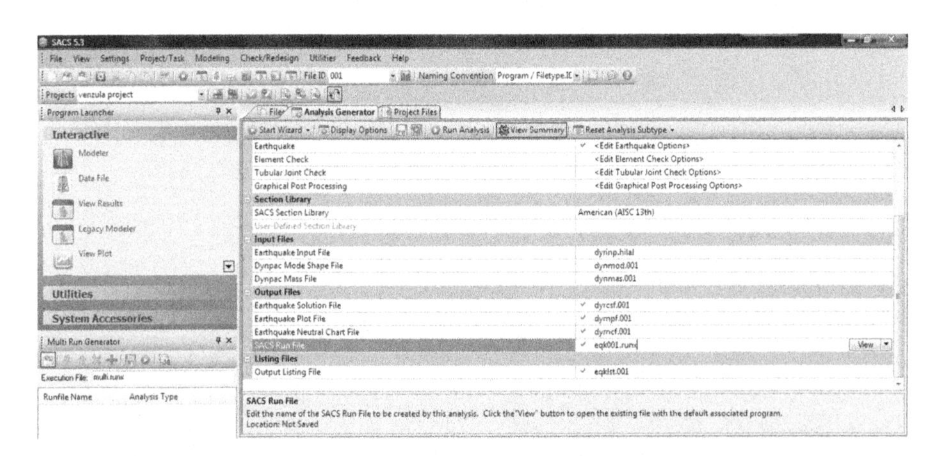

Figure 8.71 Input and output file for seismic analysis.

Start Wizard | Display Options | Run Analysis | View Summary | Reset Analysis Subtype

Analysis Type	
Type	Utilities
Subtype	Combine Solution File
Current Working Directory	
Current Working Directory	C:\Users\DR.Mohamed elreedy\Desktop\Emanuel
Section Library	
SACS Section Library	American (AISC 13th)
User-Defined Section Library	
Input Files	
Combine Input File	cmbinp.venzula
Primary SACS Common Solution File	saccsf.001
Secondary SACS Common Solution File	dyrcsf.001
Output Files	
SACS Combined Solution File	✓ cmbcsf.001
SACS Run File	✓ cmb001.runx
Listing Files	
Output Listing File	✓ cmblst.001

Figure 8.72 Combine input file.

C:\Users\DR.Mohamed elreedy\Desktop\Emanuel\cmbinp.venzula

```
CMBOPT
LCOND     1    PRSC     1.0
COMP P WAV          1.1
COMP P JNT          1.1
COMP S   1          0.9
LCOND     2    PRST     1.0
COMP P WAV          1.1
COMP P JNT          1.1
COMP S   1          0.9
LCOND     3    PRSC     1.0
COMP P WAV          0.9
COMP P JNT          0.8
COMP S   1          0.9
LCOND     4    PRST     1.0
COMP P WAV          0.9
COMP P JNT          0.8
COMP S   1          0.9
END
```

Figure 8.73 Combined loads input data.

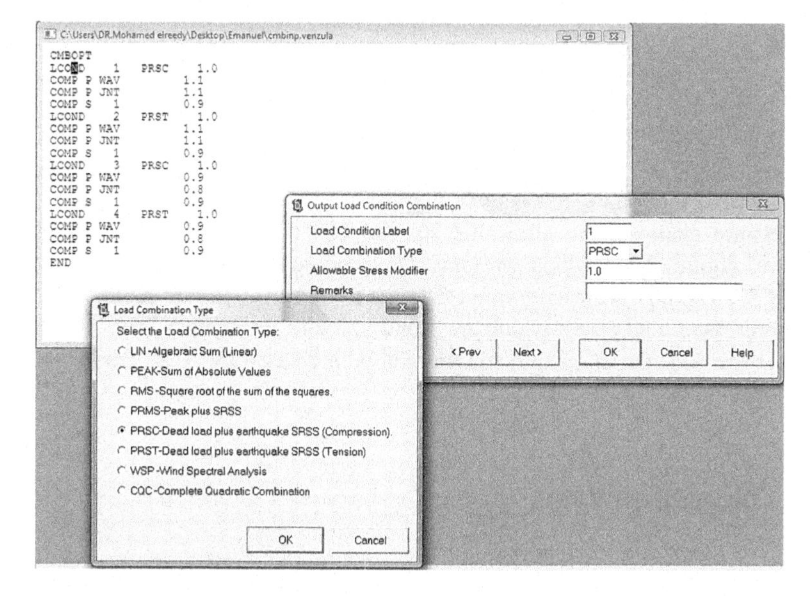

Figure 8.74 Load configuration.

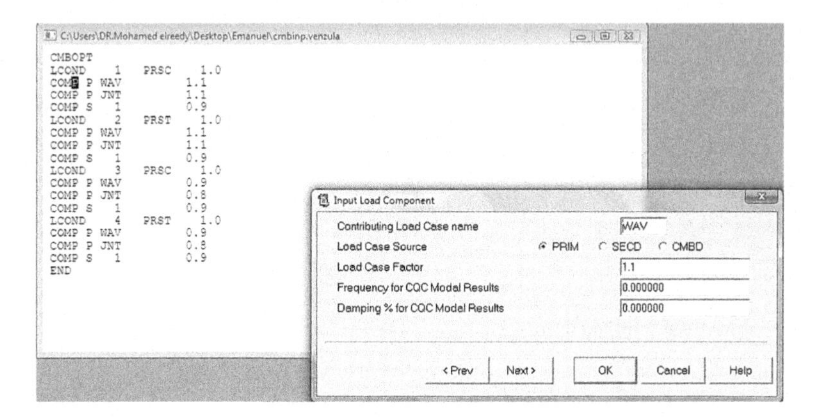

Figure 8.75 Load factor.

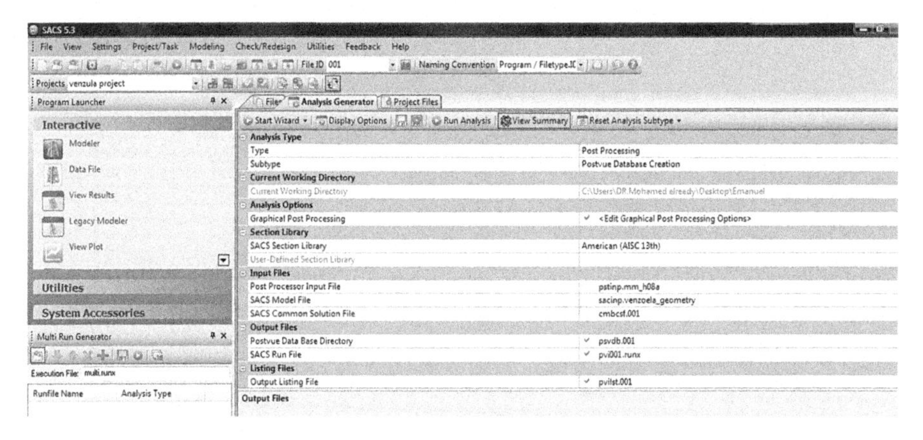

Figure 8.76 Postprocessing.

The input of the load factor is as in Figure 8.75.

In seismic analysis, the allowable stresses can be increased by 70%. This is shown in Figure 8.76, which presents the input data for postprocessing, as in Figure 8.77. The postprocessing input data are as shown in Figure 8.78.

The presented load cases, which are 1 and 2, are as in Figure 8.79.

As per the API, the allowable stresses can be increased by 70%. This is included in line AMOD, as shown in Figure 8.80.

8.8 Collapse analysis

Collapse, *pushover*, and *nonlinear analysis* are different terms with same theoretical approach. This analysis is performed by SACS under the structure type Statics and

C:\Users\DR.Mohamed elreedy\Desktop\Emanuel\pstinp.mm_h08a

```
GRAVITY LOADS + EARTQUAKE LOADING
OPTION        MN            UC   5 5  2DC  C     PTPTPTPTPT    PT      SK
LCSEL               1    2
AMOD
AMOD       1  1.70    2  1.70
END
```

Figure 8.77 Postprocessing input file.

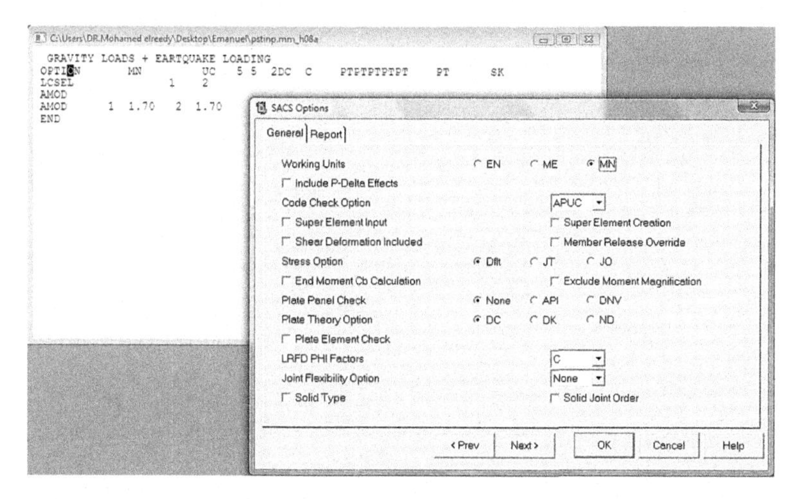

Figure 8.78 General option for postprocessing.

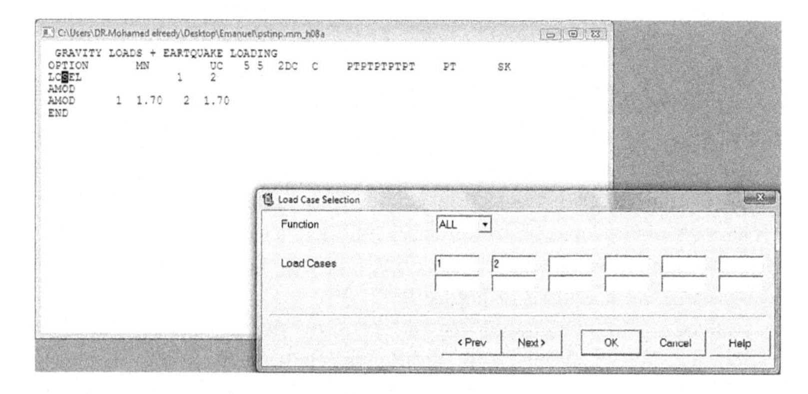

Figure 8.79 Load cases that should be considered.

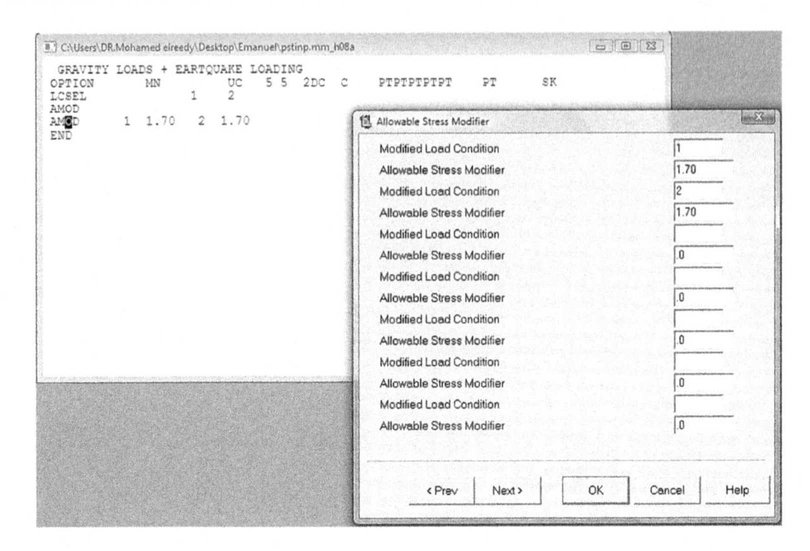

Figure 8.80 The allowable stresses in postprocessing.

Start Wizard | Display Options | Run Analysis | View Summary | Reset Analysis Subtype

Analysis Type	
Type	Statics
Subtype	Collapse Analysis (Includes Large Deflection, Plasticity, Joint Failure, etc.)
Current Working Directory	
Current Working Directory	C:\Users\DR.Mohamed elreedy\Desktop\Akhen
Analysis Options	
Environmental Loading	✓ <Edit Environmental Loading Options>
Solve	✓ <Edit Solve Options>
Foundation	<Edit Foundation Options>
Non-Linear / Plastic	✓ <Edit Non-Linear / Plastic Options>
Section Library	
SACS Section Library	American (AISC 13th)
User-Defined Section Library	
Input Files	
SACS Model File	conductor guard\sacinp.001
Collapse Input File	conductor guard\clpinp.001
Output Files	
Output Structural Data File	✓ seaoci.001

Figure 8.81 Collapse analysis.

subtype Collapse Analysis (includes large deflection, plasticity, joint failure, etc.), as in Figure 8.81.

This example studies a collapse analysis for the riser guard. The example was calculated manually in Chapter 5, here the analysis is by the software. The applied load on the riser guard is shown in Figure 8.82.

The input to the collapse data option is shown in Figure 8.83. The deflection and rotation tolerance and collapse deflection have traditional values, as shown in the figure.

The requirement of report shape is defined in the menu shown in Figure 8.84. The loading sequence is shown in Figure 8.85.

Some members are considered elastic, so they do not go through plasticity option, such as the wishbone, and should be mentioned in the list shown in Figure 8.86.

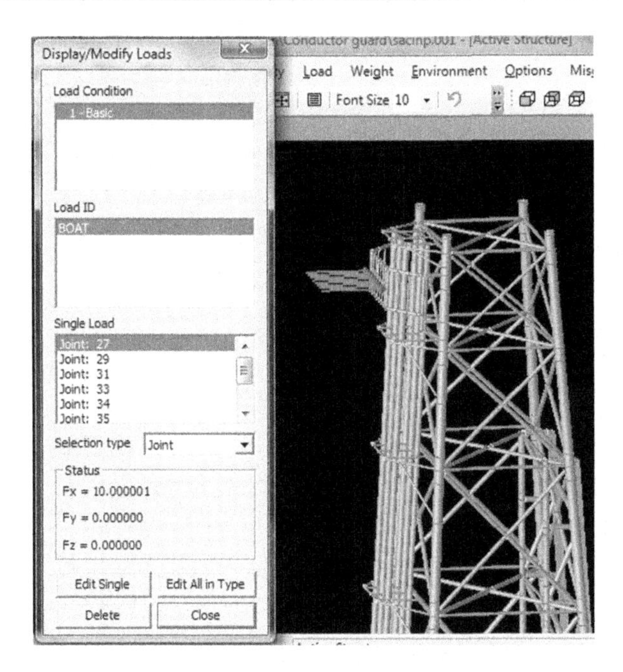

Figure 8.82 Applied loads.

COLLAPSE Options
Max.Iterations Per Load Incmnt 50
Number of Member Segments 8
Max. No. of Member Iterations 50
Treat skipped members as plastic
Continue if Max. Iterations Exceeded
Include Local Buckling Effects YES NO
Joint Flexibility Effects None JF BF
Include Pile Plasticity
Joint Strength Check Option None
All Members Elastic
Create model file w/ final defl. shape
Deflection Tolerance (CM) .01
Rotation Tolerance (RAD) 0.001
Member Deflection Tolerance (CM) .01
Collapse Deflection (CM) 100.0
Strain Hardening Ratio 0.05
< Prev Next > OK Cancel Help

Figure 8.83 Collapse option.

The numbers of the joints that will be affected by plasticity is illustrated in Figure 8.87.

After fulfilling all the requirements, the final collapse input file is as shown in Figure 8.88.

After you run the analysis, the output is illustrated as shown in Figure 8.89. The colors represent the member that fail for plasticity and another. Figure 8.89 presents the effect of a ship impact on the riser guard.

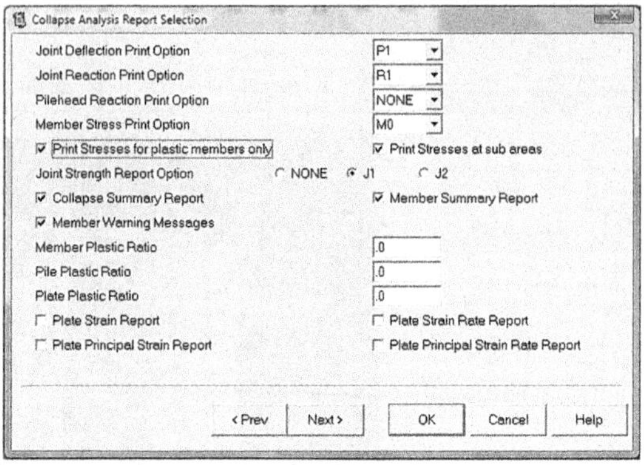

Figure 8.84 Report options.

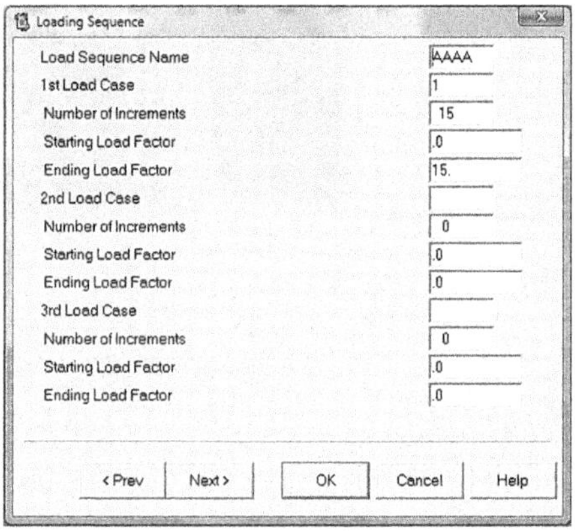

Figure 8.85 Loading sequence data.

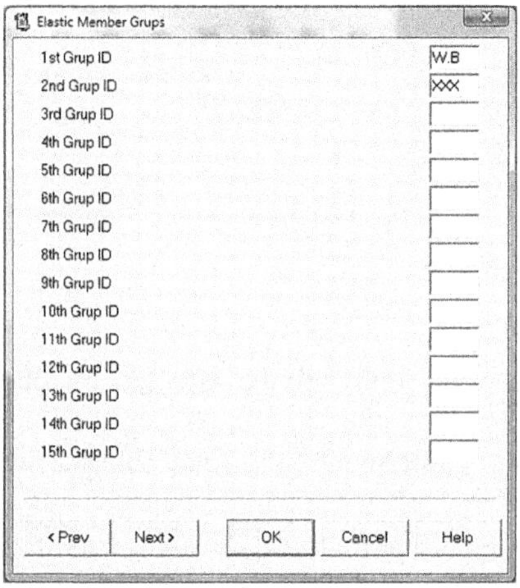

Figure 8.86 Elastic members.

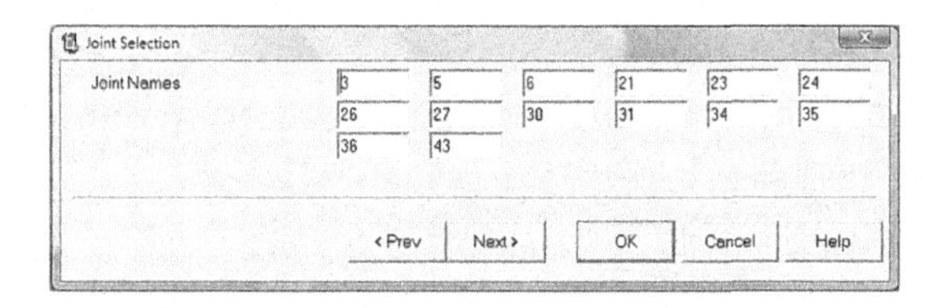

Figure 8.87 Joint selection.

```
CLPOPT      50  8 50          CN                          SF    .01  0.001.01  100.00.05
CLPRPT P1R1M0MPSPJ1SMMSPW
LDSEQ AAAA             1  15        15.
GRPELA          W.B XXX
JTSEL            3    5    6   21   23   24   26   27   30   31   34   35   36   43
JTSEL           44  104  106  112  612  613
END
```

Figure 8.88 Sample collapse input file.

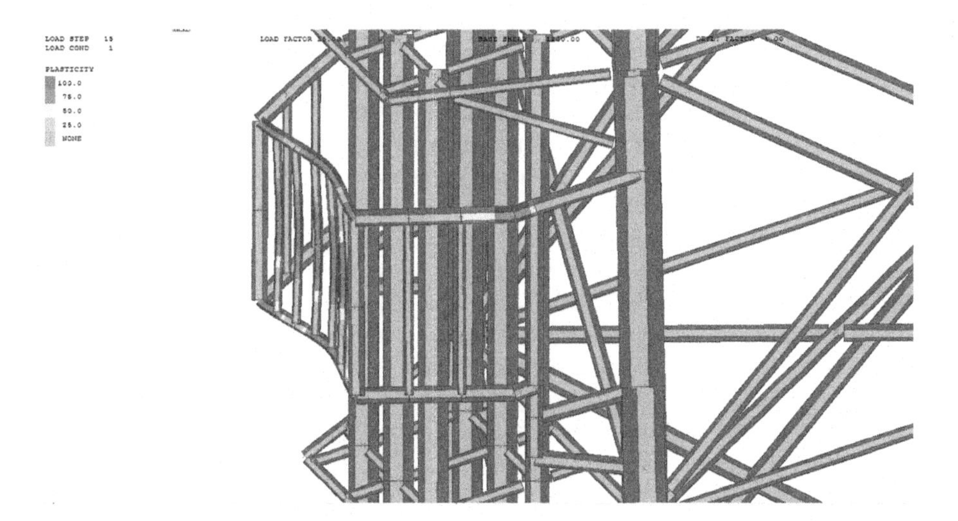

Figure 8.89 Output presentation for riser guard.

8.9 Loadout

After constructing the jacket and the topside in the construction yard, they must be transferred to the material barge then to the final location.

Figure 8.90 presents the pullout of jacket, moving it by skidding. The pulling force is applied on the slings. From this figure, you can see a four vertical supports underneath the joints. Four vertical supports underneath the joints of the jacket loadout, which are cups modeled to simulate the skid shoes. Gravity loads are completely carried by these supports under compression only, so they are defined as a gap elements.

You can add bracing members to increase the structure's strength during the load-out process. These temporary members are removed after loadout. To simulate the

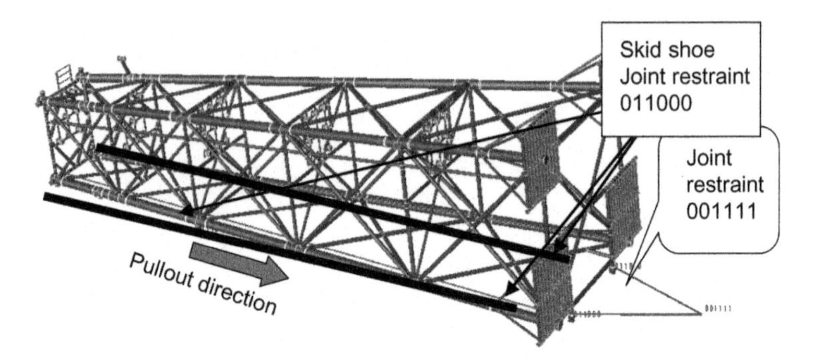

Figure 8.90 Pullout structure model.

movement of the jacket during skidding, the connection between these supported members and the skidding is released in the translation degree of freedom in the direction of the skidding except for one joint to allow calculating the stability requirement.

Two slings are modeled, connecting the front stubs to the hook point. Sling members are restrained in the three translational DOF at both ends and released in the three rotational DOF at one end; the other is restrained in torsion and released in the other two rotational DOF. To avoid numerical over flow the joint representing the hook was restrained in the vertical DOF.

For simulating the pulling load, one lateral force equals the total gravity load multiplied by the friction coefficient, which in most cases is 0.2 for the static case, applied in the direction of skidding; in addition, three forces in the opposite direction of the skidding are applied at the three skid shoes, which are released in the translational DOF to simulate the support reactions due to the pulling force.

The exact vertical reactions after applying the settlement are used in calculating the friction forces. To simulate the movement of any support, assume a vertical displacement settlement of about 25 mm at each shoe is applied at each skid shoe sequentially.

8.10 Sea fastening

After transferring the structure to the material barge, the structure analysis for the structure during transportation on the sea must be performed. The drawings in Figure 8.91 presents a simulation of the structure with additional members for sea fastening. It is traditional for this analysis to release the member ends, which are connected to the barge frames in the rotational degrees of freedom to simulate assuming a guested connection. The model uses the same pullout model but removes the slings.

As an example of the sea fastening of the jacket on the barge is presented in Figure 8.92 with the elevation and plan. Note that the location of the sea fastening and the barge layout are obtained from the installation company.

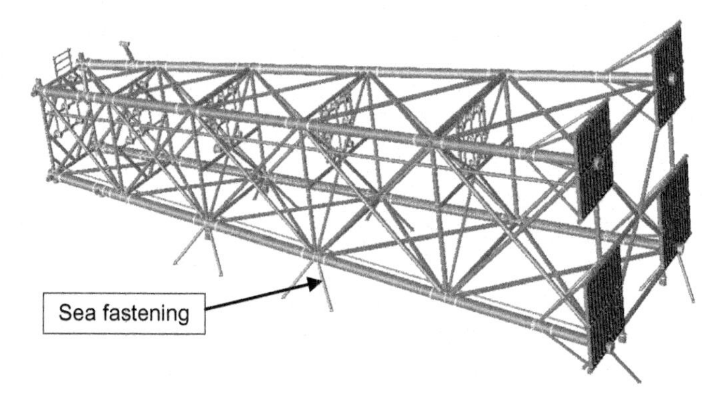

Figure 8.91 Sea fastening structure model.

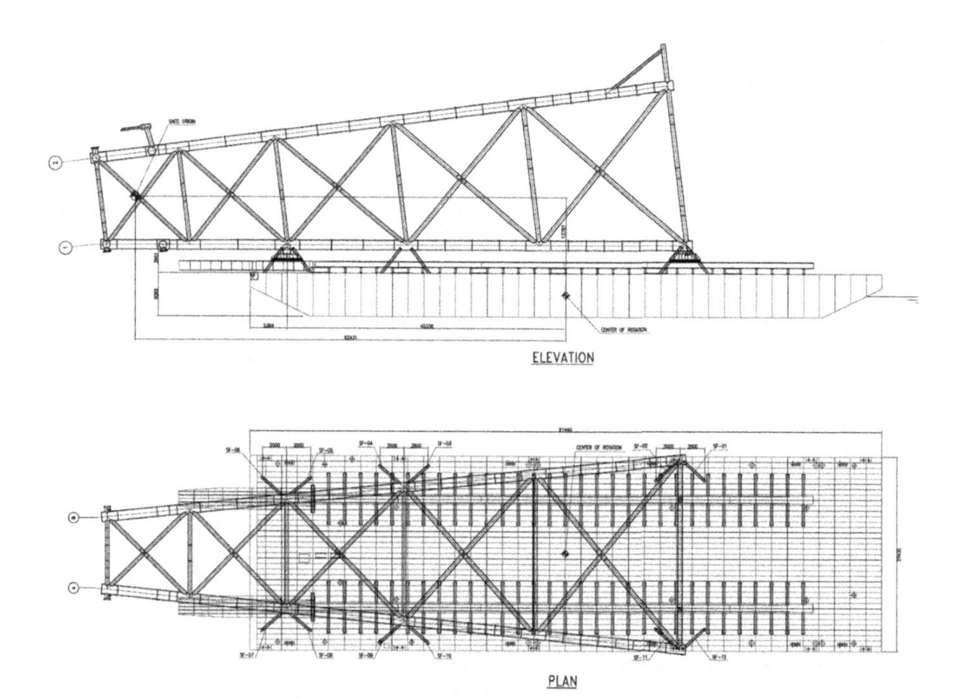

Figure 8.92 Structure model located on the barge.

The computer model is used, and the sea fastening members carry the weight from inertia loads due to rolling and pitching of the barge. In this analysis, consider the weight contingency. The material density will increase around 7%. SACS software through the Tow module considered the modeled and unmodeled element mass to generate the inertia load.

The structure location with respect to the material barge was taken into consideration. As shown in Figure 8.91, this location is defined based on the barge center of

Table 8.1 DNV relevant data

Motion	Amplitude	Period (s)
Roll	+20	10
Pitch	+12.5	10
Heave	Acceleration ± 0.2 g (ground acceleration)	

rotation, which is the barge's geometric center at sea level relative to the origin of the structure model. The orientation is modeled by aligning the barge's sway and pitch axis with the model's global Y-axis, the barge's surge and roll axis with the model's global X-axis, and the barge's heave axis with the model's global Z-axis.

As per the DNV, the data in Table 8.1 should be considered in design.

The effects of wind are assumed to be included in the preceding motion criteria. The center of motion is considered at the barge center of rotation at sea level. The motions in the table are converted by the Tow module into accelerations to generate inertia loads.

8.10.1 Load combinations

Barge motion-induced forces are combined as follows:

- +Roll + Heave
- +Roll − Heave
- −Roll + Heave
- −Roll − Heave
- +Pitch + Heave
- +Pitch − Heave
- −Pitch + Heave
- −Pitch − Heave

The effects of the wind are assumed to be included in these inertia motion criteria. The center of motion is considered to be the barge center of rotation at the sea level. These motions are converted by the Tow module into accelerations to generate the inertia load.

The combination between the gravity load and the inertia load is done by the Combine module in SACS for the following cases of loadings:

- + Gravity + Roll + Heave
- + Gravity + Roll − Heave
- + Gravity − Roll + Heave
- + Gravity − Roll − Heave
- + Gravity + Pitch + Heave
- + Gravity + Pitch − Heave
- + Gravity − Pitch + Heave
- + Gravity − Pitch − Heave

Afterward, choose Data file > Create a model.

The data generator for Tow module is chosen, as shown in Figure 8.93. Example of towing analysis for the first line input file menu as shown in Figure 8.94.

Start the analysis start by selecting Type: Loadings > Subtype: Sea State Loading (Wave, Wind, Current, Inertia, etc.), as in Figure 8.95.

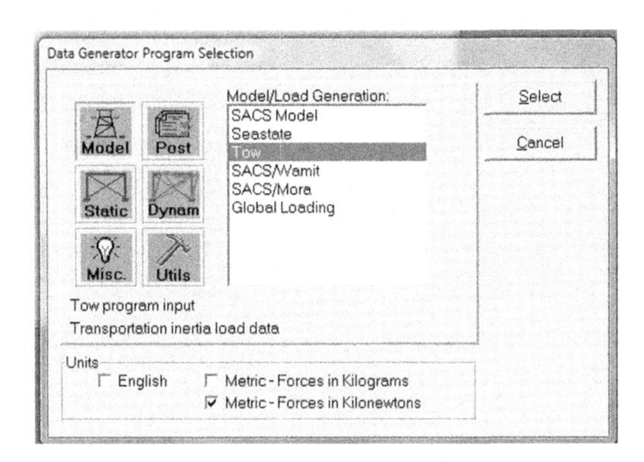

Figure 8.93 Select Tow module.

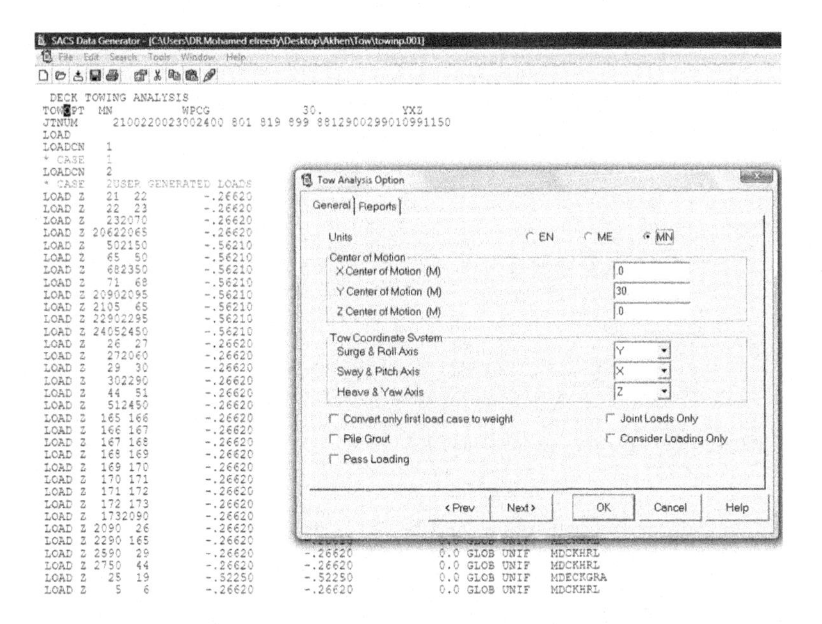

Figure 8.94 Tow module input file.

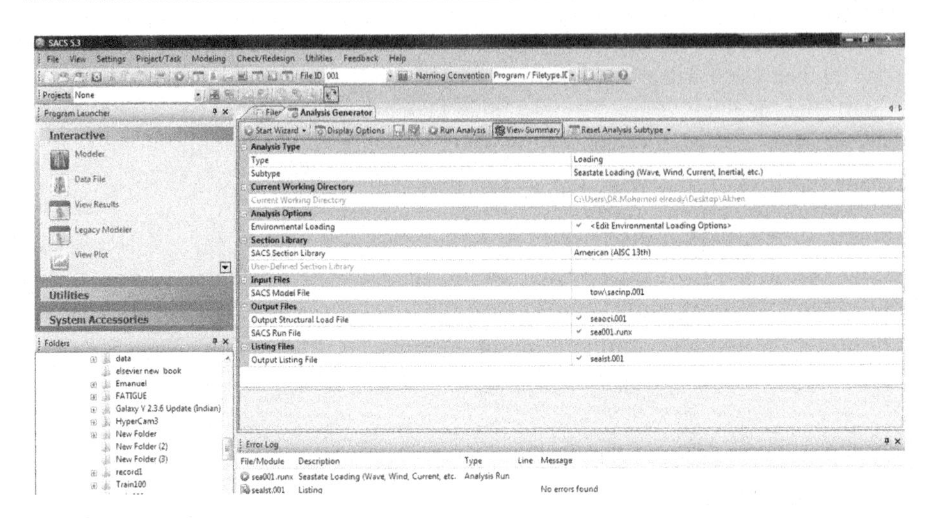

Figure 8.95 Start the Tow analysis.

Motion Data

Load Case Name	1
Roll Angle (DEG)	20.
Roll Period (SECS)	10.
Pitch Angle (DEG)	.0
Pitch Period (SECS)	.0
Yaw Angle (DEG)	.0
Yaw Period (SECS)	.0
Translational Acceleration	
Surge (G)	.0
Sway (G)	.0
Heave (G)	.2
Center of Motion	
X (M)	.0
Y (M)	.0
Z (M)	.0
Include Gravity Option	G

Exclude Structural Weight

< Prev | Next > | OK | Cancel | Help

Figure 8.96 Motion input data in load case 1.

The input data of the roll, pitch, and heave is as shown for eight load cases, as presented in Figures 8.96 through 8.103.

The transportation is done in three stages. For stage 1, the analysis under gravity load only, so the sea fastening member carries no load at this stage, as shown in Figure 8.104.

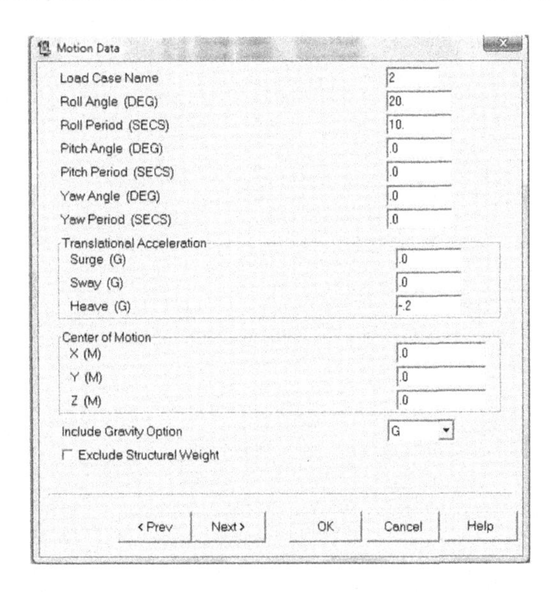

Figure 8.97 Motion input data in load case 2.

Figure 8.98 Motion input data in load case 3.

For stage 2, consider the forces due to transportation, and in this case, the sea fastening members carry the load. The last stage is to have a combination between gravity and transportation load.

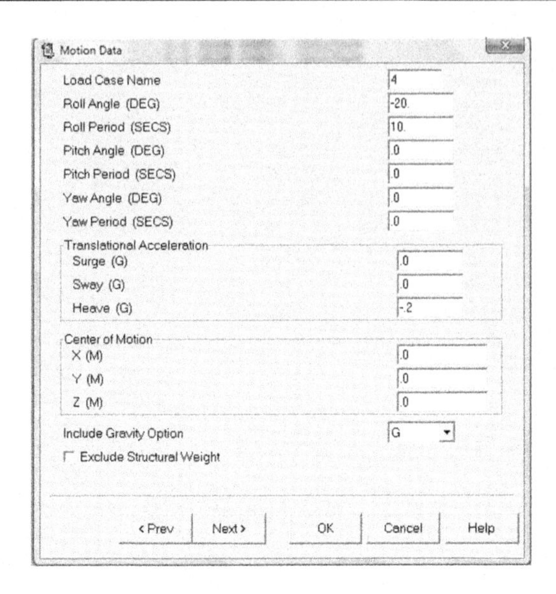

Figure 8.99 Motion input data in load case 4.

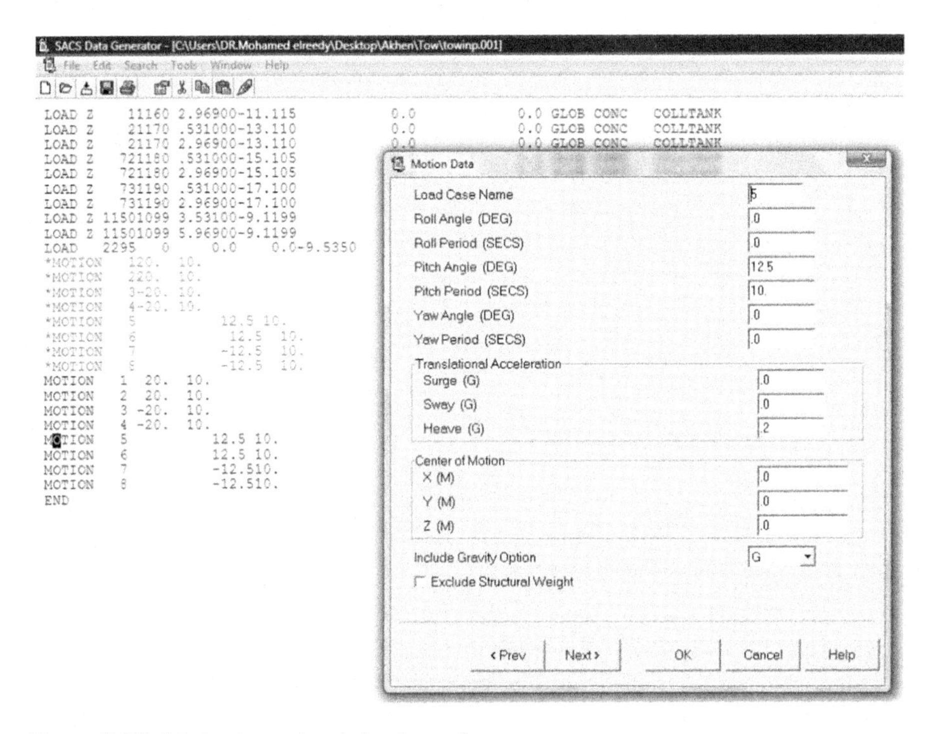

Figure 8.100 Motion input data in load case 5.

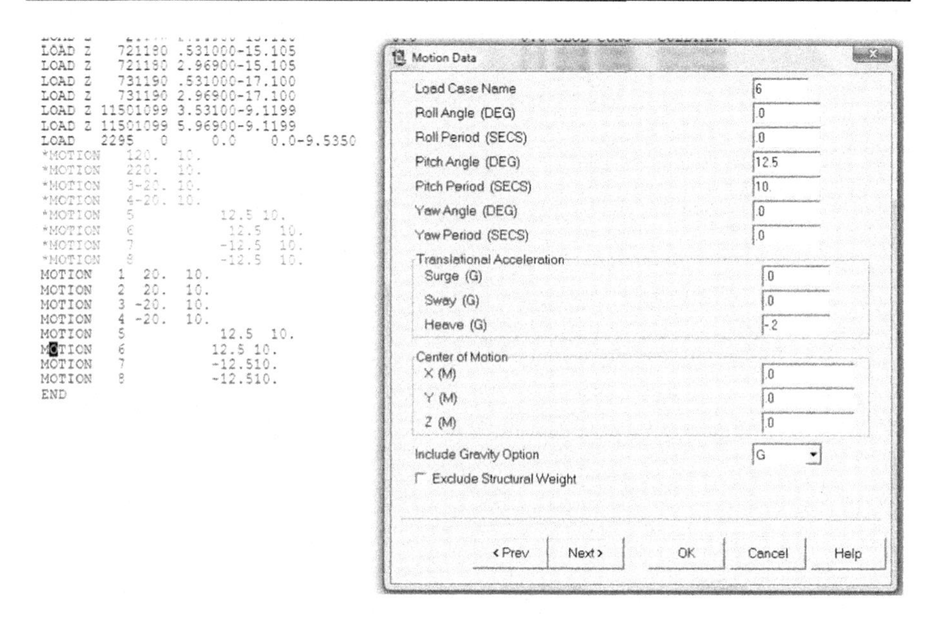

Figure 8.101 Motion input data in load case 6.

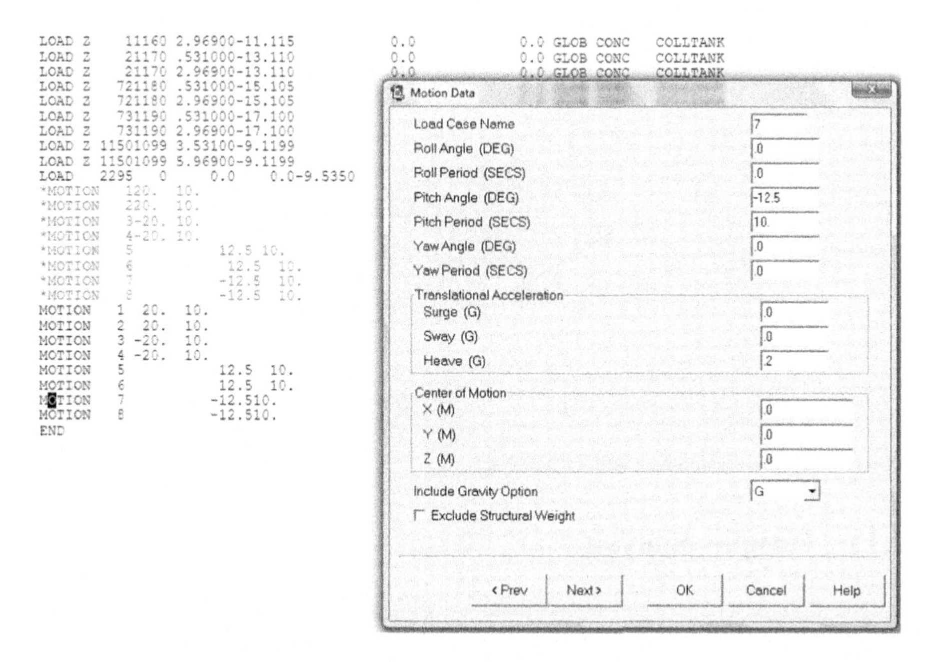

Figure 8.102 Motion input data in **load case 7**.

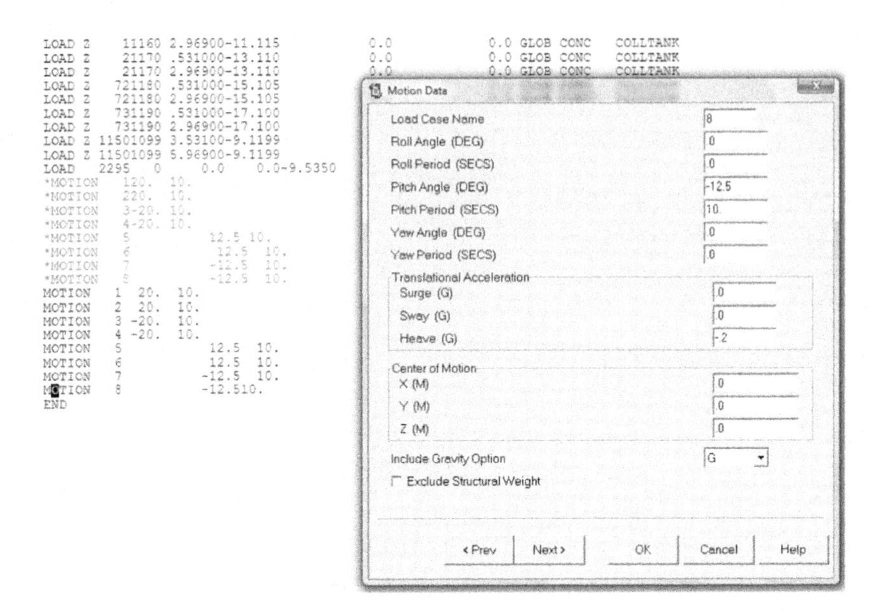

Figure 8.103 Motion input data in load case 8.

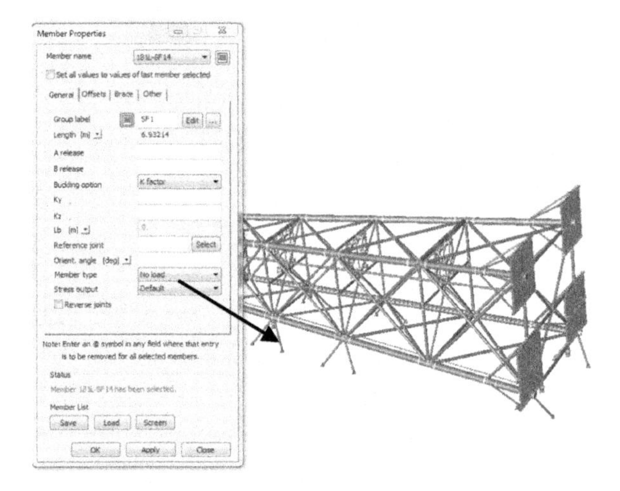

Figure 8.104 Sea fastening member properties.

8.11 Fatigue analysis

Fatigue failures occur when microcracks develop and grow until the material fractures. Such cracks are likely to occur at

- Flaws or inclusions in the material.
- Points of local nonhomogeneity.
- Points of abrupt change in the geometry of the structure.

Joint welds in welded frame structures may be sensitive to fatigue failure, because the welding process can result in microcracks or material nonhomogeneity and the local stress values can be far greater than calculated nominal values because of the relatively abrupt geometry changes.

Fatigue analysis is not easy and requires some skill for interpreting the data. This analysis is a combination between statistics and materials effect under cyclic loading with wave properties.

In most cases, we perform dynamic spectral fatigue analysis, which is a statistical approach for calculating fatigue damage in a structure. This analysis also considers the environmental data to determine the fatigue life of the jacket joints.

The spectral method assumes that there is a definable relation between wave height and stress ranges at the connections and that, at any point, the elevation of the sea above its mean value is a stationary Gaussian random process.

These assumptions are most applicable for low to moderate sea states. Since these are the sea states of interest in fatigue studies, these assumptions can reasonably be accepted.

The analysis uses a wave scatter diagram to directly represent the long-term statistics of sea state and uses wave spectra to represent the range of frequencies present in the random sea states. The effect of wave frequency on wave loading and dynamic structural response is accounted for through determination and use of dynamic transfer functions that define the relationship between stress range, wave height, and wave frequency.

Wave spectra and transfer functions are combined to develop the platform's dynamic response spectra. The expected number of occurrences at a certain stress range, and hence fatigue damage under a specific sea state, is then calculated, assuming a Rayleigh distribution of response peaks. The cumulated joint fatigue damage is equal to the summation of the calculated fatigue damages overall. The sea states are defined on the scatter diagram.

The analysis is performed using the SACS computer package through the following steps:

1. Determination of the center of damage wave parameters.
2. Generation of the foundation superelement.
3. Performance of an eigenvalue analysis to obtain the dynamic characteristics of the platform.
4. Running of the dynamic wave response analysis.
5. Performance of joint fatigue damage analysis.

8.11.1 Center of damage

The SACS center of damage calculation utility determines

Center of damage wave height $(H_{cs}) = \sum(D_i \cdot H_{si})/\sum(D_i)$

Center of damage wave height $(T_{cz}) = \sum(D_i T_{zi})/\sum(D_i)$

where

D_i = estimate damage of sea state $i = P_i H_{si}\ am/T_{zi}$

P_i = probability of occurrence of sea state i

H_{is} = significant wave height of sea state i
T_{zi} = zero crossing period of sea state i
a = taken as 1.8 for foundation linearization
m = taken as the inverse slope of S-N curve

These equation apply by using SACS software from the left menu Utilities, click on Center of damage.

The menu in Figure 8.105 is used to choose the units. Then select the fatigue input file that is to be used based on the scatter diagram data, which we obtain from metocean data. The fatigue input file is presented in sections. The example of scatter diagram is shown in Tables 8.2 through 8.9.

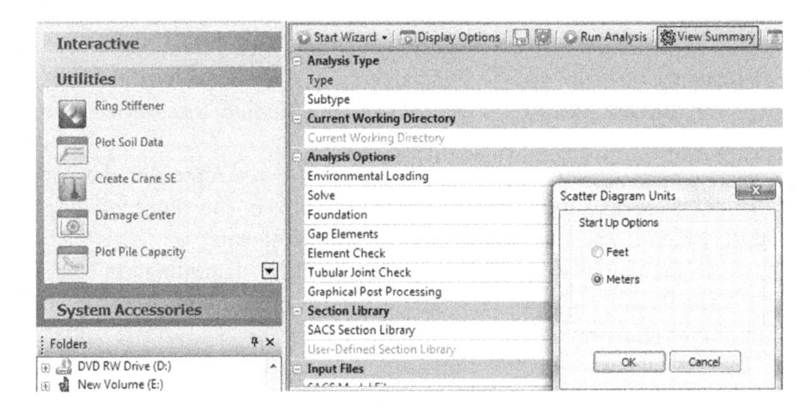

Figure 8.105 Start center of damage.

Table 8.2 Example of scatter diagram in 0° direction

Dominant period, s	Significant wave height, m		
	0.0–0.6	**0.6–1.4**	**1.4–2.6**
>1.5	0.15	0.10	0.09
1.5–2.5	0.10	0.19	0.10
2.5–3.5	0.06	0.08	0.05
>3.5	0.02	0.02	0.02

Table 8.3 Example of scatter diagram in 45° direction

Dominant period, s	Significant wave height, m		
	0.0–0.6	**0.6–1.4**	**1.4–2.6**
>1.5	0.11	0.13	0.08
1.5–2.5	0.14	0.13	0.10
2.5–3.5	0.08	0.09	0.07
>3.5	0.01	0.02	0.03

Table 8.4 **Example of scatter diagram in 90° direction**

Dominant period, s	Significant wave height, m		
	0.0–0.6	**0.6–1.4**	**1.4–2.6**
>1.5	0.10	0.12	0.08
1.5–2.5	0.11	0.15	0.11
2.5–3.5	0.06	0.09	0.08
>3.5	0.03	0.03	0.03

Table 8.5 **Example of scatter diagram in 135° direction**

Dominant period, se	Significant wave height, m		
	0.0–0.6	**0.6–1.4**	**1.4–2.6**
>1.5	0.13	0.10	0.08
1.5–2.5	0.13	0.15	0.10
2.5–3.5	0.06	0.09	0.08
>3.5	0.03	0.03	0.02

Table 8.6 **Example of scatter diagram in 180° direction**

Dominant period, s	Significant wave height, m		
	0.0–0.6	**0.6–1.4**	**1.4–2.6**
>1.5	0.11	0.10	0.08
1.5–2.5	0.12	0.12	0.09
2.5–3.5	0.07	0.12	0.09
>3.5	0.07	0.03	0.02

Table 8.7 **Example of scatter diagram in 225° direction**

Dominant period, s	Significant wave height, m		
	0.0–0.6	**0.6–1.4**	**1.4–2.6**
>1.5	0.10	0.10	0.10
1.5–2.5	0.10	0.15	0.10
2.5–3.5	0.06	0.10	0.9
>3.5	0.03	0.03	0.02

Table 8.8 **Example of scatter diagram in 270° direction**

Dominant period, s	Significant wave height, m		
	0.0–0.6	**0.6–1.4**	**1.4–2.6**
>1.5	0.11	0.10	0.08
1.5–2.5	0.11	0.15	0.10
2.5–3.5	0.08	0.07	0.08
>3.5	0.05	0.05	0.02

Table 8.9 **Example of scatter diagram in 315° direction**

Dominant period, s	Significant wave height, m		
	0.0–0.6	**0.6–1.4**	**1.4–2.6**
>1.5	0.03	0.20	0.05
1.5–2.5	0.13	0.13	0.12
2.5–3.5	0.06	0.07	0.09
>3.5	0.03	0.03	0.02

The output data for the center of damage analysis is presented in Figures 8.106 and 8.107.

8.11.2 *Generation of the foundation superelement*

As discussed previously, the nonlinear soil structure interaction with the pile foundation is calculated by a numerous iterations to form the pile stiffness matrix when working out mode shapes. SACS cannot compute this nonlinear behavior.

To overcome this the nonlinear soil-pile interaction, the analysis is linearized by creating a foundation superelement that includes a stiffness matrix at each pile head. The stiffness matrix simulates the pile-soil translational and rotational stiffness corresponding to a certain level of loading.

The foundation superelement file is produced from the static analysis of the in-place model after the drag and inertia coefficients for tubular members were taken as follows, based on API RP 2A requirements, for fatigue analysis:

$C_d = 0.50$ $C_m = 2.0$ (smooth surface)
$C_d = 0.80$ $C_m = 2.0$ (rough surface)

Drag and inertia coefficients are magnified by 5% to account for the unmodeled anodes. Member and group overrides that were entered into the in-place model to account for the environmental loads acting on the unmodeled items have been modified to account for the changes made to the drag and inertia coefficients. The modified C_d and C_m overrides used. MSL water depth is considered. Environmental loading considers wave loads only.

```
Center of Damage Utility Version 6.1.5.1

Center of Damage based on S-N Slope =  3.740 (AXP)

Scatter Diagram Wave Height Units   = METERS


Fatigue Case    1

     Foundation Linearization
      Tcz  =     1.805         Hcs  =     5.964
      Teff =     1.516         Heff =     5.308

     Most Probable Maximum Wave
      Tcz  =     1.828         Hcs  =     5.560
      Tref =     2.257         Href =    10.341


Fatigue Case    2

     Foundation Linearization
      Tcz  =     1.997         Hcs  =     5.960
      Teff =     1.678         Heff =     5.305

     Most Probable Maximum Wave
      Tcz  =     1.971         Hcs  =     5.524
      Tref =     2.433         Href =    10.275


Fatigue Case    3

     Foundation Linearization
      Tcz  =     1.986         Hcs  =     5.963
      Teff =     1.668         Heff =     5.307

     Most Probable Maximum Wave
      Tcz  =     1.985         Hcs  =     5.551
      Tref =     2.451         Href =    10.325
```

Figure 8.106 Center of damage output data.

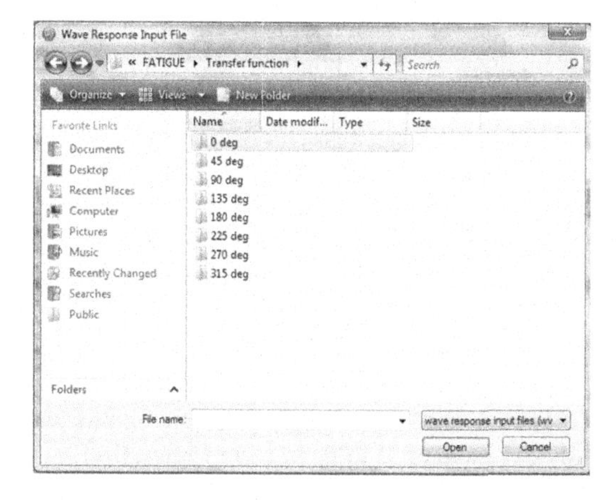

Figure 8.107 Number of analysis directions.

The environmental loads are replaced by effective waves derived from the center of damage waves using the following relations:

- Effective wave height (H_{eff}) = 0.89 H_{cs}.
- Associated wave period (T_{eff}) = 0.84 T_{cz}.
- Wave kinematics factor is taken to = 1.0.

The superelement is generated based on averaging the pile displacements resulting from combining the gravity loads with center of damage waves. X-direction stiffness is derived from averaging the four combinations that produce maximum X-direction base shear. Y-direction stiffness is derived based on the four combinations that produce maximum Y-direction base shear.

8.11.3 Dynamic wave response analysis

Dynamic wave response analysis is performed to obtain response for member stress range for a number of regular waves having different frequencies (stress range transfer functions).

The relationship between the stress ranges of unit amplitude and the corresponding wave frequency for all waves considered is the transfer function (see Figure 8.107). *For sinusoidal waves, wave height equals twice the wave amplitude.*

To generate a transfer function for a certain wave direction (fatigue load case), a number of regular waves of various heights with constant steepness are passed through the structure. The transfer function correlates the joint stress range to wave height ratio to the wave frequency.

So it can be obtain by generating a 100 waves as presented in Table 8.10 and define the location of the natural frequency with the 100 waves.

The analysis used the Seastate module to generate the static wave loads. The Wave Response module was then used to multiply such loads by the modal factors (previously obtained from the eigenvalue analysis) to obtain generalized forces. A damping of 2% of the critical damping was applied through all dynamic response analyses.

The model previously used to create the foundation superelement file was used to generate eight computer models representing the eight directions considered in fatigue loading. Since fatigue damage is due to cyclic stresses only, no gravity, buoyancy, current, or wind loads should be accounted for. Bearing that in mind, all the loads found in the superelement creation model were replaced by 100 regular Airy wave loads in each of the eight models to develop the dynamic global analysis.

To obtain the wave response analysis, Dynamic > Wave transfer/Response transfer function, as in Figures 8.108 through 8.111. The input of the damping ratio of 2% is presented in Figure 8.112. The transfer function should be obtained in eight directions, and you should have an 8 folders, as in Figure 8.113.

Figure 8.114 presents the input of different wave heighs and corresponding wave periods, as in case 75, with wave height = 0.47 m and wave period = 2.47 s.

Figure 8.115 shows an example transfer function with a frequency in the 45° direction.

Table 8.10 Variable wave steepness waves

Load case	Frequency, Hz	Wave height, H, m	Wave period, T_{app}, s	1/steepness	Relative depth, $d/(g \times T_{app}^2)$	Wave steepness, $H/(g \times T_{app}^2)$	Wave theory
1	0.082	14.10	12.20	16.48	0.056	0.010	Airy
2	0.086	12.75	11.67	16.68	0.061	0.010	Airy
3	0.090	11.40	11.10	16.87	0.068	0.009	Airy
4	0.095	10.05	10.48	17.06	0.076	0.009	Airy
5	0.102	8.70	9.80	17.24	0.087	0.009	Airy
6	0.119	5.90	8.39	18.63	0.119	0.009	Airy
7	0.130	4.82	7.68	19.09	0.142	0.008	Airy
8	0.140	4.10	7.15	19.47	0.164	0.008	Airy
9	0.150	3.50	6.67	19.85	0.188	0.008	Airy
Constant Wave Steepness Waves*							
10	0.160	3.05	6.25	20	0.214	0.008	Airy
11	0.164	2.91	6.11	20	0.224	0.008	Airy
12	0.168	2.78	5.97	20	0.235	0.008	Airy
13	0.171	2.66	5.84	20	0.245	0.008	Airy
14	0.175	2.55	5.71	20	0.256	0.008	Airy
15	0.179	2.44	5.59	20	0.267	0.008	Airy
16	0.183	2.34	5.47	20	0.279	0.008	Airy
17	0.186	2.25	5.36	20	0.291	0.008	Airy
18	0.190	2.16	5.26	20	0.302	0.008	Airy
19	0.194	2.07	5.15	20	0.315	0.008	Airy
20	0.198	2.00	5.06	20	0.327	0.008	Airy
21	0.202	1.92	4.96	20	0.340	0.008	Airy
22	0.205	1.85	4.87	20	0.352	0.008	Airy
23	0.209	1.79	4.78	20	0.366	0.008	Airy

(Continued)

Table 8.10 (Continued)

Load case	Frequency, Hz	Wave height, H, m	Wave period, T_{app}, s	1/steepness	Relative depth, $d/(g \times T_{app}^2)$	Wave steepness, $H/(g \times T_{app}^2)$	Wave theory
24	0.213	1.72	4.70	20	0.379	0.008	Airy
25	0.217	1.66	4.62	20	0.392	0.008	Airy
26	0.220	1.61	4.54	20	0.406	0.008	Airy
27	0.224	1.55	4.46	20	0.420	0.008	Airy
28	0.228	1.50	4.39	20	0.435	0.008	Airy
29	0.232	1.45	4.31	20	0.449	0.008	Airy
30	0.236	1.41	4.25	20	0.464	0.008	Airy
31	0.239	1.36	4.18	20	0.479	0.008	Airy
32	0.243	1.32	4.11	20	0.494	0.008	Airy
33	0.247	1.28	4.05	20	0.510	0.008	Airy
34	0.251	1.24	3.99	20	0.525	0.008	Airy
35	0.254	1.21	3.93	20	0.541	0.008	Airy
36	0.258	1.17	3.87	20	0.557	0.008	Airy
37	0.262	1.14	3.82	20	0.574	0.008	Airy
38	0.266	1.11	3.76	20	0.590	0.008	Airy
39	0.270	1.07	3.71	20	0.607	0.008	Airy
40	0.273	1.04	3.66	20	0.624	0.008	Airy
41	0.277	1.02	3.61	20	0.642	0.008	Airy
42	0.281	0.99	3.56	20	0.659	0.008	Airy
43	0.285	0.96	3.51	20	0.677	0.008	Airy
44	0.288	0.94	3.47	20	0.695	0.008	Airy
45	0.292	0.91	3.42	20	0.714	0.008	Airy
46	0.296	0.89	3.38	20	0.732	0.008	Airy
47	0.300	0.87	3.34	20	0.751	0.008	Airy
48	0.304	0.85	3.29	20	0.770	0.008	Airy

49	0.307	0.83	3.25	20	0.790	0.008	Airy
50	0.311	0.81	3.21	20	0.809	0.008	Airy
51	0.315	0.79	3.18	20	0.829	0.008	Airy
52	0.319	0.77	3.14	20	0.849	0.008	Airy
53	0.322	0.75	3.10	20	0.869	0.008	Airy
54	0.326	0.73	3.07	20	0.890	0.008	Airy
55	0.330	0.72	3.03	20	0.910	0.008	Airy
56	0.334	0.70	3.00	20	0.931	0.008	Airy
57	0.338	0.69	2.96	20	0.952	0.008	Airy
58	0.341	0.67	2.93	20	0.974	0.008	Airy
59	0.345	0.66	2.90	20	0.996	0.008	Airy
60**	0.349	0.64	2.87**	20	1.017	0.008	Airy
61	0.353	0.63	2.84	20	1.040	0.008	Airy
62	0.356	0.61	2.81	20	1.062	0.008	Airy
63	0.360	0.60	2.78	20	1.085	0.008	Airy
64	0.364	0.59	2.75	20	1.108	0.008	Airy
65	0.368	0.58	2.72	20	1.131	0.008	Airy
66	0.372	0.57	2.69	20	1.154	0.008	Airy
67	0.375	0.55	2.66	20	1.178	0.008	Airy
68	0.379	0.54	2.64	20	1.201	0.008	Airy
69**	0.383	0.53	2.61**	20	1.225	0.008	Airy
70	0.387	0.52	2.59	20	1.250	0.008	Airy
71	0.390	0.51	2.56	20	1.274	0.008	Airy
72	0.394	0.50	2.54	20	1.299	0.008	Airy
73	0.398	0.49	2.51	20	1.324	0.008	Airy
74	0.402	0.48	2.49	20	1.349	0.008	Airy
75	0.406	0.47	2.47	20	1.375	0.008	Airy
76	0.409	0.47	2.44	20	1.401	0.008	Airy
77	0.413	0.46	2.42	20	1.427	0.008	Airy
78	0.417	0.45	2.40	20	1.453	0.008	Airy

(Continued)

Table 8.10 (Continued)

Load case	Frequency, Hz	Wave height, H, m	Wave period, T_{app}, s	1/steepness	Relative depth, $d/(g \times T_{app}^2)$	Wave steepness, $H/(g \times T_{app}^2)$	Wave theory
79	0.421	0.44	2.38	20	1.479	0.008	Airy
80	0.424	0.43	2.36	20	1.506	0.008	Airy
81	0.428	0.43	2.34	20	1.533	0.008	Airy
82	0.432	0.42	2.31	20	1.560	0.008	Airy
83	0.436	0.41	2.29	20	1.587	0.008	Airy
84	0.440	0.40	2.28	20	1.615	0.008	Airy
85	0.443	0.40	2.26	20	1.643	0.008	Airy
86	0.447	0.39	2.24	20	1.671	0.008	Airy
87	0.451	0.38	2.22	20	1.699	0.008	Airy
88	0.455	0.38	2.20	20	1.728	0.008	Airy
89	0.458	0.37	2.18	20	1.757	0.008	Airy
90	0.462	0.37	2.16	20	1.786	0.008	Airy
91	0.466	0.36	2.15	20	1.815	0.008	Airy
92	0.470	0.35	2.13	20	1.845	0.008	Airy
93	0.474	0.35	2.11	20	1.875	0.008	Airy
94	0.477	0.34	2.09	20	1.905	0.008	Airy
95	0.481	0.34	2.08	20	1.935	0.008	Airy
96	0.485	0.33	2.06	20	1.965	0.008	Airy
97	0.489	0.33	2.05	20	1.996	0.008	Airy
98	0.492	0.32	2.03	20	2.027	0.008	Airy
99	0.496	0.32	2.02	20	2.058	0.008	Airy
100	0.500	0.31	2.00	20	2.090	0.008	Airy

*Minimum frequency of Airy waves = 0.160 Hz.
**The natural frequency period is 2.87 for first mode, and second mode of frequency is 2.61.
Maximum frequency of Airy waves = 0.500 Hz.
Number of Airy waves that follow = 91.
Wave frequency increment = 0.00378 Hz.
1/steepness of Airy waves = 20.

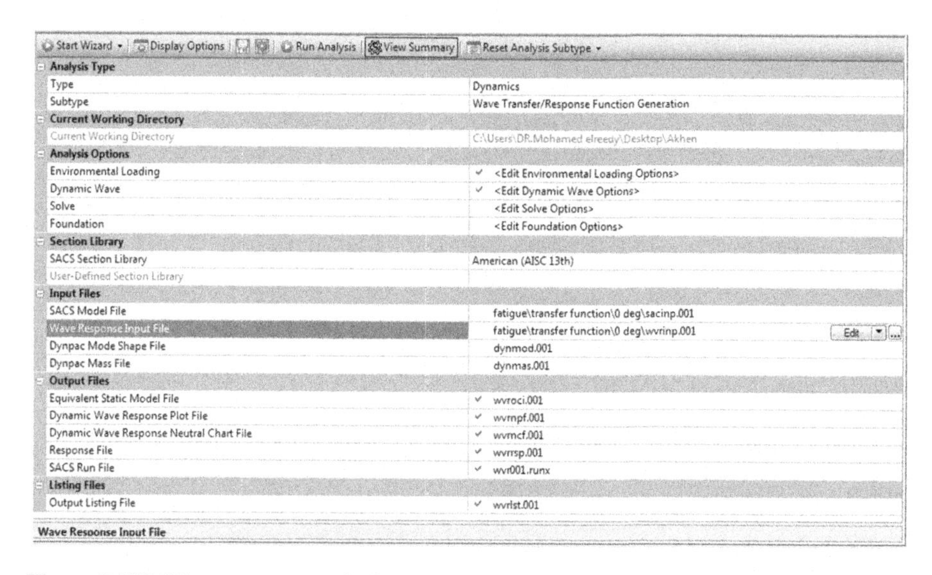

Figure 8.108 Wave response analysis.

Wave Response Options
Units selection MN
Load Case Selection ALL
Generate Plots
Iteration error tolerance(%) 0.0000
Fourier series error tolerance 0.0000
Number modes 0
Maximum allowable iterations -1
Equivalent Static Option ES
Generate Model Static Loads
Include Buoyancy Change Effects
Exclude Constant Term
Response Analysis Type DETR
Retrieve Loads
Output Option Dflt +MR Min
Elevation Cut-Off Type User Auto
Cut-Off Elevation (M) 0
LRFD Inertia Load Factor 0
<Prev Next> OK Cancel Help

Figure 8.109 Wave response data.

The last step is to run the fatigue analysis, as shown in Figure 8.116.

For each direction, you should insert the common solution file and Seastate input file with the fatigue input file.

A joint check for fatigue, fatigue damage, and joint fatigue life is evaluated using the Fatigue Damage module according to the following criteria.

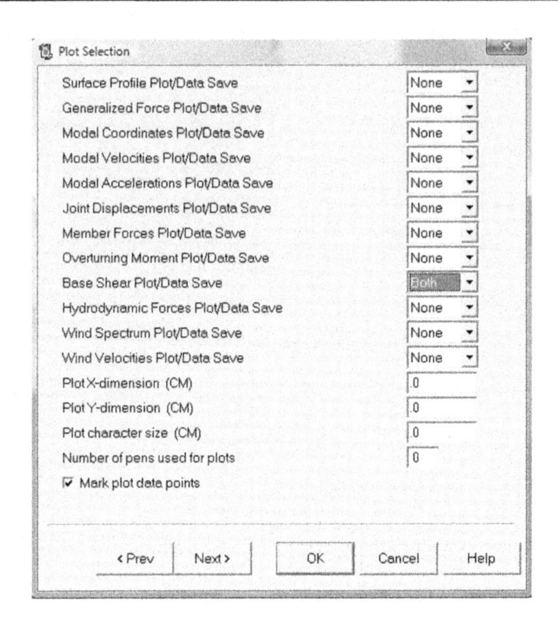

Figure 8.110 Plotting options.

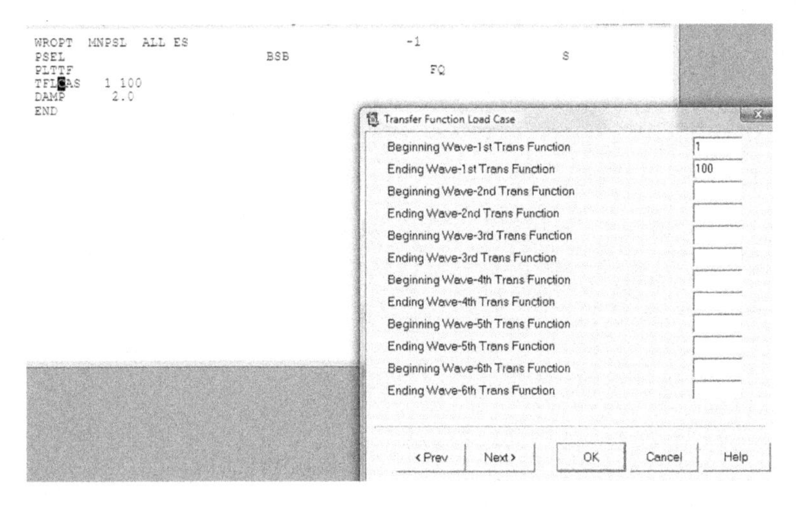

Figure 8.111 Transfer function load case.

Constant Modal Damping

Percent Modal damping (%) 2.0

< Prev Next > OK Cancel Help

Figure 8.112 Model damping ratio.

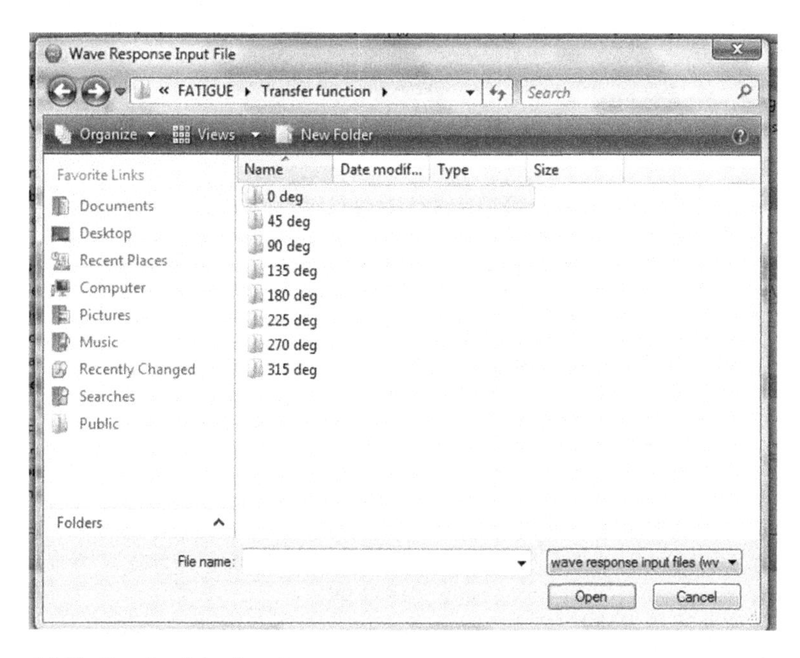

Figure 8.113 Case in eight directions.

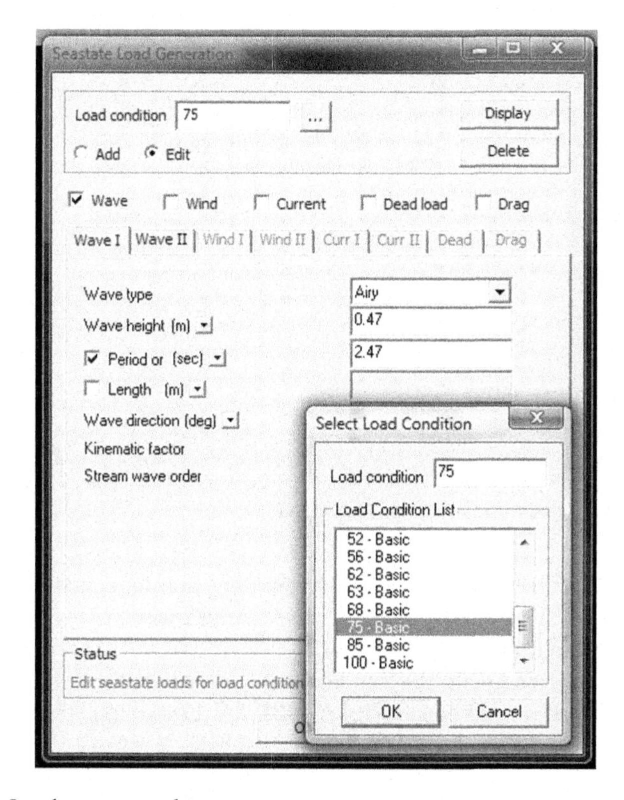

Figure 8.114 Load case wave data.

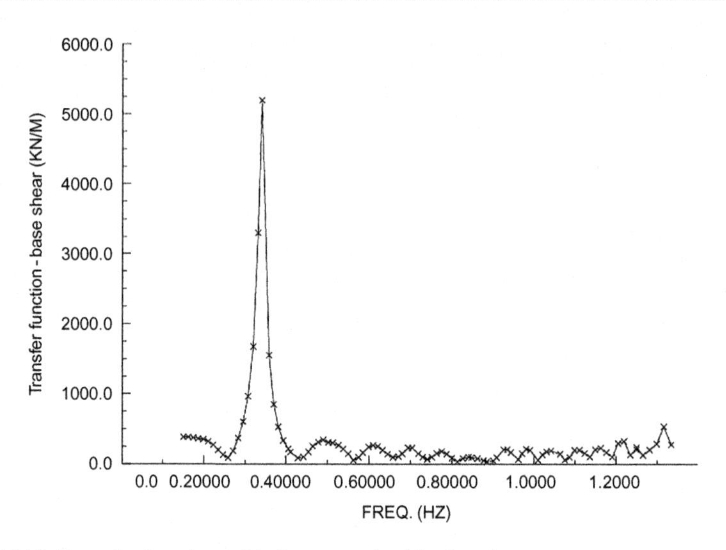

Figure 8.115 Transfer function with frequency in 45° direction.

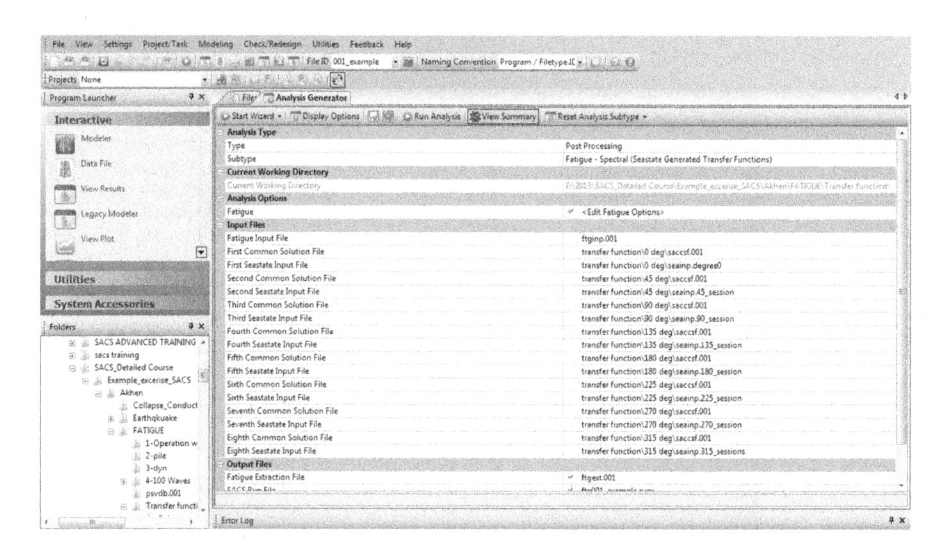

Figure 8.116 Fatigue analysis input and output data.

Eight wave spectra were developed from the environmental data using the Jonswap spectra with parameter Gamma = 3.3 and parameter C = 1.525 to compute the wave height spectral density functions.

According to the load path, the fatigue module classifies the nodal joints as T/Y, X, K, and KT joints in accordance with API RP 2A. Efthymiou stress concentration

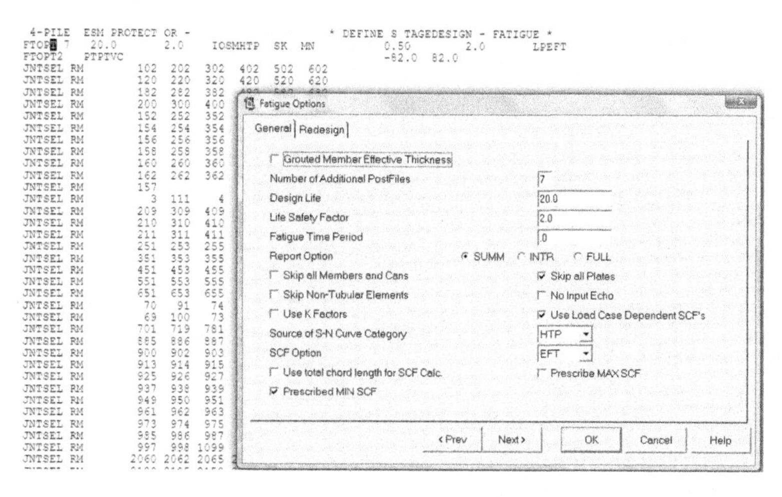

Figure 8.117 Fatigue file input data.

factors are used for simple joints. The automatic calculation is limited by setting a minimum of 1.5 SCF value.

The stress range is calculated at eight circumferential positions for each member on both the brace and chord sides of the weld.

For the pile head joints, the stress concentration factor was calculated using DNV RP-C203 and found equal to 1.0.

S-N curves (WJT curves; signal-to-noise curves), as defined in API RP 2A, are used as the default for computing the damages at all nodal joints. For the in-line pile head joints, the S-N curve for category C1, Figure 2.13 in ANSI/AWS D1.1, was used.

8.11.4 Fatigue input data

This is the last step. The input file is shown in Figure 8.117. You have many menus in which to put the fatigue data. The selected menu in Figure 8.118 is to define the input data of the members that affect the joint fatigue analysis

The input data for the wave is provided through the menu in Figure 8.119.

The limit of the stress concentration factor SCF is shown in Figure 8.120.

Spectral fatigue analysis, with the fraction of design life, is in Figure 8.121.

The data of the scatter diagram is shown in Figure 8.122. The menu in which to enter the data in Figure 8.123, which presents the frequency occurrence at the specific range of the wave period.

The output result is presented in the ftglst.filename, shown in Figure 8.124. The service live of each joint is presented in the figure.

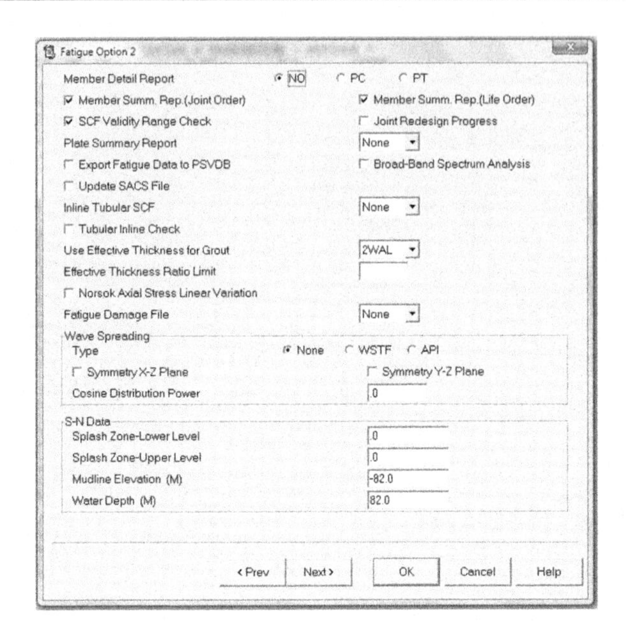

Figure 8.118 Fatigue option 2.

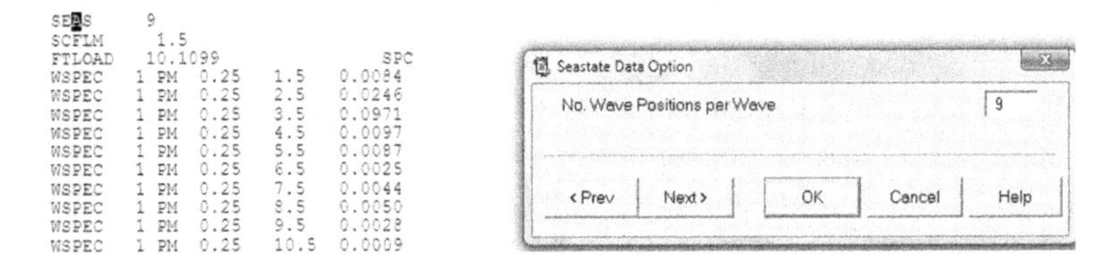

Figure 8.119 Wave positions.

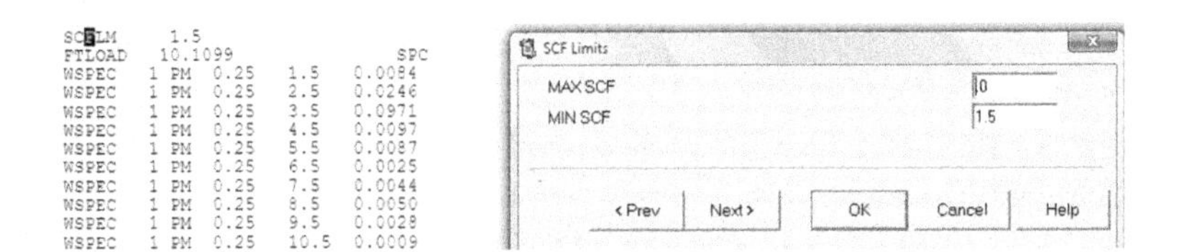

Figure 8.120 Maximum and minimum SCF.

8.12 Lifting analysis

The lifting analysis should be through a cooperation between the construction and installation company and the designer to provide the best way of lifting and to be match it with vessel's crane capacity. The lifting is designed by the Static Analysis module. You need to just enter the constraints as in Figure 8.125, as the top point is 111111 so the platform is restrained in the three translation and rotation directions.

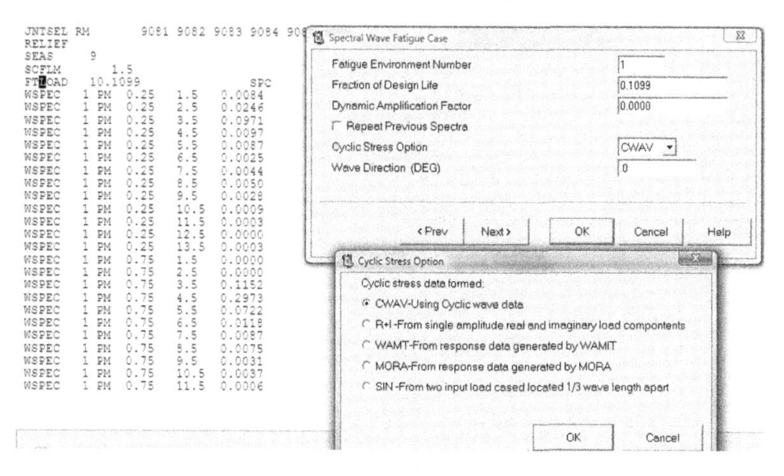

Figure 8.121 Fatigue cycle data.

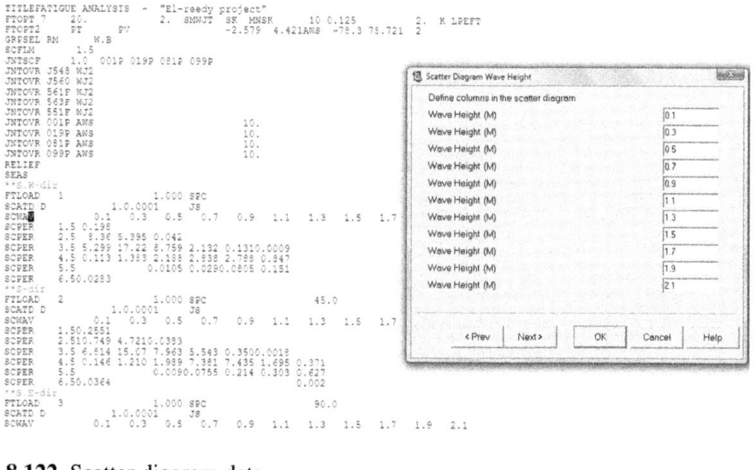

```
TITLEFATIGUE ANALYSIS  -  "El-reedy project"
FTOPT 7    20.            2.  SMWJT  SK  MNSK     10 0.125          2.  K LPEFT
FTOPT2     PT      PV                -2.579  4.421AWS  -78.3 78.721  2
GRPSEL RM      W.B
SCFLM       1.5
JNTSCF      1.0  001P 019P 081P 099P
JNTOVR J548 WJ2
JNTOVR J560 WJ2
JNTOVR 561F WJ2
JNTOVR 563F WJ2
JNTOVR 551F WJ2
JNTOVR 001P AWS                             10.
JNTOVR 019P AWS                             10.
JNTOVR 081P AWS                             10.
JNTOVR 099P AWS                             10.
RELIEF
SEAS
**S.W-dir
FTLOAD   1                  1.000 SPC
SCATD D           1.0.0001      JS
SCWAV          0.1   0.3   0.5   0.7   0.9   1.1   1.3   1.5   1.7
SCPER     1.5 0.198
SCPER     2.5  9.36 5.395 0.042
SCPER     3.5 5.299 17.22 8.759 2.132 0.1310.0009
SCPER     4.5 0.113 1.383 2.188 2.838 2.788 0.847
SCPER     5.5             0.0105 0.0290.0805 0.151
SCPER     6.50.0283
**S-dir
FTLOAD   2                  1.000 SPC                  45.0
SCATD D           1.0.0001      JS
SCWAV          0.1   0.3   0.5   0.7   0.9   1.1   1.3   1.5   1.7
SCPER     1.50.2551
SCPER     2.510.749 4.7210.0383
SCPER     3.5 6.814 15.07 7.963 5.543 0.3500.0018
SCPER     4.5 0.146 1.210 1.989 7.381 7.435 1.695 0.371
SCPER     5.5              0.0090.0755 0.214 0.303 0.627
SCPER     6.50.0364                                 0.002
**S.E-dir
FTLOAD   3                  1.000 SPC                  90.0
SCATD D           1.0.0001      JS
SCWAV          0.1   0.3   0.5   0.7   0.9   1.1   1.3   1.5   1.7   1.9   2.1
```

Scatter Diagram Wave Height
Define columns in the scatter diagram
Wave Height (M) 0.1
Wave Height (M) 0.3
Wave Height (M) 0.5
Wave Height (M) 0.7
Wave Height (M) 0.9
Wave Height (M) 1.1
Wave Height (M) 1.3
Wave Height (M) 1.5
Wave Height (M) 1.7
Wave Height (M) 1.9
Wave Height (M) 2.1
< Prev Next > OK Cancel Help

Figure 8.122 Scatter diagram data.

The bottom of the leg is to be restrained only in the translation in X and Y directions and free for another direction and rotation; so, it will be 110000, as in Figure 8.125. Figure 8.126 also illustrates the sling properties.

The spring in X and Y directions with 100 kN/m is assigned at the end of the leg, as shown in Figures 8.127 and 8.128. Select spring from Joint > Spring from the menu in Figure 8.127, and the menu in Figure 8.128 appears to receive the spring data.

The load combination is assigned as shown in Figure 8.129.

As in Figure 8.11, the dead load is inserted to calculate the self weight, as in Figure 8.130, and it becomes load case 1.

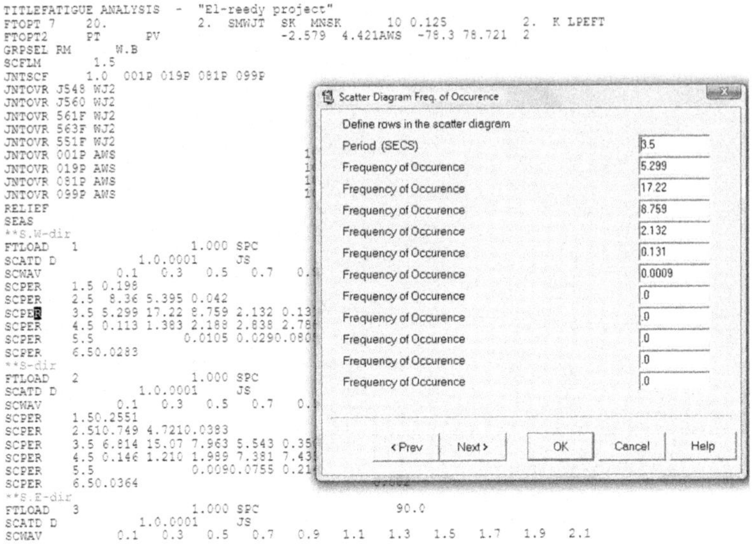

```
TITLEFATIGUE ANALYSIS  -  "El-reedy project"
FTOPT 7    20.             2.  SMWJT  SK  MNSK      10 0.125           2.  K LPEFT
FTOPT2     PT      PV                -2.579  4.421AWS  -78.3 78.721  2
GRPSEL RM      W.B
SCFLM       1.5
JNTSCF     1.0  001P 019P 081P 099P
JNTOVR J548 WJ2
JNTOVR J560 WJ2
JNTOVR 561F WJ2
JNTOVR 563F WJ2
JNTOVR 551F WJ2
JNTOVR 001P AWS
JNTOVR 019P AWS
JNTOVR 081P AWS
JNTOVR 099P AWS
RELIEF
SEAS
**S.W-dir
FTLOAD   1              1.000 SPC
SCATD D           1.0.0001     JS
SCWAV          0.1   0.3   0.5   0.7   0.
SCPER    1.5 0.198
SCPER    2.5  8.36 5.395 0.042
SCPER    3.5 5.299 17.22 8.759 2.132 0.13
SCPER    4.5 0.113 1.383 2.188 2.838 2.78
SCPER    5.5                0.0105 0.0290.080
SCPER    6.50.0283
**S-dir
FTLOAD   2              1.000 SPC
SCATD D           1.0.0001     JS
SCWAV          0.1   0.3   0.5   0.7   0.
SCPER    1.50.2551
SCPER    2.510.749 4.7210.0383
SCPER    3.5 6.814 15.07 7.963 5.543 0.35
SCPER    4.5 0.146 1.210 1.989 7.381 7.43
SCPER    5.5                 0.0090.0755 0.21
SCPER    6.50.0364
**S.E-dir
FTLOAD   3              1.000 SPC                       90.0
SCATD D           1.0.0001     JS
SCWAV          0.1   0.3   0.5   0.7   0.9   1.1   1.3   1.5   1.7   1.9   2.1
```

Scatter Diagram Freq. of Occurence

Define rows in the scatter diagram

Period (SECS)	3.5
Frequency of Occurence	5.299
Frequency of Occurence	17.22
Frequency of Occurence	8.759
Frequency of Occurence	2.132
Frequency of Occurence	0.131
Frequency of Occurence	0.0009
Frequency of Occurence	.0
Frequency of Occurence	.0
Frequency of Occurence	.0
Frequency of Occurence	.0
Frequency of Occurence	.0

< Prev | Next > | OK | Cancel | Help

Figure 8.123 Scatter diagram for wave period occurrence input data.

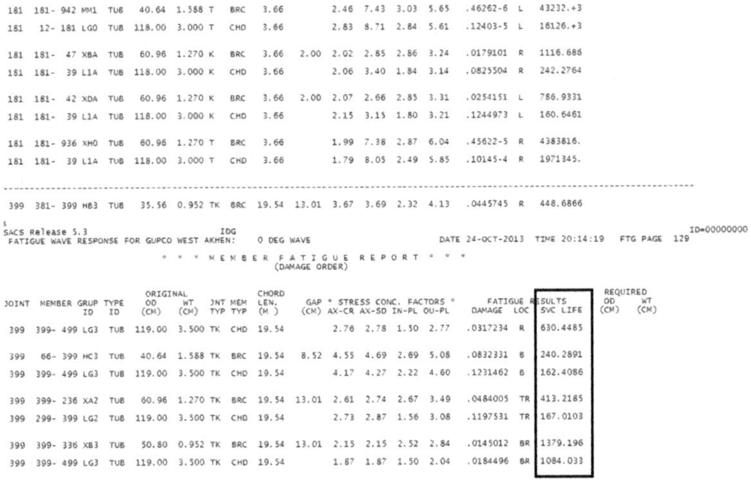

```
181  181- 942 MM1  TUB   40.64  1.588 T   BRC   3.66          2.46  7.43  3.03  5.65  .46262-6  L   43232.+3
181   12- 181 LG0  TUB  118.00  3.000 T   CHD   3.66          2.83  8.71  2.84  5.61  .12403-5  L   16126.+3

181  181-  47 XBA  TUB   60.96  1.270 K   BRC   3.66   2.00   2.02  2.85  2.86  3.24  .0179101  R   1116.686
181  181-  39 L1A  TUB  118.00  3.000 K   CHD   3.66          2.06  3.40  1.84  3.14  .0825504  R   242.2764

181  181-  42 XDA  TUB   60.96  1.270 K   BRC   3.66   2.00   2.07  2.66  2.85  3.31  .0254151  L   786.9331
181  181-  39 L1A  TUB  118.00  3.000 K   CHD   3.66          2.15  3.15  1.80  3.21  .1244973  L   160.6461

181  181- 936 XH0  TUB   60.96  1.270 T   BRC   3.66          1.99  7.38  2.87  6.04  .45622-5  R   4383816.
181  181-  39 L1A  TUB  118.00  3.000 T   CHD   3.66          1.79  8.05  2.49  5.85  .10145-4  R   1971345.

--------------------------------------------------------------------------------------------------------------

399  381- 399 HB3  TUB   35.56  0.952 TK  BRC  19.54  13.01   3.67  3.69  2.32  4.13  .0445745  R   448.6866

1
SACS Release 5.3                      IDG                                                                ID=00000000
 FATIGUE WAVE RESPONSE FOR GUPCO WEST AKHEN:    0 DEG WAVE           DATE 24-OCT-2013  TIME 20:14:19   FTG PAGE  129

                          * * *  M E M B E R   F A T I G U E   R E P O R T  * * *
                                               (DAMAGE ORDER)

                             ORIGINAL              CHORD                                                              REQUIRED
JOINT  MEMBER GRUP TYPE    OD       WT    JNT MEM  LEN.     GAP * STRESS CONC. FACTORS *     FATIGUE RESULTS          OD     WT
                ID   ID   (CM)     (CM)   TYP TYP  (M )    (CM) AX-CR AX-SD IN-PL OU-PL   DAMAGE   LOC  SVC LIFE   (CM)   (CM)
399  399- 499 LG3  TUB  119.00  3.500 TK  CHD  19.54          2.76  2.78  1.50  2.77  .0317234  R   630.4485

399   66- 399 HC3  TUB   40.64  1.588 TK  BRC  19.54   8.52   4.55  4.69  2.69  5.08  .0832331  B   240.2891
399  399- 499 LG3  TUB  119.00  3.500 TK  CHD  19.54          4.17  4.27  2.22  4.60  .1231462  B   162.4086

399  399- 236 XA2  TUB   60.96  1.270 TK  BRC  19.54  13.01   2.61  2.74  2.67  3.49  .0484005  TR  413.2185
399  299- 399 LG2  TUB  119.00  3.500 TK  CHD  19.54          2.73  2.87  1.56  3.08  .1197531  TR  167.0103

399  399- 336 XB3  TUB   50.80  0.952 TK  BRC  19.54  13.01   2.15  2.15  2.52  2.84  .0145012  BR  1379.196
399  399- 499 LG3  TUB  119.00  3.500 TK  CHD  19.54          1.87  1.87  1.50  2.04  .0184496  BR  1084.033
```

Figure 8.124 Fatigue output data.

The load combination cases are identified for the loading scenario, as in Figure 8.131.

The load menu and its location of the topside during lifting are presented in Figure 8.132.

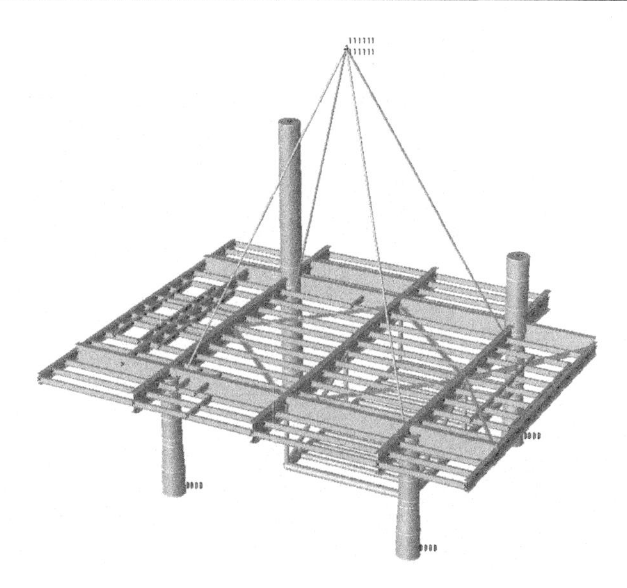

Figure 8.125 Lifting analysis with end release.

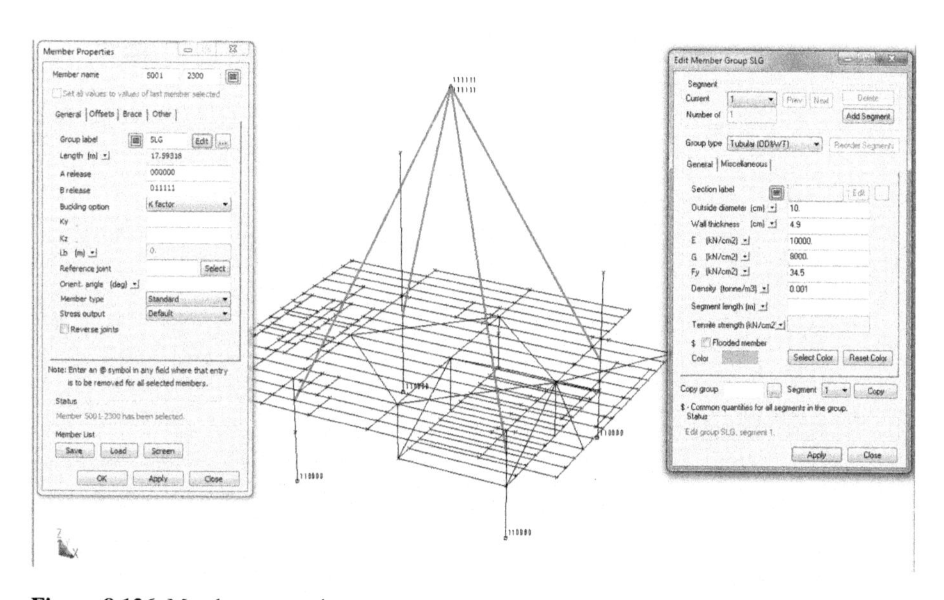

Figure 8.126 Member properties menu.

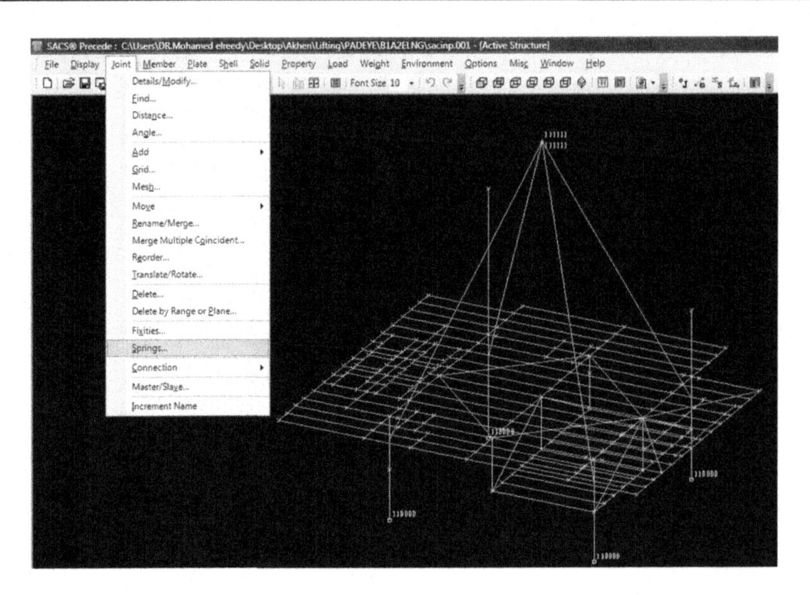

Figure 8.127 Choose Spring.

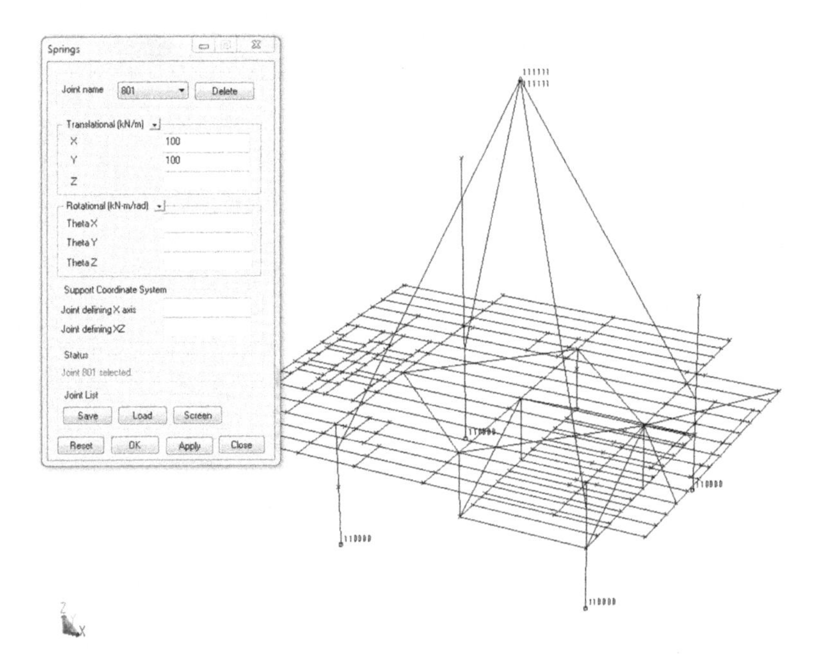

Figure 8.128 Spring data.

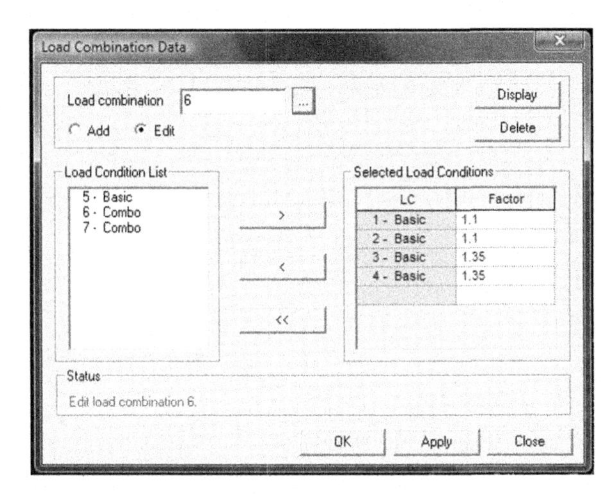

Figure 8.129 Load combination.

Figure 8.130 Self weight input data.

8.13 Flotation and upbending

The Flotation and Upbending module allows us to calculate the maximum hook load and the steps of flotation and upbending of the jacket. It is accessed via Marine application > Flotation and Upbending. Figure 8.133 presents the input data of upbending and flotation through the sacinp.filename and fltinp.file name.

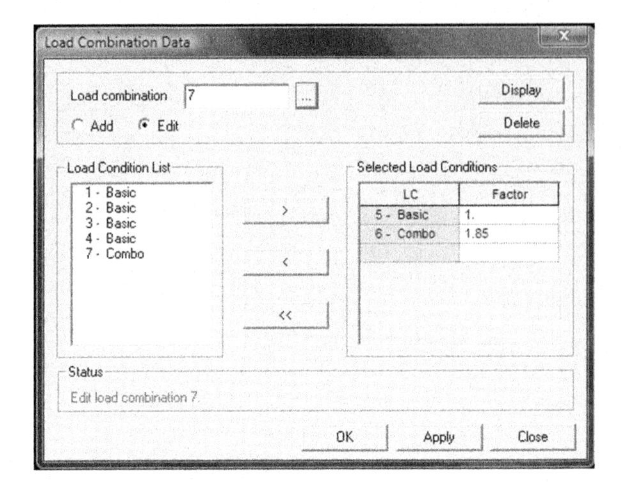

Figure 8.131 Load combination for lifting.

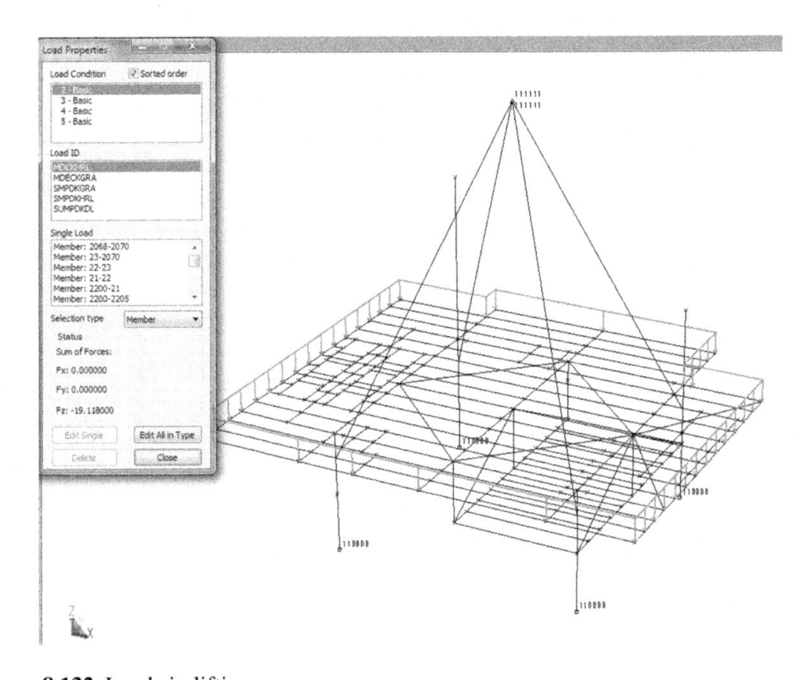

Figure 8.132 Loads in lifting.

The output is shown in the following curves. Figure 8.134 presents the hook load at every step. Figure 8.135 shows the mud clearance distance. Figure 8.136 shows the sling load.

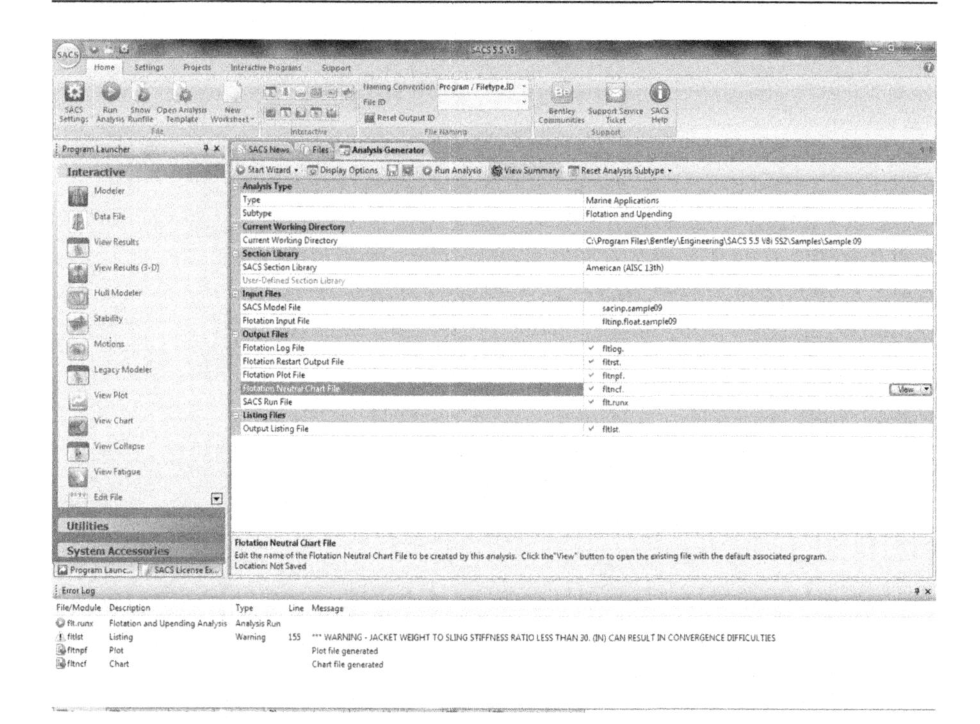

Figure 8.133 Flotation and upending analysis.

8.14 On-bottom stability

The analysis of this design condition is carried out to ensure that the jacket can be placed on the seabed, before stabbing any piles, without stability risks, as in Figure 8.137. The computer model used for this analysis is based on in-place analysis design except for the following modifications:

- The in-place model was modified by deleting the topsides, piles, boat landing, barge bumper, and conductors.
- The water depth considered for the computation of the environmental loads and buoyancy was the MSL + (0.5 m installation tolerance).
- Corrosion allowance was disregarded in the model as per the design basis.
- Marine growth was eliminated from the model as per the design basis.

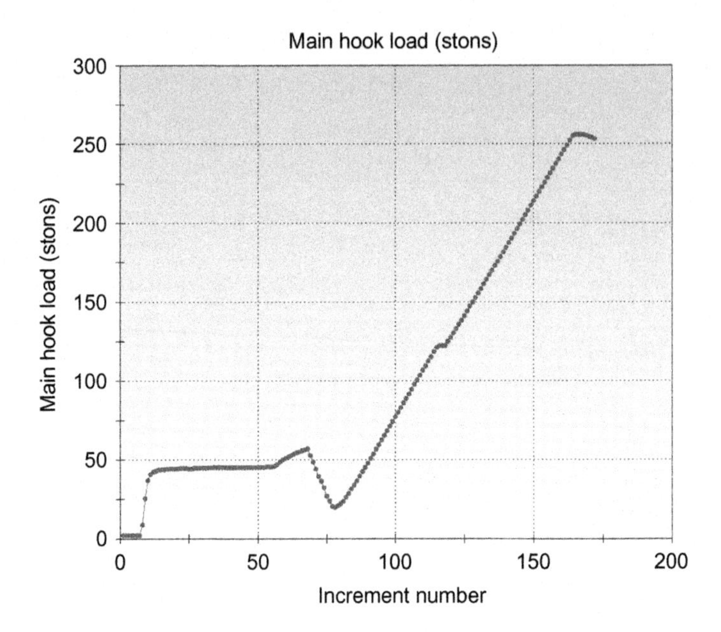

Figure 8.134 The maximum hook load.

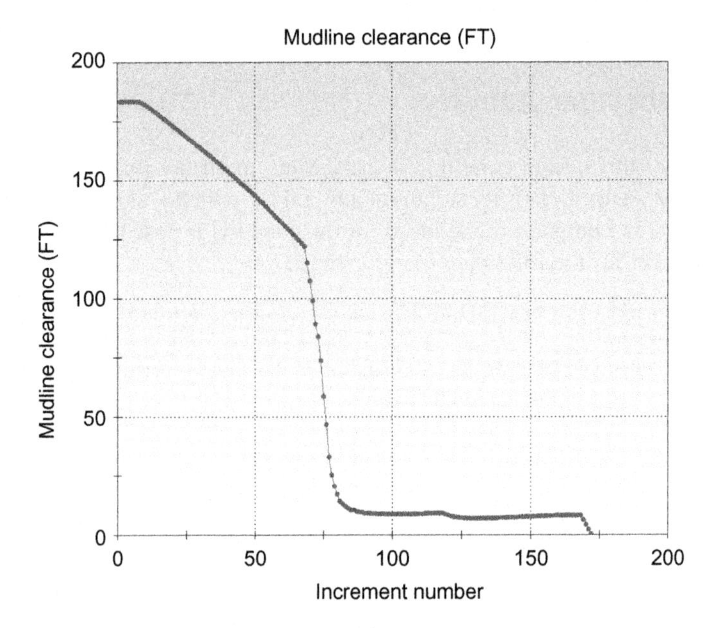

Figure 8.135 Mud clearance distance.

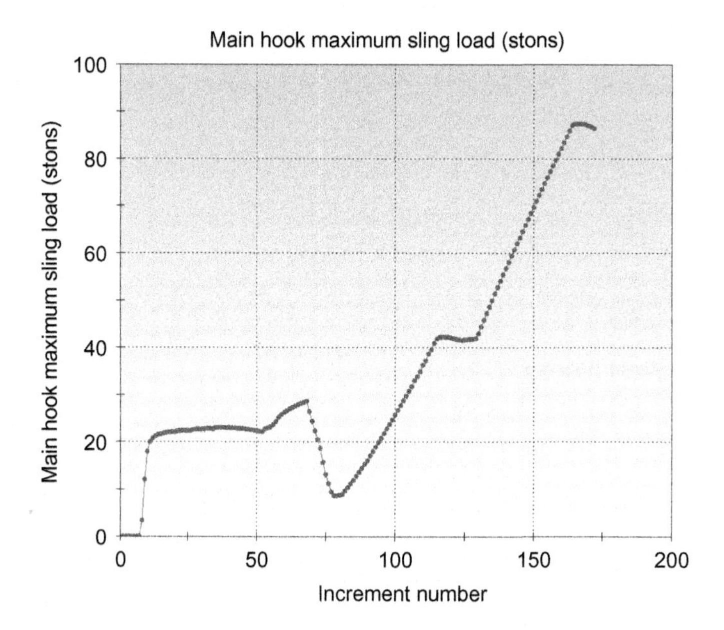

Figure 8.136 Sling load.

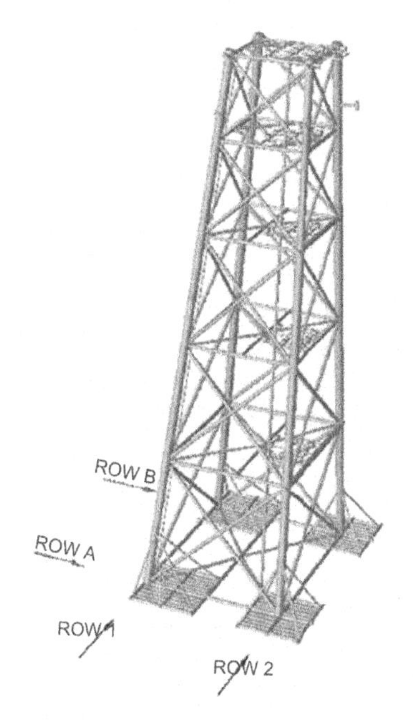

Figure 8.137 Bottom stability structure model.

Analysis Type	
Type	Marine Applications
Subtype	Launch
Current Working Directory	
Current Working Directory	E:\Sample 09
Section Library	
SACS Section Library	American (AISC 13th)
User-Defined Section Library	
Input Files	
SACS Model File	sacinp.sample09
Launch Input File	lnhinp.launch.sample09
Output Files	
Launch Plot File	✓ lnhnpf.
Launch Neutral Chart File	✓ lnhncf.
Launch Restart Output File	✓ lnhrst.
SACS Run File	✓ lnh.runx
Listing Files	
Output Listing File	✓ lnhlst.

Figure 8.138 Launch analysis.

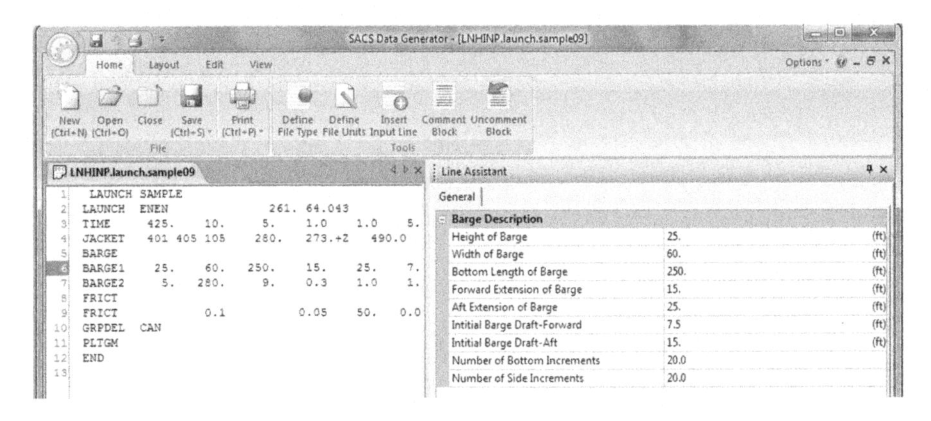

Figure 8.139 Launch data file.

8.15 Launch analysis

In launching the jacket, the Launch analysis can be reached by going to Marine application > launch for the input and output data, as shown in Figure 8.138.

The file LNHinp.project contains data on the barge, as shown in Figure 8.139. Note that this view is from SACS revision 5.0, as the edit file presents two menu as in the figure. Lines 6 and 7 define the barge data, as in Figure 8.139.

8.16 Summary

8.16.1 Static analysis

1. Build the geometric model.
2. Enter the load cases..
3. Perform static analysis. If everything is correct,

4. Enter PSI data in the data file datagen.
5. Run the static information with the nonlinear pile-soil interaction.
6. Review the data.
7. Obtain the superelement as generating the file dysef.

8.16.2 Dynamic analysis

1. Define the master joint 222000 in the model.
2. Open the Dynpac file by Data File, datagen.
3. Run the model to define the natural frequency.
4. You have two files, dynmod and dynmass.

8.16.3 Seismic analysis

1. Run the model for the gravity load only.
2. Run Dynamic with the subtitle Spectral Earthquake.
3. Enter the Earthquake file by Data File datagen.
4. Run the model to obtain the seismic stresses.
5. Start the combination between the gravity and seismic loads by
6. Data file "datagen" to have a combination file define the primary and secondary loads.
7. Run the combination.
8. From Postvue, run this module by specifying the model and the output from the combination with Type: Postprocess and Subtype: Postvue Database Creation.
9. Read the data on the model.

8.17 Fatigue analysis

1. Obtain the scatter diagram.
2. Build the fatigue file ftginp"by Data file datagen.
3. Run the Center of Damage.
4. Define the wave in eight directions.
5. Obtain 100 waves from the Excel sheet.
6. Run the static analysis in eight directions to obtain the transfer function in eight directions.
7. Check the maximum shear in the range of the waves.
8. Run tFatigue analysis by Type: Postprocessing and Subtype: Fatigue—spectral.
9. Enter the common solution file in the eight directions.
10. From the ftglist file, check that the joint has a service life longer than assigned.

8.17.1 Collapse analysis

1. Choose Type: Static and Subtitle: Collapse analysis.
2. Run the collapse file clpinp by the Data File datagen.
3. Run the model to see the collapse view.

8.17.2 Lifting analysis

Run a static analysis, but give the jointe a special constraint in the lifting point and leg on the deck.

8.17.3 On-bottom stability

1. Run a static analysis but the constraint on the leg is 111000.
2. Obtain the overturning moment and maximum shear.

8.17.4 Tow analysis

1. Run the mode by selecting Type: Loading and Subtype: Tow.
2. Perform the file datagen for the Tow.
3. Run the model.
4. Obtain the data.
5. See the results on the geometric model by Postprocessing.

Appendix: Assignment

More than one answer may be correct. A reasonable time for completing this is 45 minutes.

1. The accepted scour depth besides the pile
   - **a.** 1.5 from the pile diameter
   - **b.** 1.25 from leg diameter
   - **c.** 1.5 from leg diameter
   - **d.** 1.25 from pile diameter
2. For the offshore structure, welding is checked if
   - **a.** Welding is no less than 5 mm
   - **b.** allowable, but not if less than 0.2 Fu
   - **c.** No check is made
3. In obtaining the reserve strength ratio (RSR), what type of analysis do you use?
   - **a.** In-place analysis
   - **b.** Pushover analysis
   - **c.** Collapse analysis
   - **d.** Nonlinear analysis
4. To increase the jacket capacity, what is the best mitigation?
   - **a.** Decrease conductors
   - **b.** Decrease the riser
   - **c.** Remove marine growth
5. Which type of these structure configuration has the highest RSR
   - **a.** X-bracing
   - **b.** K- Bracing
   - **c.** X-bracing with horizontal bracing
   - **d.** A diagonal member with horizontal bracing
6. The sleeve and master joint should be defined in the
   - **a.** Seismic analysis
   - **b.** Tow analysis
   - **c.** Eigenvalue analysis
   - **d.** Supperelement
7. What is the conductor shielding factor for conductors of 24 in. spaced 4 ft apart?
   - **a.** 1.0
   - **b.** 0.75
   - **c.** 0.5
   - **d.** 0.85
8. We can neglect the current by increase the wave height by
   - **a.** 1.05
   - **b.** 1.10
   - **c.** 1.15
   - **d.** 1.20

9. Please write the name of each type of motion in Figure A.1.
10. With a water depth of 40 m, a wave height of 6 m, and a wave period of 6 s, what is the best wave theory to apply?
    a. Strokes
    b. Stream function3
    c. Stream function 5
    d. Airy
11. After obtaining the eigenvalue that the natural frequency for the platform is 2.9 s,
    a. Consider the DAF
    b. Do not consider the DAF
12. In doing the dynamic analysis Dynpac, it is important to use non linear soil interactions.
    a. True
    b. False
13. In doing in-place analysis, it is important to use non linear soil interactions.
    a. True
    b. False
14. In doing bottom stability analysis, the boat landing is modeled.
    a. True
    b. False

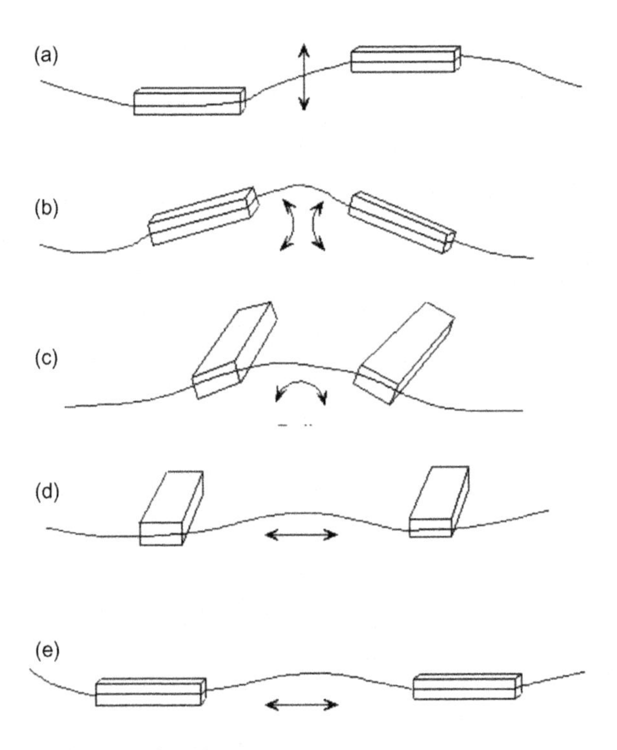

Figure A.1 Motions for Exercise 10.

15. What is the reserve strength ratio (RSR)?
 a. Maximum stress/minimum stress
 b. Failure load/design load
 c. Load at the start for first plastic joint
 d. Load value of last plastic joint

16. Remote Operating Vehicle (ROV) presents four cutting bracing members with a K- bracing configuration, using structure analysis of many members over the UC value. What is the first step?
 a. Perform pushover analysis
 b. Repair 50% of the flooded members
 c. Repair 100% of the members
 d. Perform nonlinear structure analysis

17. What is the traditional safety factor of fatigue analysis for jacket joints?
 a. 1.5
 b. 2.0
 c. 2.5
 d. 10.0

18. The fatigue analysis wave spectral diagram is obtained from SACS.
 a. True
 b. False

19. A T-Z curve shows
 a. Axial pile deflection versus load
 b. Lateral pile deflection versus load
 c. End bearing deflection versus load

20. If the displacement of the last deck to the pile head is 25 cm for a platform 40 m high, it is acceptable in a water depth of 25 m.
 a. True
 b. False

21. In the design of a new platform to withstand operating storm conditions, the applied maximum compression load is 723 tons and the pile capacity in compression is 1360 tons, so it will be safe.
 a. True
 b. False

22. In the design of a new platform to withstand extreme storm conditions, the applied maximum compression load is 723 tons and the pile capacity in compression is 1360 tons, so it will be safe.
 a. True
 b. False

23. The engineering cost percentage for the total project is equal to
 a. 3%
 b. 5%
 c. 10%
 d. 1%

24. The percentage of design and drafting cost for the total project is around
 a. 60%
 b. 10%
 c. 30%
 d. 90%

25. For seismic analysis, the dynamic damping ratio is
    a. 1%
    b. 3%
    c. 5%
    d. 10%
26. For the strength requirement of a seismic analysis, the basic AISC allowable value may be increased by
    a. 70 %
    b. 33%
    c. 20%
    d. 30%
27. The pile diameter is 48 in. and the scour depth, by ROV, is 180 cm, so there is a scour problem.
    a. True
    b. False
28. To start the Center of Damage module on SACS, you should have the file
    a. ftginp
    b. sacinp
    c. eqlst
    d. dyrinp
29. The results of the pushover analysis are seen through
    a. postvue
    b. precede
    c. collapseview
30. You use the mean yield strength rather than the nominal yield strength in which analysis?
    a. In-place
    b. Dynamic
    c. Collapse
    d. Fatigue
31. In a weighed weight jacket, the contingency weight factor is
    a. 3%
    b. 4%
    c. 5%
    d. 10%
32. In an unweighed jacket, the weight of the topside the contingency weight factor is
    a. 3%
    b. 4%
    c. 5%
    d. 10%
33. In spectral fatigue analysis, the method assumes that there is no relationship between wave height and stress ranges at the connections.
    a. True
    b. False
34. Stress Range is difference between maximum and minimum values of stress over one complete cycle of loading
    a. True
    b. False

35. The transfer function describes the relationship between stress range and frequency.
  a. True
  b. False
36. For seismic analysis, the modal responses in each of the three orthogonal directions (X, Y, and Z) are combined by
  a. CQC
  b. SRSS
37. For riser guard design, the member is designed to reach 75% plasticity.
  a. True
  b. False
38. For a boat landing design, the vessel absorbs 70% of the impact energy.
  a. True
  b. False
39. The impact energy of collision force is based on a 100-ton boat's impact at 0.5 meter per second velocity is
  a. 13.8×10^7 N.m
  b. 12.75×10^7 Nm
  c. 13.8×10^6 Nm
40. The wave theory used in fatigue analysis is
  a. Airy
  b. Stroke
  c. Stream
41. In fatigue analysis, the following percentage of the live load is considered:
  a. 100%
  b. 75%
  c. 50%
  d. 0%
42. A helideck safety net is designed for a load and distance of
  a. 50 kg, distance 1 m
  b. 100 kg, distance 500 mm
  c. 100 kg, distance 1 m
  d. 1 ton, distance 1000 mm
43. If the CoG is shifted 500 mm from the original CoG, it is acceptable.
  a. True
  b. False
44. In takeoff loading, the wind load at operation is considered based on CAP437.
  a. True
  b. False
45. For a helicopter at rest under loading conditions, the wind load at operation is considered based on CAP437.
  a. True
  b. False
46. After conceptual design, the cost estimate accuracy is
  a. 10
  b. 15
  c. 5%
  d. 20%
47. Issuing the drawing draft to the client, the progress is considered to be
  a. 0% finished
  b. 5% finished

c. 15% finished
d. 30% finished

48. The cost of preparing the proposal for a tender is less than how many percentage of the total project cost?
a. 1%
b. 5%
c. 10%
d. 3%

49. In the engineering phase, the cost of engineering and drafting should be about how much of a maximum percentage of the project cost?
a. 60%
b. 50%
c. 30%
d. 25%

50. Who is responsible to deliver the plot plan of the platform?
a. Piping designer
b. Structural engineer
c. Members of all disciplines

Index

Note: Page numbers followed by "*f*" and "*t*" refer to figures and tables, respectively.

A

Abnormal level earthquake (ALE), 62–63
  requirements, 65
  structural and foundation modeling, 65–68
    topside appurtenances and equipment, 67–68
Add Member Group dialog box, 348–350
Add/Edit Conductor Data button, 346–348
Airy and Stokes theories, 47
Allowable joint capacity, 146–148
Allowance
  conceptual design, 311
  and contingencies of weight, 311–312
  defined, 316*t*
  detailed design, 311
  fabrication, 311–312
  in weight, 311*t*
Aluminum helideck, 205
API RP2A, 52, 253, 258–259, 300, 314
  cylinder member strength calculation by, 125–135
    axial compression, 125
    axial tension, 125
    axial tension and hydrostatic pressure, 130–132
    bending, 126
    combined axial compression and bending, 129
    combined axial tension and bending, 129
    design hydrostatic head, 127
    hoop buckling stress, 128
    local buckling, 125–126
    member slenderness, 129
    pressure, 127
    safety factors, 132–135
    shear, 126–127
    torsional shear, 127
  joint calculation from, 136–149
    allowable joint capacity, 146–148
    punching shear, 145–146
    tubular joint punching failure, 148–149
  joint classification and detailing, 136–139
  lifting forces, 313
  pile-capacity factor of safety in, 238*t*
  safety factor based on, 133*t*
  shape coefficient by, 44*t*
  tubular joint calculation, 139–144
    chord load factor, 141
    grouted joints, 143–144
    joints with thickened cans, 142–143
    overlapping joints, 143
    strength check, 143
    strength factor, 140–141
Appraise phase, 16
ASTM C109, 79
Auxiliary platforms, 6
Axial load-deflection (*t-z* and *Q-z*) data, of pile foundations, 241–244
Axial pile capacity, 244–246
Axial pile performance, 240–253
  axial load-deflection data, 241–244
  axial pile capacity, 244–246
  changes in axial capacity in clay with time, 252–253
  cyclic response, 240–241
  lateral bearing capacity
    for sand, 249–252
    for soft clay, 247–248
    for stiff clay, 248–249
  laterally loaded piles reaction, 246–247
  static load-deflection behavior, 240

B

Barge, 303–306
  bumpers, 206
  defined, 316*t*
  installing the topside on, 314*f*
  structure model located on, 389*f*
  transportation, 313
  transportation of the topside to, 315*f*
Base weight, defined, 316*t*
Basis of design (BOD) documents, 16, 35

Bearing capacity, lateral. *See* Lateral bearing capacity
Bending efficiency factor, 331, 331*t*
Bending reduction factor, defined, 316*t*
Boat landing design, 205–213
  boat impact methods, 210–211
  calculation, 206–209
  cases of impact load, 209
  tubular member denting analysis, 211–213
  using nonlinear analysis method, 209–210
Brace stresses, 153, 361*f*
Bracing input data, 348*f*
Bracing system, 92–94
Breaking load, 330–331
  defined, 316*t*
Bridges, 6, 7*f*, 8*f*, 178
Brinch Hansen formula, 228
Bulk materials, 22, 309–310

C

Cable-laid sling, defined, 316*t*
Campos Basin, 8
Cast joint (CJ) curves, 155
Catwalk, 6
Cement grout, 79–82
Certificate of approval, defined, 316*t*
Checklist
  for jacket and topsides transportation analysis, 29*t*
  for jacket in-place analysis, 24*t*
  for jacket on-bottom stability analysis, 28*t*
  report, 32*t*
  for topside in-place analysis, 27*t*
  for topside lift analysis, 31*t*
Circulational currents, 57
Clay, 227, 243*t*, 245
  sand, lateral bearing capacity for, 249–252
  soft clay, lateral bearing capacity for, 247–248
  stiff clay, lateral bearing capacity for, 248–249
Clearances, 335–340
  around crane vessel, 335–340
  around lifted object, 335
  example, 338–340
Cohesionless soils, 253
  degree of compactness for, 229*t*
  evaluating soil resistance drive, 261
  shaft friction and end bearing in, 236–239
Cohesive soil
  consistency of, 229*t*
  evaluating soil resistance drive, 261
  skin friction and end bearing in, 234–236
Collapse analysis, 382–386, 384*f*, 423
Collapse option, 385*f*
Collision events, 76–78
Combine module in SACS, 390
Complete quadratic combination (CQC) method, 61, 380
Composite pile, 270–274
  allowable axial force, computation of, 271–274
Concept stage, 15–16
Conceptual design allowance, 311
Conceptual design package, 16
Concrete gravity platform, 1, 6, 9–10
Conductor guides, 104*f*
  and piles tolerances, 292
Conductor shield factor (SF), 56–57
Conductors input data, 347*f*
Conductors tab, 346–348
Cone penetration test (CPT), 220–224, 221*f*
  arrangement for, 222*f*
  CPTU testing, 220
  equipment requirements, 222
  pile capacity calculation based on, 253–256
  results, 223–224
Cone penetrometer (Dutch cone), 219, 221–222
Conidial wave theory, 47
Connectivity tab, 348
Conoco Phillips, 9–10
Consequence factors, 332–333, 332*t*
  defined, 316*t*
Construction process, lifting process during, 284–295
  conductor guides and piles tolerances, 292
  dimensional control, 293
  fabrication tolerances, 285–289
    leg spacing tolerances, 286
    tolerances of leg alignment and straightness, 288
    tubular joint tolerances, 289
    tubular member tolerances, 287–288

vertical level tolerances, 287
jacket assembly and erection, 293–295
stiffener tolerances, 290–291
Cost control, engineering, 22
Cost time and resources (CTR) sheets, 19
sample form, 20*t*
CPT. *See* Cone penetration test (CPT)
Crane lift factors, 330
Crane support structures, 39–41
Crane vessel, 341
clearances around, 335–340
defined, 316*t*
Critical hoop buckling stress, 128
Critical movement, 246
Current data, 353*f*
Current load, 57–59
Cyclic loadings effects, on piles, 240–241
Cylinder member strength, 112–135
calculation by API RP2A, 125–135
axial compression, 125
axial tension, 125
axial tension and hydrostatic pressure, 130–132
bending, 126
combined axial tension and bending, 129
combined stresses for cylindrical members, 128–129
design hydrostatic head, 127
hoop buckling stress, 128
local buckling, 125–126
pressure, 127
safety factors, 132–135
shear, 126–127
torsional shear, 127
calculation by ISO19902, 113–124
axial tension, 113
bending, 115–116
column buckling, 114
effective lengths and moment reduction factors, 123–124
hoop buckling, 118–119
hydrostatic pressure, 117–118
local buckling, 114–115
shear, 116
torsional shear, 117
tubular members combined forces without hydrostatic pressure, 119–120
tubular members combined forces with hydrostatic pressure, 121–123

D

Dead load, 33–34
Deck areas, 36–37
functional loads on, 37*t*
Deck Girder tab, 348
Deck wizard, 349*f*
Design management, engineering, 18–31
cost control, 22
engineering interfaces, 22–23
structural engineering quality control, 23–31
time schedule control, 19–22
Det Norske Veritas (DNV), 198
Detail engineering phase, 17–18
Detailed design allowance, 311
Detailed Report button, 368
Determinate lift, defined, 316*t*
Dimensional control, 293
Drag coefficient tab, 352
Drag effect direction, 353*f*
Drag effect in Y direction, 352, 353*f*, 354*f*, 355*f*
Drilling loads, 310
Drilling mode, 220
Drilling/well protected platforms, 5
Dry weight, 37, 310
Ductility level (DLE) earthquake, 63
Dutch cone, 219
Dynamic amplification factors (DAFs), 315, 373–375
calculations, 373–375
defined, 316*t*
in different locations, 318*t*
Dynamic analysis, 61, 373–377, 375*f*, 423
special fixity for, 377*f*
Dynamic structure analysis, 106–112
Dynamic wave response analysis, 402–411
Dynpac input file, 376, 376*f*
Dynpac module, 373, 375–376

E

Earthquake load, 59–68
abnormal level earthquake (ALE) requirements, 65
ALE structural and foundation modeling, 65–68
extreme level earthquake (ELE) requirements, 64–65
Edit Model Extraction Option, 376

Efthymiou stress concentration factors, 410–411
Eigenvalue analysis, 375–377
output of, 378f
Elastic hoop buckling stress, 118–119, 128
Elastic local buckling stress, 126
Elastic members, 386f
Elevations tab, 343–346
End-bearing capacity, 239, 246
Engineering management for marine structure, 13
design management, 18–31
cost control, 22
engineering interfaces, 22–23
structural engineering quality control, 23–31
time schedule control, 19–22
detail engineering phase, 17–18
FEED engineering phase, 16–17
field development, 13–16
concept stage, 15–16
project cost and the life cycle, 13–15
screening process, 15–16
Engineering manager, challenge for, 19
Engineering phase costs, breakdown of, 23t
Engineering report progress measurement guideline, 21t
Entrapped mass, 376
EPCI contract, 8
Equation of motion, 107
Espirito Santo FPSO, 8
Extract Mode Shapes, 375
Extreme level earthquake (ELE), 62–63
requirements, 64–65
Exxon, 1
Exxon Mobil, 6

F

Fabrication change allowance, 311–312
Fabrication tolerances, 285–289
leg alignment and straightness, tolerances of, 288
leg spacing tolerances, 286
tubular joint tolerances, 289
tubular member tolerances, 287–288
vertical level tolerances, 287
Factored weight, defined, 312, 316t
Fatigue analysis, 150–172, 171t, 396–411, 423–424
center of damage, 397–400
collapse analysis, 423
dynamic wave response analysis, 402–411
fatigue input data, 411
foundation superelement, generation of, 400–402
lifting analysis, 423
on-bottom stability, 424
S-N curves
for all members and connections, except tubular connections, 154–155
jacket fatigue design, 168–172
for tubular connections, 155–168
stress concentration factors, 151–154
in cast nodes, 154
in grouted joints, 154
Tow analysis, 424
Field development, 13–16
concept stage, 15–16
overview of, 2–4
project cost and the life cycle, 13–15
screening process, 15–16
Field vane test, 225–226
Fixed offshore platform, 4–5, 33, 217
construction of, 283
design procedure, 100f
extreme environmental situation for, 72–73
Fixed platforms
categories of, 1
operating environmental situations for, 73–74
Flare jackets and flare towers, 6
Floating production, storage, and offloading (FPSO) vessels, 7–8
Flotation and upbending module, 417–418
Foundation model, 362–367
Foundation piling model, 98f
Foundation superelement, generation of, 400–402
Friction sleeve, 221
Front end engineering design (FEED), 14–15, 309
engineering phase, 14, 16–17

G

Geotechnical investigation, 217–218
performing offshore investigation, 218

Golar LNG, 7
Gravity load, 33–41
  crane support structures, 39–41
  dead load, 33–34
  design for serviceability limit state, 38–39
  impact load, 38
  live load, 35–38
Grommets, 331, 332*f*
  defined, 316*t*
Gross weight, defined, 316*t*
Grouted joints, 143–144
  stress concentration factor (SCF) in, 154
Grouting, of piles platform loads, 270–271
Gulf of Mexico (GoM), 258
  hurricanes in, 2
  steel template type development in, 1

H
Hammer types, efficiency for, 259*t*
Hammering data, pile, 259*t*
Hammock effect, 193–194
Handrails, 175
Helicopter landing loads, 189–193
  area load, 192
  helicopter static loads, 192
  helicopter tie-down loads, 192, 193*f*
  installation motion, 193
  loads for helicopter landings, 190–192
    dead load of structural members, 190
    dynamic loads due to impact landing, 190
    lateral load on landing platform supports, 190
    overall superimposed load on landing platform, 190
    punching shear, 192
    sympathetic response of landing platform, 190
    wind loading, 190–191
  loads for helicopters at rest, 192
  single main rotor helicopter, 189
  tandem main rotor helicopter, 189
  weights, dimensions and D value for helicopters, 191*t*
  wind loading, 192
Helideck design
  aluminum helideck, 205
  design load conditions, 198–205
  helicopter landing loads, 189–193
  helideck layout design steps, 201–203
  parameters, 195*t*
  plate thickness calculation, 204–205
  safety net arms and framing, 193–198
Helideck wind area, 356*f*
Helidecks, 6, 8*f*
Hook load, 319
  defined, 316*t*
Horizontal tolerance, 286–287, 286*f*
Hydrocarbon reserves, 2
Hydrostatic collapse data, 355, 357*f*
Hydrotest, 310

I
Ice loads, 68
Impact load, 38, 38*t*, 205
  cases of, 209, 216
Indeterminate lift, defined, 316*t*
Inelastic local buckling stress, 126
In-place analysis, 343–348
  by ISO19902, 71–72
In-plane bending, 159, 161, 161*f*
  for three braces, 164
  unbalanced, 162–163
In-plane joint detailing, 131*f*
In-plane tubular joint, 152*f*
Input the load data, 350–367
  foundation model, 362–367
  joint can options, 357–361
In-situ testing, 220–226
  cone penetration test, 220–224
  field vane test, 225–226
Installation lifting analysis, 296–308
  barges, 303–305
  launching and upending forces, 305–308
  loadout process, 297
  transportation process, 297–303
ISO 19901-1, 47, 58
ISO 19901-2, 62–63, 65
ISO 19901-4, 275
ISO19902, 11–12
  axial tension, 113
  bending, 115–116
  column buckling, 114
  cylinder member strength calculation by, 113–124
  effective lengths and moment reduction factors, 123–124

ISO19902 (*Continued*)
for member strength checking, 124*t*
hoop buckling, 118–119
hydrostatic pressure, 117–118
in-place analysis by, 71–72
local buckling, 114–115
shear, 116
torsional shear, 117
tubular members combined forces without hydrostatic pressure, 119–120
axial compression and bending, 120
axial tension and bending, 120
tubular members combined forces with hydrostatic pressure, 121–123
axial tension, bending, and hydrostatic pressure, 121–123

J
Jackets, 5, 16
assembly and erection, 293–295
for deeper water, 283–284
deformation of, 374*f*
design, 87*f*, 88*f*, 94
fatigue design, 168–172
in-place analysis, checklist for, 24*t*
loadout stages, 298*f*
major secondary steelwork, 104
on-bottom stability analysis, checklist for, 28*t*
primary steelwork, 104
structure system, 85
and topsides transportation analysis, checklist for, 29*t*
views, 87*f*
Joint can
data generator, 358*f*
design, 374*f*
editor, 357, 358*f*
input, 363*f*
option, 357–361, 359*f*
output data, 373*f*
redesign option, 360*f*
reports, 359*f*
second method, 363*f*
Joint coordinates, 102–104
Joint detailing, 138
Joint eccentricities, 105–106
Joint selection, 387*f*

K
K-brace pattern system, 92–93
K-bracing geometry, 96*f*
Keppel shipyard, 7–8

L
Ladders, 175–177
Lateral bearing capacity
for sand, 249–252
for soft clay, 247–248
for stiff clay, 248–249
Laterally loaded piles reaction, 246–247
Launch analysis, 422, 422*f*
Leg alignment and straightness, tolerances in, 288
Leg spacing tolerances, 286
Legs tab, 346
Lift point
defined, 316*t*
design, 334–335
Lift weight, 310
defined, 316*t*
Lifting analysis, 283, 308–341, 412–416, 423
clearances, 335–340
around crane vessel, 335–340
around lifted object, 335
example, 338–340
construction procedure, 283–284
construction process, 284–295
conductor guides and piles tolerances, 292
dimensional control, 293
fabrication tolerances, 285–289
jacket assembly and erection, 293–295
stiffener tolerances, 290–291
execution methods, 284
installation process, 296–308
barges, 303–305
launching and upending forces, 305–308
loadout process, 297
transportation process, 297–303
lift point design, 334–335
lifting calculation report, 340–341
crane vessel, 341
lifting procedure and calculation, 313–333
bending efficiency factor, 331

calculated weight, 319
consequence factors, 332–333
crane lift factors, 330
example, 321–329
grommets, 331
hook load, 319
part sling factor, 330
resolved padeye load, 320–329
shackle safety factors, 331–332
skew load factor, 319
sling force, 330
termination efficiency factor, 330–331
loads from transportation, launch, and lifting operations, 313
structural calculations, 334
weight accuracy, classification of, 310–312
allowances and contingencies, 311–312
weight engineering procedures, 312
weight calculation, 309–310
weight control, 308–309
Littoral currents. *See* Longshore currents
Live load, 35–38, 45*t*, 310
Living accommodations platform. *See* Quarters platforms
Lloyd's Register of Shipping (LRS), 198, 204
LNG FPSOs, 7–8
Load cases options, 360*f*
Load configuration, 381*f*
Load factor, 348, 382*f*
Loading sequence data, 386*f*
Loadout, 387–388
defined, 316*t*
process, 297, 310, 387–388
Loadout, lifted
defined, 316*t*
Loads, 16, 28*t*
for helicopter landings, 190–192
for helicopters at rest, 192
Loads on padeye, calculating, 337*f*
Local axis technique, 105*f*
Local member axes, 105
Longshore currents, 57

M

Manufacturing delivery progress measuring guideline, 22*t*
Marine growth, 69, 376
data, 356*f*
Material strength, 78–79
Materials takeoffs (MTOs), 22
Maximum takeoff weight (MTOW), 189–190
Mean water line (MWL), 89
Member Diagram, 369, 372*f*
Member effective lengths, 105
Member Group Manage dialog box, 348
Member properties, defining, 348–349
Member Review, 373
Members properties, 186, 349*f*
Minimum required breaking load, defined, 316*t*
Minimum wall thickness, of pile, 258–260
Mode of deflection, 110*f*
Modes of deformation, 110*f*
Morison equation, 51–52
Motion input data, 392*f*, 393*f*, 394*f*, 395*f*, 396*f*
Mud mat design, 274–277

N

Natural frequency, 107–112
Net weight, defined, 315, 316*t*
Nonlinear analysis, 209–210, 382–384

O

Offshore fixed platform design
approximate dimensions, 89–92
bracing system, 92–94
bridges, 178
cylinder member strength, 112–135
calculation by API RP2A, 125–135
calculation by ISO19902, 113–124
dynamic structure analysis, 106–112
natural frequency, 107–112
fatigue analysis, 150–172
S-N curves
stress concentration factors, 151–154
jacket design, 94
preliminary design, guide for, 85–94
structure analysis, 95–106
global structure analysis, 96–99
loads on the piles, 99–101
modeling techniques, 102–106
topside design, 172–177
deck design to support vibrating machines, 173–174

Offshore fixed platform design (*Continued*)
grating design, 174–175
handrails, walkways, stairways, and ladders, 175–177
topside structure analysis, 172–173
tubular joint design, 135–149
simple joint calculation from API RP2A, 136–144
vortex-induced vibrations (VIVs), 179–186
Offshore lifting terminology, 316*t*
Offshore platforms, types of, 4–6
auxiliary platforms, 6
bridges, 6, 7*f*, 8*f*
drilling/well protected platforms, 5
flare jackets and flare towers, 6
helidecks, 6, 8*f*
production platforms, 5
quarters platforms, 6
self-contained platforms, 5
tender platforms, 5
Offshore structures, 13
cement grout, 79–82
collision events, 76–78
design for ultimate limit state, 69–76
extreme environmental situation for fixed offshore platforms, 72–73
in-place analysis by ISO19902, 71–72
load factors, 70–74
operating environmental situations for fixed platforms, 73–74
partial action factors, 74–76
earthquake load, 59–68
abnormal level earthquake requirements, 65
ALE structural and foundation modeling, 65–68
extreme level earthquake requirements, 64–65
field development, 2–4
gravity load, 33–41
crane support structures, 39–41
dead load, 33–34
design for serviceability limit state, 38–39
impact load, 38
live load, 35–38
history of, 1–2
ice loads, 68
loads and strength, 33
marine growth, 69
material strength, 78–79
offshore loads, 46–59
current load, 57–59
wave load, 47–57
platforms types, 4–6
scour, 69
types of, 6–12, 12*f*
wind load, 41–45
Oil and gas platforms
facing problem of aging, 4
field development, 3
in offshore structures, 1
On-bottom stability, 274–275, 419–421, 424
Operating condition, 310
Organization of Petroleum Exporting Countries (OPEC), 2
Out-of-plane bending, 159, 160*f*, 161–162, 161*f*
balanced, 161–162
unbalanced, 164–165, 164*f*
Out-of-plane joint detailing, 138*f*
Overconsolidation ratio (OCR), 261–262
Overlapping joints, 143

P

P&ID, engineering preparation for, 17, 21*t*
Padear, defined, 316*t*
Padeye, 308, 319
defined, 316*t*, 319
dimensions, 326
loads on, calculating, 337*f*
Part sling factor, 330
Piezocone test, 220
Pile drivability analysis, 261–267
drivability calculations results, 265–266
recommendations for pile installation, 266–267
soil resistance drive evaluation, 261
unit shaft resistance and unit end bearing for uncemented materials, 261–262
upper- and lower-bound SRD, 262–263
wave equation analysis results, 263–265
Pile foundations, 231–256
axial pile performance. *See* Axial pile performance
foundation size, 239–240

pile penetration, 240
pile capacity
for axial loads, 232–239
calculation, 237*t*
calculation methods, 253–256
safety factors, 238*t*
shaft friction and end bearing in cohesionless soils, 236–239
skin friction and end bearing in cohesive soils, 234–236
Pile load tests, 234, 245
Pile properties input data, 364*f*
Pile wall thickness, 256–260
design pile stresses, 256–257
driving shoe and head, 260
minimum wall thickness, 258–260
pile section lengths, 260
stresses due to weight of hammer during hammer placement, 257–258
Pile-soil interaction menu, 364*f*
Pile-to-jacket leg annulus, 106
Plate thickness calculation, 204–205
Platform configuration input data, 347*f*
Plot soil data, 367
Plotting options, 408*f*
Point loads, 37
Postprocessing, 382*f*, 383*f*, 384*f*
Pressure meter, 219
Process flow diagrams (PFDs), 15
Procurement progress measurement, 21*t*
Production platforms, 5
Progress measuring percentage, 21*t*
Project cost and life cycle, 13–15
Project management, goals of, 22–23
Project manager, 23
challenge for, 19
Project progress measuring guideline, 22*t*
Pullout structure model, 388*f*
Pushover analysis, 66, 382–384
P-Y curve, 247–249, 251–252
for different strata, 370*f*
input data, 369*f*
number of point on, 368*f*
number of soil strata for, 368*f*

Q
Quality control (QC), 293
Quarters platforms, 6

R
Rayleigh method, 107
Remote vane. *See* Vane shear device
Report checklist, 32*t*
Reservoir management plan, 3
Resolved padeye load, 320–329
Review Joint Can, 373
Review Member, 368
Review Member data, 371*f*
Rigging, defined, 316*t*
Rigging facilities, checking, 339*f*
Rigging weight, defined, 316*t*
Riser guard, 214–216, 384–385, 387*f*
cases of impact load, 216
design calculation, 214–216
Rope, defined, 316*t*

S
SACS Combine module, 380
SACS software, 343
bracing input data, 348*f*
collapse analysis, 382–386
conductors input data, 347*f*
current data, 353*f*
Deck wizard, 349*f*
drag effect direction, 353*f*
drag effect in Y direction, 355*f*
dynamic analysis, 373–377, 375*f*, 423
eigenvalue analysis, 375–377
fatigue analysis, 396–411
center of damage, 397–400
dynamic wave response analysis, 402–411
fatigue input data, 411
foundation superelement, generation of, 400–402
fatigue analysis, 423–424
collapse analysis, 423
lifting analysis, 423
on-bottom stability, 424
Tow analysis, 424
flotation and upbending module, 417–418
helideck wind area, 356*f*
hydrostatic collapse data, 357*f*
in-place analysis, 343–348
input the load data, 350–367
foundation model, 362–367
joint can, 357–361
Launch analysis, 422

SACS software (*Continued*)
  lifting analysis, 412–416
  loadout process, 387–388
  marine growth data, 356*f*
  member properties, defining, 348–349, 349*f*
  on-bottom stability, 419–421
  output data, 368–373
  overview, 344*f*
  platform configuration input data, 347*f*
  project start menu, 344*f*
  sea fastening, 388–395
    load combinations, 390–395
  seismic analysis, 378–382, 423
    combination of seismic and gravity loads, 380–382
  start, options to, 345*f*
  start structure wizard, 346*f*
  static analysis, 422–423
  units menu, 345*f*
  wave input data, 350*f*
    continuation of, 351*f*
  wave load, 354*f*
  wind and drag effect, 354*f*
  wind area, 352*f*
  wind input data, 351*f*
Safe working load, defined, 316*t*
Safety net arms and framing, 193–198
Sand, 227, 228*t*, 243*t*, 246, 252*t*, 255
  lateral bearing capacity for, 249–252
SAP2000, 343
Satellite platform, 7, 9*f*
Scatter diagram, example of, 398*t*, 399*t*, 400*t*
Scour, 69
Screening process, 15–16
Sea fastening, 388–395
  defined, 316*t*
  load combinations, 390–395
  member properties, 396*f*
  structure model, 389*f*
Seabed mode, 220
Seastate module, 402
Seismic analysis, 423
  input and output file for, 380*f*
  input data, 379*f*
Seismic and gravity loads, combination of, 380–382
Seismic load, 61
Seismic risk category (SRC), 63
Select phase, 16
Self-contained platforms, 5
Shackle safety factors, 331–332
Shaft friction, 234, 236–239
Shell, 6
Shell Castellon, 7
Shock cell, 205, 207*f*
Shoes, pile, 260
Simple harmonic motion, 107
Sinusoidal progressive curve, 49*f*
Skew load factor (SKL), 319
  defined, 316*t*
Sling breaking load, defined, 316*t*
Sling eye, defined, 316*t*
Sling force, 330
Sling load, 330, 418, 421*f*
S-N curves, 411
  for all members and connections, except tubular connections, 154–155
  jacket fatigue design, 168–172
  for tubular connections, 155–168
    axial load, balanced, 159–161
    axial load, chord ends fixed, 157–158
    axial load, general fixity conditions, 158–159
    balanced axial load, 162
    balanced axial load for three braces, 164
    in-plane bending, 159, 161
    in-plane bending for three braces, 164
    out-of-plane bending, 159, 161–162
    thickness effect, 156–165
    unbalanced in-plane bending, 162–163
    unbalanced out-of plane bending OPB, 163–164
    unbalanced out-of-plane bending for three braces, 164–165
    weld toe position, effect of, 165–168
Software in design
  global structure analysis of, 96–99
  structural analysis of, 95–106
Soil data, 364, 365*f*
Soil investigation, 217. *See also* Cone penetration test (CPT); Soil tests
  onshore and offshore, difference between, 217
Soil investigation report, 265–266
Soil properties, 226–231

soil characterization, 230–231
strength, 227–229
Soil resistance drive (SRD)
evaluation of, 261
upper- and lower-bound, 262–263
Soil strata, number of
for P-Y data, 368*f*
Soil tests, 218–220
Soliton currents, 57
SOR (statement of requirements) document, 18–19
Splice, defined, 316*t*
Spreader bar (frame), defined, 316*t*
Spring, Choosing, 416*f*
Square root of the sum of the squares (SRSS) method, 61, 380
Staad, 343
Stairways, 175–177
Statement of requirements (SOR), 35
Static analysis, 422–423
Static load-deflection behavior, 240
Steel, strength of, 78–79
Steel template platform, 1, 11–12, 85
jacket, 85
piles, 85
topside facilities, 85
Stiffener tolerances, 290–291
Storm surges, 47
Straining action, 372*f*
Stream function theory, 47, 52
Strength level (SLE) earthquake, 63
Stress concentration factor (SCF), 151–154
in cast nodes, 154
in grouted joints, 154
Stress range, 411
Stresses, pile, 256–257
Structural engineering quality control, 23–31
1-Structural steel, 309
Structural steel pipes, mechanical properties for, 82*t*
Structural steel shapes, mechanical properties for, 82*t*
Structure Definition dialog box, 343–346
Study reports, progress measured in, 21*t*

T

Tender platforms, 5
Tension leg platform (TLP), 1, 9
Termination efficiency factor, 330–331
defined, 316*t*
Three-legged platform, 7, 10*f*, 11*f*
Tidal currents, 57
Tie-down loads, helicopter, 87*f*, 99–101
Time schedule control, 19–22
Tolerances
conductor guides and piles tolerances, 292
fabrication tolerances. *See* Fabrication tolerances
stiffener tolerances, 290–291
Topside design, 172–177
deck design to support vibrating machines, 173–174
grating design, 174–175
handrails, walkways, stairways, and ladders, 175–177
topside structure analysis, 172–173
Topside facilities, 85
cellar deck, 85
drilling deck, 85
wellhead/production deck, 85
Topside in-place analysis, checklist for, 27*t*
Topside lift analysis, checklist for, 31*t*
Topside major secondary steelwork, 104
Topside primary steelwork, 104
Topside structure analysis, 95–106
global structure analysis, 96–99
loads on the piles, 99–101
modeling techniques, 102–106
joint coordinates, 102–104
joint eccentricities, 105–106
local member axes, 105
member effective lengths, 105
software programs on, 95–96
Torsion adhesion data, 367*f*
Torsion head data, 367*f*
Tow analysis, 392*f*, 424
Tow module, 389, 391, 391*f*
Transfer function load case, 408*f*
Trunnion, 320, 321*f*
defined, 316*t*
Tsunamis, 57
Tubular joint check, 359
Tubular joint design, 135–149
joint calculation from API RP2A, 145–149
simple joint calculation from API RP2A, 136–144

Tubular joint tolerances, 289
Tubular member denting analysis, 211–213
simplified method for denting limit calculation, 212–213
Tubular member tolerances, 287–288
Turbidity currents, 57
T-Z axial bearing, 366*f*
soil data, 366*f*
strata, 366*f*
T-Z axial curve, 370*f*
T-Z axial input data, 365*f*
T-Z axial stratum soil data, 365*f*
T-Z bearing curve, 369*f*
T-Z soil axial data input, 365*f*

U

Undefined Groups window, 348
Unit end-bearing component, of SRD, 261
Unit shaft resistance component, of SRD, 261
Upbending module, 417–418

V

Vane shear device, 219
Vane test, 225–226, 225*f*
Vertical level tolerances, 287
Vortex-induced vibrations (VIVs), 179–186

W

Walkways, 175–177
Wall thickness, of pile, 256–260
Wave equation analysis, 231, 261
results, 263–265
Wave input data, 350*f*
continuation of, 351*f*
Wave load, 47–57, 354*f*
calculation, 52
conductor shield factor (SF), 56–57
drag force, 51
example, 53–57
inertia force, 52
wind and wave calculation, comparison between, 53–55
Wave Response module, 402
Weight accuracy, classification of, 310–312
allowances and contingencies, 311–312
weight engineering procedures, 312
Weight calculation, 309–310
Weight control, 308–309
Welded joint (WJ) curves, 155
Well-protectors (well jackets), 5
Wheel loads, 37
Wind and drag effect, 354*f*
Wind and wave calculation, comparison between, 53–55
Wind area, 352*f*
Wind input data, 351*f*
Wind loading, 41–45
on helideck structure, 190–192
Wind-generated currents, 57–59
Workhours percentage in engineering phase, 20*t*

X

X-bracing, 94, 95*f*

Printed in the United States
By Bookmasters